The International Library of Environmental, Agricultural and Food Ethics

Volume 26

Series editors

Michiel Korthals, Wageningen, The Netherlands
Paul B. Thompson, Michigan, USA

The ethics of food and agriculture is confronted with enormous challenges. Scientific developments in the food sciences promise to be dramatic; the concept of life sciences, that comprises the integral connection between the biological sciences, the medical sciences and the agricultural sciences, got a broad start with the genetic revolution. In the mean time, society, i.e., consumers, producers, farmers, policymakers, etc, raised lots of intriguing questions about the implications and presuppositions of this revolution, taking into account not only scientific developments, but societal as well. If so many things with respect to food and our food diet will change, will our food still be safe? Will it be produced under animal friendly conditions of husbandry and what will our definition of animal welfare be under these conditions? Will food production be sustainable and environmentally healthy? Will production consider the interest of the worst off and the small farmers? How will globalisation and liberalization of markets influence local and regional food production and consumption patterns? How will all these developments influence the rural areas and what values and policies are ethically sound?

All these questions raise fundamental and broad ethical issues and require enormous ethical theorizing to be approached fruitfully. Ethical reflection on criteria of animal welfare, sustainability, liveability of the rural areas, biotechnology, policies and all the interconnections is inevitable.

Library of Environmental, Agricultural and Food Ethics contributes to a sound, pluralistic and argumentative food and agricultural ethics. It brings together the most important and relevant voices in the field; by providing a platform for theoretical and practical contributors with respect to research and education on all levels.

More information about this series at http://www.springer.com/series/6215

Sahotra Sarkar · Ben A. Minteer
Editors

A Sustainable Philosophy—The Work of Bryan Norton

Springer

Editors
Sahotra Sarkar
Departments of Philosophy
and Integrative Biology
University of Texas
Austin, TX
USA

Ben A. Minteer
School of Life Sciences
Arizona State University
Tempe, AZ
USA

ISSN 1570-3010 ISSN 2215-1737 (electronic)
The International Library of Environmental, Agricultural and Food Ethics
ISBN 978-3-030-06470-9 ISBN 978-3-319-92597-4 (eBook)
https://doi.org/10.1007/978-3-319-92597-4

Printed on acid-free paper

This Springer imprint is published by the registered company Springer International Publishing AG part of Springer Nature
The registered company address is: Gewerbestrasse 11, 6330 Cham, Switzerland

Foreword

Rudolf Carnap and *External Questions*: My Path to Sustainability

I am sometimes asked: What philosopher has most affected your work? When I respond, "Rudolf Carnap," those not associated with philosophy give me a blank look, while those inquirers who know the story of philosophy in the twentieth century are even more nonplussed: How could Carnap's work, best known for its attack on the meaningfulness of metaphysics and most of traditional philosophy, support my efforts to use philosophy to address contemporary social problems, especially environmental problems?

The key to this puzzle requires that we distinguish between two sometimes-clashing themes in Carnap's work: his *empiricist (verificationist) view, associated with the Vienna Circle*, and his *principle of tolerance* (PT), *associated with his conventional notion of language*. In my dissertation and an early book, I showed that these two themes co-existed throughout Carnap's career, with the former view being more dominant in the early decades, while the latter view emerged as more central in his more mature philosophy.

In a University of Michigan seminar (1968) led by Prof. Jaegwon Kim, we read Carnap's 1950 piece, "Empiricism, Semantics, and Ontology" (ESO); as I was struggling to understand how to separate "meaningless metaphysics" from what seemed to me to be valuable philosophical insights, I came to see that Carnap's attitude toward traditional philosophy in 1950 embraced a shift away from his early emphasis on empiricism and the denigration of philosophical utterances as unverifiable (see Appendix). By the time he wrote "Empiricism, Semantics, and Ontology," empiricism appears not as an essential criterion of meaningfulness but rather as posing problems for his ongoing work in semantics, in which he constructed languages that formed sentences referring to entities, some of which were not observable and would, hence, be unverifiable and meaningless according to the strict empiricism of the Vienna Circle. In particular, he was stung by criticisms from empiricists who charged him with falling into metaphysics when he introduced conventions that allowed reference to unverifiable (abstract) entities such as numbers, classes, properties, and relations because he included such entities in the ontologies he introduced to provide semantic content for his constructed languages, which he fashioned to allow creative improvements on the language of science.

To respond to these uncomfortable criticisms, Carnap returned in "Empiricism, Semantics, and Ontology" to the second theme, which had been formulated in his early masterwork, The Logical Syntax of Language (1937). There, Carnap rejected attempts of philosophers to exclude some linguistic forms based on "negative requirements" and instead proposed his *Principal of Tolerance: It is not our business to set up prohibitions, but to arrive at conventions* (Carnap 1937: 51). Elaborating on this point, he said: "*In logic there are no morals*. Everyone is at liberty to build up his own logic, i.e. his own form of language, as he wishes. All that is required of him is that, if he wishes to discuss it, he must state his methods clearly, and give syntactical rules instead of philosophical arguments" (Carnap 1937: 52).

PT merely dramatizes the position more commonly referred to as "linguistic conventionalism," which asserts that language forms are tools for communication and that choosing linguistic forms necessarily involves choices not fully dictated by the structure and content of reality; they are better understood as proposals to adopt specific languages in particular situations which, in turn, give shape to the reality humans experience.

Whereas the strict empiricism of the Vienna Circle positivists aimed to destroy and put to rest philosophical problems, PT aims more at a reconstruction of traditional philosophy. This reconstruction was accomplished by arguing that philosophical assertions such as "There are material objects" or "There are numbers" can be rehabilitated by restating them in the metalanguage as assertions about the usefulness, in particular situations, of languages that refer to material objects or to numbers, respectively.

Without going into great detail here (see Norton 1977, 2005, 2015, Appendix for more detail), we can see that Carnap was advocating a new mode and purpose for philosophy. The mode was to use linguistic analysis in cooperation with active empirical scientists to improve communication of scientific findings and, in the process, to create "ontologies," which represent useful interpretations of scientific observations. Chosen linguistic categories can thus reflect models of reality through semantic choices justified not by appeals to "essences" existing in the real world, but with a method of appealing to the practical usefulness of proposed language forms. This apparently innocuous viewpoint has deep and consequential implications for philosophy because choices legitimized by PT can only be resolved or justified by appeal to *human purposes* and thus require appeals to human interests and values. The linguistic choices that Carnap defends in ESO—based on PT and a human interest in language reformation—are "pragmatic" decisions.

In The Logical Syntax of Language, Carnap describes his view of the nature of philosophy as follows:

> The supposed peculiarly philosophical point of view from which the objects of science are to be investigated proves to be illusory.... Apart from the questions of the individual sciences, only the questions of the logical analysis of science, of its sentences, terms, concepts, theories, etc, are left as genuine scientific questions. We shall call this complex of questions the *logic of science*....
>
> According to this view, then, once philosophy is purified of all pseudoscientific elements, only the logic of science remains. In the majority of philosophical investigations, however, a sharp division into scientific and unscientific elements is quite impossible. For this reason, we prefer to say: *the logic of science takes the place of the inextricable tangle of problems which is known as philosophy* (Carnap 1937: 279).

Carnap's alternative view of philosophy, explicated in ESO, which sees the role of philosophy as limited to the study of the language of science, however, offers a powerful alternative. He says, "Our thesis that the logic of science is syntax must therefore not be misunderstood to mean that the task of the logic of science could be carried out independently of empirical science and without regard to its empirical results."[1] (Carnap 1937: 332). So, by 1950 and the publication of "Empiricism, Semantics, and Ontology," Carnap clarified the role of philosophy as the study of choices made regarding the language by which the empirical results of science can be perspicuously stated, discussed, and resolved, including syntactics, semantics, and pragmatics.[2] The main innovation in ESO was to emphasize that, properly understood, philosophical discourse mixes together two quite different types of questions he designated as "internal" and "external" in nature. Internal questions are asked within a linguistic framework already developed, and their answers are usually "analytic"—e.g., the sentence "There are numbers" would be true for anyone who accepts and uses normal arithmetic. The question whether to adopt such a linguistic framework involves an external question which must be understood as a "proposal" made in the metalanguage; as such, it involves a choice that will depend on which linguistic rules will foster communication and improved understanding.

The key point here is that Carnap saw philosophy as an activity understood as making linguistic choices at the boundary between philosophy and empirical science. This point has two further implications. First, it entails—and supports—the

---

[1] Actually, this formulation, published in The Logical Syntax of Language, was altered in an important way before 1950, when these ideas were used in "Empiricism, Semantics, and Ontology." In the 1930s, Carnap thought all linguistic work can be formalized into discussions of syntax (limited to the physical form of words and sentences), but subsequent work by logicians convinced him that the logic of science must include semantics, so in 1950, he would expand the domain of the logic of science to include metalinguistic discussions of the semantics and pragmatics of language.

[2] The pragmatics of language studies the use of language in communication and human deliberation. Again, his later study of pragmatics, like his embrace of semantics, broadened his conception of the "logic of science." In my dissertation, I formulated this point simply as a claim that, properly understood, philosophical discourse should be carried out in a "metalanguage." Philosophers' role, so understood, is to work closely with scientists and to help them construct languages that are both empirically grounded and useful in communicating important ideas, including ideas with empirical content.

controversial argument that philosophy is intimately tied to language development and the evolution of scientific concepts. Second, this view should encourage philosophers to work closely with empirical scientists, as new observations can lead to, or even demand, conceptual/linguistic change.

In my work subsequent to my examination of Carnap's later work, I built upon this creative and rehabilitative view of philosophy, and at first emphasizing work in epistemology based on pragmatic principles. During this period, I continued to learn from many philosophers other than Carnap—especially, Willard Van Orman Quine and other conventionalists, but it was my early exposure to Carnap and ESO that supported my work in pragmatic thought.

Then, in the late 1970s, acting on my long-standing interest in environmental problems and issues, I initiated a class, "Humanities and the Environment" which, in turn, led to an opportunity to do research on the question, "What is the rationale for the US Endangered Species Act of 1973?" This question occupied me from 1979 until the late 1980s as I prepared two books designed to bring the methods of philosophy to bear upon the policy questions provoked by the growing concern for protection of biological resources—what we came to call "biodiversity" (Norton 1986, 1987).

With those books behind me, and having learned from my experience in working with policy-oriented groups and individuals in Washington, D.C., I recognized that talk of "sustainability" was gaining prominence in most environmental discussions and I realized that the approach and methodology I used to clarify obligations to protect species could also be usefully employed in articulating a discourse on the meaning of "sustainability." In particular, I noticed that authors discussing sustainability make very different linguistic choices, and their advocated policies tended to fall in two camps: one group embraced economic methods and set out to define sustainability in those terms, while others—often ecologists by training—urged that sustainability can only be achieved if it includes protection of ecological characteristics and systems, such as "integrity" and "resilience." Building on this difference in disciplinary approaches to understanding sustainability, I emphasized a distinction to characterize these two approaches, respectively: "weak" and "strong" sustainability. I hope it is clear by now that this extremely important distinction can be usefully explored within the broadly pragmatist/conventionalist view of language and its role in public discourse.

That role requires that value concepts, as well as scientific explorations, must unavoidably be involved in choices regarding how to talk and how to make policy arguments in the melees in which public values are pursued. All of my policy work has been supported on these ideas, which have coalesced in my books on sustainability. Those books articulate the argument that, while weak sustainability theorists interpret environmental problems within a single paradigm—that of the economics of cost-benefit—and seek algorithmic solutions to problems, strong sustainability will require richer concepts and methods than can be developed on the assumptions and definitions of economics.

Two important consequences follow. First, richer concepts will require that strong sustainability theorists work closely with natural scientists such as ecologists and foresters to develop a vocabulary richer in describing physical features associated with sustainability. Second, strong sustainability will never be defined and operationalized so definitively that problems can be solved algorithmically, once and for all. Consequently, problems encountered as we move forward toward a more sustainable future will not be resolved by deductive or academic arguments. Solutions must be sought by engaging public and interested parties in discussion and deliberation: the route to sustainability will, therefore, involve both experience and scientific learning, on the one hand, and articulation of the values and aspirations of human communities, on the other hand.

My debt to Carnap has thus guided my work throughout my career: his work established for me the key premise that "external questions" like "How should we define 'sustainability'?" will only be resolved in forums in which science, and the broadest understanding of human values, can all be brought to bear upon how we talk, how we think, and how we choose.

Atlanta, Georgia Bryan G. Norton

References

Carnap, R. 1937. *The logical syntax of language*. London: K. Paul, Trench, Trubner & Co.

Carnap, R. 1950. Empiricism, Semantics, and Ontology. *Revue Internationale de Philosophie* 4: 20–40. Reprinted in the Supplement to R. Carnap, *Meaning and necessity: A study in semantics and modal logic,* enlarged edition (University of Chicago Press, 1956).

Norton, B.G. 1977. *Linguistic frameworks and ontology: A re-examination of carnap's metaphilosophy*. The Hague: Mouton Publishers, Janua Linguarum Series, #145.

Norton, B.G. 1986. *The preservation of species,* ed. Princeton, NJ: Princeton University Press.

Norton, B.G. 1987. *Why preserve natural variety?* Princeton, NJ: Princeton University Press.

Norton, B.G. 2005. *Sustainability: A Philosophy of Adaptive Ecosystem Management*. Chicago, IL: University of Chicago Press.

Norton, B.G. 2015. *Sustainable Values, Sustainable Change: A Guide to Environmental Decision Making*. Chicago, IL: University of Chicago Press.

Contents

Contributors

Dean Anderson Manaaki Whenua-Landcare Research, Lincoln, New Zealand

Evelyn Brister Philosophy Department, Rochester Institute of Technology, Rochester, NY, USA

Daniel W. Bromley Department of Agricultural and Applied Economics, University of Wisconsin-Madison, Madison, WI, USA

J. Baird Callicott University Distinguished Research Professor Emeritus, University of North Texas, Denton, USA

Jennifer M. Chirico Sustainable Pacific Consulting, Paia, HI, USA

Paul D. Hirsch Department of Environmental Studies, State University of New York College of Environmental Science and Forestry, Syracuse, NY, USA

R. Bruce Hull College of Natural Resources and Environment, Virginia Tech (0324), Blacksburg, VA, USA

Jozef Keulartz Institute for Science in Society, Radboud University, Nijmegen, The Netherlands

Curt Meine Aldo Leopold Foundation and Center for Humans and Nature, Prairie du Sac, WI, USA

Ben A. Minteer School of Life Sciences, Arizona State University, Tempe, AZ, USA

Bryan G. Norton Distinguished Professor Emeritus in Philosophy and Policy, Georgia Institute of Technology, Atlanta, GA, USA

Christopher Rojas School of Life Sciences, Arizona State University, Tempe, AZ, USA

Sahotra Sarkar Departments of Philosophy and Integrative Biology, University of Texas at Austin, Austin, TX, USA; Presidency University, Kolkata, India

Piers H. G. Stephens Philosophy Department, University of Georgia, Athens, GA, USA

Paul B. Thompson Department of Philosophy, Michigan State University, East Lansing, MI, USA

Bruce Warburton Manaaki Whenua-Landcare Research, Lincoln, New Zealand

Clark Wolf Departments of Philosophy, Political Science, and Sustainable Agriculture, Iowa State University, Ames, IA, USA

Asim Zia Department of Community Development & Applied Economics, University of Vermont, Burlington, VT, USA

Chapter 1
Introduction

Sahotra Sarkar and Ben A. Minteer

For the last thirty years Bryan Norton has been one of the most important figures in environmental philosophy and probably the most influential proponent of the perspective that has come to be called environmental pragmatism. He was one of the very first philosophers to bridge the gap between philosophy and environmental public policy discussions in the 1980s through his work on the practical ethical foundations of the Endangered Species Act (1973) and the emerging scientific and policy attention to biodiversity (Norton 1987). In the 1990s, Norton's work at the interface of philosophy and environmental policy widened and deepened. The result was a series of important papers and books, including most notably *Toward Unity among Environmentalists* (1991), which remains one of the most influential discussions of the policy implications of environmental ethics. Since then his impact has been felt not only on a new generation of environmental philosophers but across the much wider field of environmental decision making, including environmental and ecological economics, environmental management and, especially, the emerging field of sustainability science (Norton 2003, 2005, 2015).

This volume is intended as a celebration of Norton's life and work. Reflecting the multidisciplinarity of Norton's scholarship, the contributors to this book range from philosophers and historians to biologists and economists who have all been deeply engaged with some aspect of Norton's oeuvre. The contributions cover the

S. Sarkar (✉)
Departments of Philosophy and Integrative Biology, University of Texas,
2210 Speedway, Austin, TX 78712, USA
e-mail: sarkar@austin.utexas.edu

S. Sarkar
Presidency University, Kolkata, India

B. A. Minteer
School of Life Sciences, Arizona State University, P.O. Box 874501,
Tempe, AZ 85287-4501, USA
e-mail: ben.minteer@asu.edu

S. Sarkar and B. A. Minteer (eds.), *A Sustainable Philosophy—The Work of Bryan Norton*, The International Library of Environmental, Agricultural and Food Ethics 26, https://doi.org/10.1007/978-3-319-92597-4_1

spectrum of that work, from Norton's early contributions to the ethics of biodiversity conservation to his later work on sustainability, adaptive management, and procedural rationality. The breadth of the contributors and topics is a testament to Norton's wide influence, while the recurring themes of the volume—the vexed meaning of sustainability, the promise and limits of deliberative and rationalistic approaches to decision making, the inescapable pluralism and dynamism of environmental values—speak to his scholarly curiosity and philosophical coherence over a long and distinguished career.

Norton provides a Foreword that traces his pragmatism to the early influence of the later work of Rudolf Carnap and logical empiricism. The Appendix includes his correspondence with Carnap when Norton was still a graduate student at the University of Michigan specializing in the philosophy of language. Norton's comments on that correspondence show how Carnap's pluralism about linguistic frameworks (encapsulated in his well-known Principle of Tolerance) forms the basis for the pluralism about values that environmental pragmatism embraces. The comments also explicitly record his acknowledgement of his debt to Carnap and his work.

The first part of this volume following this introduction (Chaps. 2–7) engages with Norton's work on characterizing and achieving sustainability, a project that has occupied him since the 1990s. Paul Thompson argues that Norton takes the debate between strong and weak sustainability as being central to a theoretical understanding of sustainability. In response Thompson tries to shift attention away from developmentalism of this sort to a characterization of sustainability that draws on a system's robustness and resilience that influence its capacity for adaptive management (another theme running through Norton's work that will be explored in the next section). Baird Callicott, one of the founders of environmental ethics and a scholar who has been engaged in a decades-long argument with Norton over the field's philosophical commitments and purposes, also returns to the debate between strong and weak sustainability and finds it wanting. For him sustainability must be achieved by a human economy patterning itself at all levels of organization on the economy of nature. It is a departure from Norton's own approach, which focuses more on the pragmatics of communication and the method of community-level problem-solving. Clark Wolf compares Norton's concept of sustainability to those inherited from Solow, Rawls, Mill, and Sen and argues that none of them are entirely free from problems. He also shows how adequate concepts of sustainability must fit into practical procedures for adaptive management.

To be truly useful, philosophical considerations of sustainability must ultimately be grounded in particular geographies and operationalized within specific social, political, and professional contexts. Accordingly, Jennifer Chirico analyzes the case of Hawaii and how attempts to achieve sustainability there can take advantage of Norton's multiscalar approach to adaptive management. Paul Hirsch explores the relevance of Norton's approach to sustainability embracing the public interest in the context of the political and social divisiveness that marks the present. According to him, addressing these divisions while framing public policy will require sensitivity and efforts beyond what Norton envisioned for a less fractured social context. Finally, Bruce Hull focuses on how Norton's work on deliberative decisions and procedural rationality can be used to identify leadership practices to help achieve sustainable solutions to environmental problems. Together, these six contributions provide a very useful overview of the philosophical moorings and social currency of Norton's contributions to sustainability.

The second part of this volume (Chaps. 9–15) pivots toward Norton's work on environmental ethics and decisions. Piers Stephens reviews the debate between Callicott and Norton in interpreting the work and thought of the American conservationist Aldo Leopold and his influental "land ethic." He provides a critical endorsement of Norton's claim of the influence of William James on Leopold's work and goes on to argue that the American pragmatist tradition has far more radical potential than Callicott allows. Daniel Bromley also urges environmental ethicists to grasp the promise of pragmatism but with a focus on Charles Sanders Peirce. He understands this view to be a core component of Norton's lifelong project of environmental scholarship.

Evelyn Brister turns to democratic decision procedures in which context she takes Norton's pragmatism to be influenced by Dewey and Habermas. She argues that Norton's advocacy of decisions based on inclusive, democratic, and public inquiry may not be achievable in many local decision contexts and, at best, provide an ideal for less rigid decision making procedures. Asim Zia's chapter focuses on the adaptive management framework that Norton has developed based on hierarchy theory and drawing on decision science and planning theory. He uses that framework to analyse problems of deforestation in Peru, Tanzania, and Vietnam and argues that the intractability of problems arises due to power asymmetries, conflicting values, and scale-dependent politics.

Jozef Keulartz's contribution considers Norton's convergence hypothesis about policy-making, i.e., that metaethical disagreements over the source of value and other foundational issues disappear in the context of practical policy formulation. Perhaps again reflecting on the divisiveness of the times, he finds Norton's expectations to be overly optimistic and agues that Norton consciously ignores power relationships and strategic moves during group policy formulation. Sahotra Sarkar turns to Norton's early interest in the conservation of biodiversity. He argues that Norton has been too hasty in his dismissal of formal decision analysis which, when used with due flexibility, can incorporate Norton's emphases on value pluralism, multiple objectives, and deliberative decision making.

The next pair of essays explore the continuing relevance of Norton's seminal work on the ethics of animal conservation. Bruce Warburton and Dean Anderson show how, following Norton's emphasis on adaptive mangement, ecology, economics, and ethics can be jointly incorporated into a program for sustainable wildlife management through probabilistic modeling. Ben Minteer and Christopher Rojas draw on Norton's notion of "transformative value" as they consider zoos' ability to inspire their visitors to care for nature, a discussion that provides a bridge between Norton's early writing in value theory and newer work on the zoo experience coming out of the field of conservation psychology.

The final contribution of this book before the Appendix is a revealing and inspiring interview of Norton by Curt Meine recorded in July 2016 during the North

American Society for Conservation Biology Meeting in Madison, Wisconsin. The interview traces Norton's intellectual biography, from his early work in philosophy of language and science to his career in helping found and subsequently develop the new field of environmental philosophy.

As noted earlier, Norton was a pioneer in moving environmental philosophy to engagement with the practical world of decisions and policy formulation. Indeed, one of the most valuable and distinctive aspects of his work over the years has been its ability to cross the academy's borders and absorb himself into the thickets of environmental practice. It is one of the reasons why his work has become so well-known and respected across diverse communities of practice, especially (but not only) conservation biologists and wildlife scientists. This reputation is reinforced by his extensive collaborative ventures over the decades (see, e.g., Norton 1986; Costanza et al. 1992; Takacs 1996; Norton et al. 1995; Sharpe et al. 2001; Luck et al. 2012; Chan et al. 2016). The interview records how Norton's early attempts at boundary crossing, however, were derisively dismissed by professional philsophers more interested in abstract arguments about absolute foundations than in the messy world of environmental problem solving. Perhaps the greatest testimonial to Norton's transformative role in the discpline is that, from today's perspective, the insularity and resistance to multidisciplinarity on the 1980s seems perverse.

As he notes in Meine's interview, Norton was the only philosopher of science then interested in working on environmental issues in that early period. The others were all ethicists, perhaps explaining why the field now known as environmental philosophy was long called environmental ethics even though it was never in practice so limited in scope. His background in the philosophy of science probably also explains why Norton was comfortable in crossing disiplinary boundaries which was common practice among philosophers of science since the heyday of logical empiricism. It also helps explain why so much of Norton's later work on sustainabilty incorporated insights from decision theory (now considered to be central to formal epistemology) and hierarchy theory (now part of complexity theory).

The interview also records how participation in policy groups consisting mainly of ecologists helped move Norton to environmental pragmatism because it showed both that these decision makers did not have much time or sympathy for philosophical foundations and, irrespective of their deeper value commitments, that they could achieve agreement on policy. The second of these features later became the basis for Norton's well-known convergence principle. Finally, the interview records where Norton sees the field of environmental philosophy going and how he reads the current political situation in the light of increasing divisiveness. As always, Norton is ultimately optimistic.

As editors, our hope is that readers already familiar with Norton's many contributions to environmental philosophy will find new ideas and possibilities in his corpus, especially regarding the field's potential to improve our understanding of environmental practice and develop novel opportunities to inform environmental research, decision making, and the quest for sustainability. Those new to Norton's work will find this collection to be a reliable companion and guide, but also a provocation; a syllabus for exploring some of the most significant and challenging questions in environmental philosophy and its application. Most of all, we hope this collection serves as a fitting tribute to the life and work of a trailblazing philosopher, one whose ideas and arguments have influenced so many of us, including the usual cohort of fellow-travelers, collaborators, and former students, but also skeptical interlocutors and even longtime critics. Our collective indebtedness to Norton and to his work can be measured not only by the depth of engagement with his thought in the discussions that follow, but in the scholarly gratitude and warmth issuing from these pages.

References

Chan, K.M.A., P. Balvanera, K. Benessaiah, M. Chapman, S. Díaz, E. Gómez-Baggethun, R. Gould, N. Hannahs, K. Jax, S. Klain, G.W. Luck, B. Martín-López, B. Muraca, B. Norton, K. Ott, U. Pascual, T. Satterfield, M. Tadaki, J. Taggart, and N. Turner. 2016. Opinion: Why protect nature? Rethinking values and the environment. *PNAS* 113: 1462–1465.

Costanza, R., B.G. Norton, and B.J. Haskell (eds.). 1992. *Ecosystem health: New goals for environmental management*. Washington, DC: Island Press.

Luck, G.W., K.M.A. Chan, U. Eser, E. Gómez-Baggethun, B. Matzdorf, B. Norton, and M.B. Potschin. 2012. Ethical considerations in on-ground applications of the ecosystem services concept. *BioScience* 62: 1020–1029.

Norton, B.G. (ed.). 1986. *The preservation of species: The value of biodiversity*. Princeton, NJ: Princeton University Press.

Norton, B.G. 1987. *Why preserve natural variety?* Princeton, NJ: Princeton University Press.

Norton, B.G. 1991. *Toward unity among environmentalists*. New York: Oxford University Press.

Norton, B.G. 2003. *Searching for sustainability: Interdisciplinary essays in the philosophy of conservation biology*. New York: Cambridge University Press.

Norton, B.G. 2005. *Sustainability: A philosophy of adaptive ecosystem management*. Chicago: University of Chicago Press.

Norton, B.G. 2015. *Sustainable values, sustainable change: A guide to environmental decision making*. Chicago: University of Chicago Press.

Norton, B.G., M. Hutchins, E.F. Stevens, and T.L. Maple (eds.). 1995. *Ethics on the ark: Zoos, animal welfare and wildlife conservation*. Washington DC: Smithsonian Institution Press.

Sharpe, V.A., B.G. Norton, and S. Donnelley. 2001. *Wolves and human communities: Biology, politics, and ethics*. Washington, D.C.: Island Press.

Takacs, D. 1996. *The idea of biodiversity: Philosophies of paradise*. Baltimore, MD: Johns Hopkins University Press.

Chapter 2
Norton and Sustainability as Such

Paul B. Thompson

Abstract Bryan Norton takes the debate over weak and strong sustainability to characterize the key conceptual disagreement among attempts to elaborate a theoretical approach to sustainability. In contrast, I argue that this debate is mired within assumptions of economic development theory that fail to recognize how elements of fragility, stability, resilience and adaptive capability within system design or organization have been material to the way that sustainability has been conceptualized in many domains. Two paradigmatic conceptualizations of sustainability compete within the scholarly literature, one stressing the availability of key resource stocks, the other stressing the robustness and resilience of system organization. A better approach would move beyond weak and strong sustainability to acknowledge a deeper and more philosophically potent set of paradigmatic features within sustainability scholarship.

Keywords Resources · Natural capital · Gross Domestic Product (GDP)
Resilience · Systems thinking

2.1 Introduction

Following path-breaking work on biodiversity and a failed attempt to forge consensus among environmental philosophers, Bryan Norton has devoted the last 25 years of his career to the theme of sustainability. His 2005 book, *Sustainability: A Philosophy of Adaptive Ecosystem Management*, argues that the environmental crisis is, in large part, a failure to integrate diverse strands of knowledge *about*

In addition to helpful comments from Minteer and Sarkar, I would like to thank Evelyn Brister and Zachary Piso for suggestions that I have implemented in the final version of this chapter.

P. B. Thompson (✉)
Department of Philosophy, Michigan State University,
503 S. Kedzie Hall, East Lansing, MI 48922, USA
e-mail: thomp649@msu.edu

S. Sarkar and B. A. Minteer (eds.), *A Sustainable Philosophy—The Work of Bryan Norton*, The International Library of Environmental, Agricultural and Food Ethics 26, https://doi.org/10.1007/978-3-319-92597-4_2

ecosystems (ecology, toxicology, systems biology, and the diverse fields we have come to call "climate science") but also a failure to recognize the inevitability of value judgments and a corresponding incapacity to discuss or negotiate value conflicts in planning, policy, and environmental management. The examples discussed in the book reflect Norton's long-running interest in conservation biology, the protection of habitat for threatened species, and the challenges to these goals posed by urbanization and human incursion on the ecosystems that support all forms of life on Earth. A succinct statement of Norton's prescription for change is that humanity must learn how to discuss these problems: environmental science cannot be expected to save the day (Norton 2005). Sustainability is, in this respect, a project in building collective communicative capacity around the various environmental challenges that human beings face on a worldwide basis.

On the basis of this work, Norton is far more influential in shaping the discourse on sustainability than any other academic philosopher. Yet, as I have noted before, analysis of sustainability as a concept or as a heuristic for shaping theory and practice is largely missing from his magnum opus (Thompson 2007). Norton has admitted that he treats sustainability as a general name for an overall philosophical approach that is more precisely characterized in terms of the values-cognizant procedures for adaptive ecosystem management that he describes in some detail (Norton 2007). While conceding that Norton's use of the word 'sustainability' is consistent with ordinary usage circa 2015, this chapter will nag him a bit on some points that were contested in our 2007 exchange. Norton and I both identify ourselves as "environmental pragmatists." I will begin by summarizing how Norton's views on sustainability are grounded in his philosophical commitments to pragmatism, and will move from there to a discussion of how those views have and have not changed in his 2015 book *Sustainable Values, Sustainable Change.* I will conclude by arguing that given the current state of public discourse on sustainability, Norton's approach to sustainability as such—that is, to the way in which the idea of sustainability does and might function in public discourse—fails to engage key contested elements within the public discussion. In this, Norton fails to hold himself accountable to the standard of public engagement that he has articulated and defended throughout his work.

2.2 Norton's Pragmatism

The first task is to situate both Norton and myself within the American philosophical tradition. Norton makes explicit links to philosophical pragmatism in *Sustainability: A Philosophy of Adaptive Ecosystem Management,* especially in the lengthy appendix where he describes his epistemological approach as an extension of work by Peirce and Dewey. Norton argues that debates carried on in an exchange of letters between W. V. O. Quine and Rudolf Carnap conclude with Carnap—the arch Vienna Circle positivist—becoming convinced that a pragmatist approach to meaning and truth is inescapable. Carnap's move from positivism to pragmatism can be summarized as a reinterpretation of the meta-theory for propositions. Succinctly, the positivist program had been committed to the claim that propositions are fully analyzed in terms of "sense" (or semantic content) and "reference"—a relationship to states of affairs that (according to positivists) were fully amenable to empirical verification. The positivist program of verificationism began to face an endless regress of meta-languages for specification of semantic content. At the same time, performative contradictions in its own specification of rules for verification and the inescapable theory-ladenness or reliance on background assumptions became increasingly apparent to Carnap. Propositions presuppose value judgments in the very specificity with which states-of-affairs are characterized. What is more, failed attempts at verification are always open to modifications in these value-informed background assumptions. Thus, rather than simply rejecting a claim as either false or "meaningless", as the logical-positivists had hoped, scientists are more inclined to reinterpret the nature of their overall project (a practice that Thomas Kuhn characterized as paradigm change). Norton argues that these philosophical revisions to the positivist program put both Quine and Carnap much closer to Dewey than to Bertrand Russell or the early Wittgenstein. Dewey had always interpreted propositions as implicit proposals whose ultimate value (including their truth value) is only seen when one attempts to act on them.

Norton applies this linguistic pragmatism to his environmental philosophy in numerous ways. At the outset of the book he characterizes the general inapplicability of environmental science as arising from the illusion of positivist, value-free science. Separate communities of environmental scientists are able to persevere in developing ever more refined empirical methods, but they overlook the way in which theoretical constructs fail to connect up with each other, much less with the terms in which policy decisions are inevitably discussed. Norton counters this with the claim that "language is our environment", an argument intended to show how perception, judgment and even the ability to communicate are inextricably shaped by history, by quotidian practice and by ordinary language. Environmental value debates, he argues, must be conducted in the vernacular, rather than the specialized terminology of technical elites. The argument here is pragmatist. In the first place, the terminology of technical elites is itself dependent on background commitments reflective of ordinary language use. Scientist's attention may be focused on the stipulative definition of their technical terms, but this definition-giving practice is only possible because they are already proficient speakers of ordinary language.

Second, ordinary speech is grounded in practical communicative activity or *use,* as Wittgenstein and even Russell came to see, (Misak 2016). A given individual's ability to use language implies interaction with other human beings who are engaged in various purposeful endeavors. Individuals rely upon a presumption of quotidian common purposes to invest their speech acts with meaning, or to put it more simply, meaningful speech becomes possible only within a social context. Ordinary non-technical language is well adapted to a very wide variety of contexts, and communities of practice develop more specific vocabularies that reflect the narrower range of purposes that bind them. For example, 'aglet,' 'welt,' 'vamp' and 'shank' are terms that indicate specific parts of a shoe, and their definitions can be found in any English dictionary. These are not terms that will be familiar to all speakers of English who also wear shoes, but they will certainly be known by cobblers and others in the community of practice that is the shoe trade. Science is, in this respect, a community of practice in which similarly specific vocabularies evolve. Norton's point is that the environmental sciences are in fact many communities of practice each of which has evolved distinct vocabularies.

Finally, the argument is pragmatist in holding that the success of technical elites in refining or developing their particular vocabulary depends far more on their practice as a community of inquiry than it does on putatively scientific methods. As Peirce or Dewey would have argued, inquiry is the general practice that all sciences share, as well as being the specific practice that distinguishes them from cobblers or other communities that develop more nuanced vocabularies. However, inquiry is always a response to some question, disturbance or felt need, and although one question often gives rise to another, a community of inquiry will only emerge when some impetus to inquire is shared. As Peirce argued most sagaciously, science is also distinguished by shared commitments to the means for fixation of belief in response to these driving questions. Science is thus shaped by values reflective of the way in which questions, methods and received opinions determine community bonds, in turn giving rise to an emergent technical vocabulary. *These* values may or may not coincide with the purposes that drive and define a political community. This is in no way to reject the importance of environmental science for policy-making. It is, instead, a plea for scientific communities of inquiry to be first self-conscious and second forthcoming in recognizing the values that shape their discourse.

In *Sustainability* Norton relies heavily on a metaphor introduced by Otto Neurath to convey how tests of coherence and practical fruitfulness can satisfy the demand for justifiability that is traditionally associated with science-based claims. This is the famous ship that is repaired, plank-by-plank, while at sea. It is an expression of the general pragmatist principle that while one neither can nor needs to doubt every possible proposition at once, it is nonetheless possible to regard any specific claim as fallible and open to potentially falsifying tests. However, Norton goes well beyond coherentists in analytic philosophy who use this approach merely to develop a theory of justified assertability. Norton's view of adaptive management applies fallibility criteria to policy contexts. This means that applications of environmental science to environmental problems must be accompanied by systematic monitoring of results and potential revision or modification in light of experience. It is not just that the planks *can* be replaced while the boat is at sea; they *should* be.[1] In the context of making and implementing policy, Norton's pragmatism relies on the way that Dewey understands experimentalism in epistemology as presupposing a commitment to democratic governance and participatory decision making. In short, Norton's entire philosophy, from his analysis of scientific method and

---

[1]Thanks to Evelyn Brister for this felicitous turn of phrase.

justification to his view of policy making, participation and governance is deeply informed by pragmatist principles.

2.3 Norton on Sustainability

Sustainability figures in Norton's philosophy in both a broad and a narrow sense. As noted already Norton uses the word to name his general approach to environmental ethics. This begins with his volume of collected papers *Searching for Sustainability* in 2003 and is most strongly reflected in the title of *Sustainability: A Philosophy of Adaptive Ecosystem Management*. However, Norton does discuss debates from the scholarly literature on sustainability. A 1995 paper discusses what he identifies as "competing paradigms" in the sustainability literature, though he does not name these paradigms or offer succinct discussions of them. One paradigm is illustrated by the work of Robert Solow, an economist known for his work on economic growth. Norton points out that Solow assumes total fungibility among the various forms of capital that play a role in the growth process, committing himself to what is widely referred to as "weak sustainability." This view defines general sustainability in terms of non-declining amounts of total capital. But given the fungibility assumption, capital in this approach is simply a term that generalizes or universalizes all those resources critical for economic growth. Relying especially on the work of Talbot Page, Norton discusses a number of reasons to question this view, stressing especially the importance of temporal and spatial scale as developed in ecological hierarchy theory. He concludes, "The movement to introduce health and integrity into policy analysis and implementation is best understood as a movement to define a structured and appropriate bequest from the current to future generations, especially as that package affects the functional characteristics of whole ecological systems," (Norton 1995, p. 126).

Ten years later, *Sustainability: A Philosophy of Adaptive Ecosystem Management* continues to discuss the concept of sustainability through the debate over the fungible nature of capital. Although Norton has dropped the theme of competing paradigms, he again summarizes Solow's position as an exemplar of weak sustainability and contrasts it with that of economist Herman Daly, who defends the view that there are limits to the extent that technology, knowledge and other forms of manufactured or human capital can substitute for natural capital—a concept that is increasingly tied to planetary ecosystems that regulate climate, and supply clean air or water. Norton characterizes Daly's defense of strong sustainability as the claim that humanity must continue to protect more than some generalized notion of capital, it must protect "stuff", that is, at least some natural ecosystems and ecosystem processes.

It should be noted that the dichotomy between weak and strong sustainability leaves room for many intermediate positions. For example, one can admit that advances in technology or social cooperation could, for example, reduce the amount of water wasted by replacing inefficient irrigation systems, on the one hand, or by convincing people to take shorter showers, on the other. One might also admit that getting to a situation where these efficiencies and social practices are possible will require expenditure of natural capital over the short run. Yet such an admission need not also compel one to permit any and all declines of natural capital on the grounds that some unimagined new future technology will permit our children's grandchildren to manufacture water out of thin air. Norton's new book recapitulates the strong versus weak sustainability debate by explicitly advocating for this kind of middle-ground position. While in *Sustainability,* Norton's commitment to democratic process at times appears to override his commitment to strong sustainability, in *Sustainable Values, Sustainable Change* he claims that maintaining some forms of natural capital is an absolute necessity. The problem lies in knowing *which* natural capital must be preserved: Is it natural habitats, the systems that render ecosystem services for clean air and water, climate regulation, or even some stock of minerals? Consistent with the position he has argued since *Toward Unity,* Norton denies that putatively "value-free" environmental sciences can answer this question. Some process for the elicitation and construction of a community value consensus is essential (Norton 2015).

Sustainable Values, Sustainable Change revisits the philosophy of adaptive ecosystem management within the framework of a political debate that is being conducted at scales ranging from the very local micro-politics of neighborhood associations up to that of international organizations that attempt to govern human-environment interactions at a planetary scale. In this book Norton recognizes that the idea of sustainability is the umbrella under which these intense political debates are presently occurring. He sees his own adaptive ecosystem management approach as a way for communities to accommodate the various factors that are forcing them to change their institutions as well as their practices of production, distribution and consumption of resources. The strategy of adaptive management stresses the value-laden (one might better say "value-saturated") nature of resource conflicts, while proposing a way to conceptualize resource management in a manner that is sensitive to existing community values. Norton usefully contrasts his approach with expert-driven management styles that pivot around some conception of optimization: allocating resources to their "most valued" use (including, of course, the flow of provisioning, regulatory, supportive and cultural services). The optimizing approach relies upon environmental science (including environmental economics) both to characterize ecosystem services and to assess the relative value of these services, (Norton 2015).

In effect, then, Norton's current conception of sustainability is an expansion and replacement of the "convergence hypothesis" advanced in his 1991 book, *Towards Unity among Environmentalists.* Norton was there reacting to the way that anthropocentrically justified environmental policies were being attacked on philosophical grounds by those who advocated direct moral status for non-human animals, in some cases for all living organisms (biocentrism), or for seemingly abstract entities such as ecosystems or species (ecocentrism). Norton argued that as one expands one's horizons both spatially and temporally, the implications of these alternative philosophical positions converge on the same prescriptions. Thus as the scope of the anthropocentrist's concern comes to recognize humanity's dependence on global ecosystems and extends to future generations, defense of non-humans, conservation of species, and protection of ecosystems are easily defended on anthropocentric grounds. In his shift to sustainability, Norton has maintained his interest in unifying diverse and seemingly antagonistic philosophical commitments around a common set of social goals and policy prescriptions (Norton 2009). Much of the effort in *Sustainable Values, Sustainable Change* is devoted to developing a decision-oriented account of how consensus on action-oriented policies can be achieved.

One might trace the trajectory of Norton's current position as originating with his pragmatic approach to the question "Why preserve natural variety?" Norton's work in the 1980s showed why instrumental arguments for the conservation of biodiversity must acknowledge an ethical commitment to the needs of future generations. Once these arguments are examined in their details, one can see how narrowly technical and putatively "science-based" forms of argumentation actually presuppose a community-based commitment to the claim that preserving biodiversity follows from duties to future generations. Norton also argued that there were no substantive policy differences between those who defended the preservation of habitats and natural variety as a duty to future generations and those who asserted a human-independent intrinsic value for these goods (Norton 1987). This work was followed by an unsuccessful attempt to convince other environmental philosophers to adopt a pragmatic orientation (e.g. *Toward Unity*), with Norton then moving on to a version of adaptive management that might be generally applied to any problem in environmental policy. Although 'sustainability' might have been thought of as a technical term in 2005, by 2015 Norton is certainly correct to point out that it has a general use that somewhat vaguely indicates the goals being pursued in environmental politics.

In sum, the connections between pragmatism and sustainability in Norton's work are primarily associated with his use of the word 'sustainability' as the name for his general environmental philosophy. That philosophy is indisputably a particular form of environmental pragmatism, while environmental pragmatism can be seen as a contemporary school of thought being advanced by Andrew Light, Anthony Weston, Ben Minteer, Jennifer Welchman and others, including myself. Norton is, I think, by far the most influential figure in this school. However, Norton's pragmatist commitments are not especially in evidence when he comes to specific discussions of the concept of sustainability or of ongoing debates over how the concept should be operationalized. Norton is content to accept the weak and strong framing as it has been developed within environmental and ecological economics, and in his later work has focused rather narrowly on the fungibility of various types of capital. While this is not necessarily inconsistent with pragmatist principles, it is a construal of the sustainability debate that could be accepted by people with no commitment to pragmatism as an approach to epistemology and social thought.

2.4 Sustainability: The Current Discourse

Stated in brief, my disagreement with Norton turns on my belief that his status as the most exemplary philosopher of sustainability notwithstanding, he pays too little heed to the intellectual history of sustainability as a concept. As Norton is certainly aware, sustainable yield has a long history, and has been well characterized for managing renewable resources in forestry and fisheries for at least a century. Ulrich Grober argues that the definitive formulation can be traced to Hans Carl von Carlowitz's text *Silvicultura Oeconomica* published in 1713, (Grober 2012). The precise historical origins notwithstanding, by the 1970s complaints about unsustainable practices were being voiced widely, and theorists were beginning to explore what sustainable alternatives would require in many policy domains. The weak versus strong sustainability debate represents a significant narrowing of theoretical perspective when compared to discussions that were taking place in conservation biology, operations research, agriculture and urban planning prior to Norton's entrée into the field. While redefining sustainability as an umbrella term for a social movement or general philosophical approach does indeed broaden the concept, it also sacrifices much that might have been gleaned from paying closer attention to the specific points that were being debated within these subfields of environmental science and policy. The dominance of Norton's influence within environmental philosophy circles has, in effect, skewed the discourse away from any engagement with this more complex intellectual tradition. I will argue that the philosophical discourse has not only been less rich than it might have been, but that it has also been less capable of productive engagement with other disciplines in environmental science.

It will be obvious to everyone that a plethora of environmental, economic and social objectives have been advocated under the banner of sustainability since the World Commission on Environment and Development (henceforth "the Brundtland commission") issued its report on sustainable development in 1987. My book *The Agrarian Vision: Sustainability and Environmental Ethics* offers a somewhat selective history of this discourse, emphasizing how "sustainable agriculture" had developed a constituency and a defined discourse well before 1987. Grober's *Sustainability: A Cultural History* can be read as supporting my analysis with a complementary emphasis on forestry, though Grober is less interested than I am in understanding the tensions within that discourse, (Grober 2012). Concisely stated, the debate within food systems was conducted between two contending perspectives. One understood sustainability in terms of whether planetary agricultural production would be sufficient to feed the global population, given projected trends for population growth. The other focused on agricultural ecosystems, on the one hand, and agriculture's key social institutions, especially the household farm, on the other. The concern was that patterns of industrialization within agriculture were threatening the metabolic and social mechanisms that reproduce this socio-ecological infrastructure over time (Thompson 2010).

Within the context of farm policy and agricultural science, the debate was significant because those who defined sustainability in terms of feeding the world were quite willing to accept mechanization, chemical use and yield increasing biological transformations on the grounds that they increased the total productive potential of global agriculture. Those who measured sustainability in terms of the resilience of agro-ecosystems and rural institutions saw industrialization as a threat. While the first view emphasized agriculture's ability to sustain a growing global population, the second questioned whether the systemic integration of agriculture itself could be sustained in the face of profit-driven changes that were threatening soil fertility and the livelihoods of farming communities. In *The Agrarian Vision,* I was trying to generalize this tension to other domains in which sustainability was being advanced as a guiding philosophy for environmental or social policy. I sharpen the contrast between one view of sustainability that emphasizes the availability of critical resources, and another view that is focused on the integrity and reproducibility of critical systems, (Thompson 2010).

Already by the mid-1990s this debate over sustainable agriculture had been institutionalized within the agricultural sciences. As research and education programs dedicated to sustainability within food systems (and journals with the word 'sustainability' in their title) began to appear it became clear that there were both philosophical and theoretical orientations that were shaping the way that scientists were conceptualizing sustainability. I believe that the conceptual outline of these orientations can be seen in quite a few fields, not just agricultural science. For these audiences, the competing paradigms approach is highly appropriate. These audiences are composed of experts who often work from the perspective of specific scientific disciplines, but who are increasingly sensitized to the need for an interdisciplinary approach. Nevertheless, being committed to interdisciplinary integration does not necessarily imply that one is also sensitive to the extent to which one's approach is limited by a specific paradigm, in the sense that Thomas Kuhn initially proposed (Kuhn 1962) and that Norton uses in his 1995 paper.

Norton abandoned the terminology of "competing paradigms" in his discussion of weak and strong sustainability for the 2005 book, and I think he was correct to have done so. I will first explain why the weak/strong sustainability debate does *not* reflect incompatible paradigms before explaining how competing paradigms do pervade the discourse on sustainability. We can begin by reminding ourselves that Solow is a leading theorist of the role that capital plays in economic growth, rather than a resource economist. The point of disagreement in the weak versus strong debate concerned Solow's willingness to condone growth of Gross Domestic Product (GDP) as the leading indicator of economic development and his assumption that the various forms of capital are fully fungible. Economists differ among themselves about the latter point, and their debates pick apart distinctions that can be drawn between human, social and financial capital, and the durable capital that is sometimes referred to as "the means of production." These disagreements typify normal science, not incompatible paradigms. Daly was contesting both the emphasis on GDP *and* the assumption that human, social or financial capital could substitute perfectly for durable capital, especially when productive capacity depends upon a natural (as opposed to manufactured) endowment. But *everyone* was talking about economic development in this debate. It is clear today that the multitude of goals and practices now sought under the aegis of sustainability cannot be reduced to any measure of economic growth. As such, Norton's return to the weak/strong interpretation of "the sustainability debate" in *Sustainable Values, Sustainable Change* is actually a problem.

Although Solow and Daly probably shared a common paradigm,[2] I *do* think that many environmental scientists (including some economists) who frame their work in terms of sustainability suffer from an inability to understand one another that can be diagnosed in terms of competing paradigms. How might we generalize the diverging paradigmatic commitments that separate scientifically grounded individuals who are developing research and education programs in sustainability? In *The Agrarian Vision,* I introduced the terms "resource sufficiency" and "functional integrity" to characterize the two conceptually divergent approaches. Here I will shift to words that have more consonance with ordinary discourse. Clearly, many scholars continue to be very strongly influenced by the 1987 Brundtland report. They may not realize the extent to which the theoretical underpinnings of this approach depended on prior theories of economic development. I will refer to this as "the Brundtland paradigm." Other scholars have flirted with the Brundtland ideas, but they have increasingly become aware that their systems orientation is poorly supported by the core ideas of this approach. Using language that was introduced into the sustainability literature by Ludwig et al. (1997), this group of scholars has increasingly stressed the idea of resilience. Indeed, the Resilience Institute has become an important spokesperson for this way of approaching sustainability, and I refer to their approach as "the resilience paradigm."

2.5 The Brundtland Paradigm

As Norton (2015, p. 28) and the Resilience Institute group (Biermann et al. 2012) point out, many of the key ideas in the Brundtland report surfaced at a 1972 conference in Stockholm. What these accounts do not emphasize was the fact that the Brundtland commission was *not* focused on sustainability as such, but was dealing primarily with long recognized issues of inequality in global development. Robert Solow also elided the direct reference to development in his influential essay "An Almost Practical Step toward Sustainability," but as Solow was one of the world's leading authorities on economic growth, his audience was justified in assuming that he meant sustainable economic development (Solow 1992). What was obvious to the specialized audience of economists but unclear to many was that

[2]Daly's broader work is sometimes described as foundation for ecological economics, which may indeed reflect a revolution in economic thought. I think one would be in error to interpret his remarks on capital in this light.

however one evaluates the sustainability of the development process, economic development itself has traditionally been measured in terms of year-to-year increases in GDP. Although large sections of Daly and Cobb's *For the Common Good,* are devoted to the critique of GDP, this aspect of Daly's critique would not be clear to anyone who was focused solely on the fungibility of various types of capital. As such a brief historical exegesis may prove helpful.

When Franklin Delano Roosevelt became President in 1933 the United States was in the throes of the Great Depression. At the request of Roosevelt's Secretary of Commerce Daniel Roper, Simon Kuznets developed a new approach to estimating the health of the national economy that divided it into sectors: mining, manufacturing, agriculture and the like. In the method eventually deployed by the Department of Commerce, each sector's contribution to the overall national economy is taken to be the sum of value added by production in that sector. This sum can be understood as a "stock" that is determined by inflows and outflows. It is increased by the sale of goods from that sector, but decreased by the sector's purchase of materials (including labor) needed for the production process. When each sector is represented in a spreadsheet, these relationships among various sectors yield a system of equations that can be used to calculate the net sum of value added by productive activity to the economy as a whole. The stunning result is a number (in fact *the* number) that is now widely taken to serve as an overall indicator of economic health, e.g. GDP. Kuznets himself was somewhat critical of the way that national accounts were being calculated because he felt that the spreadsheet approach ignored crucial philosophical questions about *which* kinds of economic activity actually contribute to human well-being (Kapuria-Foreman and Perlman 1995). The concerns that Kuznets voiced in the 1930s are in fact many of the same ones that Daly emphasizes in his collaboration with theologian John Cobb (Daly and Cobb 1989).

In 1962 Harold J. Barnett and Chandler Morse followed Kuznet's general approach in the book that arguably launched the field of environmental economics. Barnett and Morse examined how scarcity in natural resources would affect economic growth by estimating the impact that declining availability of raw materials would have on inflows and outflows for each sector of the economy. Although the initial expectation was that as nonrenewable resources were depleted, they would become more expensive with subsequent negative impacts on productivity adumbrating through the economy, Barnett and Morse found that the higher cost of increasingly scarce resources triggered compensating substitutions that significantly dampened the overall impact of resource scarcity on GDP. In fact, the most serious impacts were from defensive expenditures needed to deal with pollution and declines in potentially renewable resources like fertile soil, clean water or (as we now know) a stable climate. Since these goods are not traded in any conventional sense, damage to them does not readily appear in the form of increased costs that would, in turn, trigger change in measures that account for scarcity. This way of thinking lies at the root of policies that identify ways for the costs to be more effectively reflected in the market prices being paid by those who use (or misuse) natural resources. The thought that people will economize on scarce or costly inputs is the driving idea behind "market approaches" to increasing the sustainability of productive activities (Barnett and Morse 1962).

More generally, however, any approach based on GDP is inherently tied to an accounting mentality, and the shift to more comprehensive or sensitive indicators does not alter this basic orientation. A practice or process will be sustainable to the extent that the resources needed to carry it out are foreseeably available. Barnett and Morse recognized that for a process as broadly defined as "economic growth", the ability to continue it indefinitely can be achieved in many ways: increasing the efficiency of a production process, or shifting from relatively scarce to relatively plentiful resources. Furthermore, while the roots of this approach to measuring stocks and flows are in economics, it can be applied to other goods, as well. By the 1970s, matrix models for predator-prey relationships and for energy flows in ecosystems were using the same general quantitative approach that Kuznets had applied in developing national income accounts. As Kuznets had modeled the economy through a complex linkage of stocks and flows, ecologists such as Bernard Patten were modeling biodiversity and ecosystem health in very similar ways (Patten 1971). If it is the number of organisms (or species, or trophic flows) existing in an ecosystem that matter for conservation, accounting for the key stocks and flows in a manner much like Kuznets devised for a national economy can be a quantitatively sophisticated way to understand how well an ecosystem is faring. This is what I have called a *resource sufficiency* understanding of sustainability (Thompson 2007).

There are, of course, many things to debate within this paradigmatic way of approaching sustainability. One might start with Kuznets' warning about GDP. There are many bad things that contribute to economic growth. When people pay doctor and hospital bills for treatment of diabetes or heart disease, it is a positive flow into the healthcare sector of the economy. The profits of oil companies boost GDP, even while combustion of their product pollutes the atmosphere. When we make defensive expenditures to develop new weapons systems they register as contributions to the economy in numerous sectors: research and development, education and manufacturing. Even paying to clean up toxic spills makes a positive contribution to GDP. Yet it is far from clear that we want to think of all these types of economic activity as making our society more sustainable. As already noted, the critique of GDP was a centerpiece in Herman Daly's writings on sustainability, and the additional fact that forms of natural capital were not being priced in markets was a further basis for his concern about Solow's weak sustainability (Daly and Cobb 1989; Daly 2005). Yet it is crucial to see that these debates take place *within* the accounting approach implied by the Brundtland paradigm. They are debates about what to count and how to count it. They do not challenge the basic idea that meeting the needs of the present (without compromising future generation's ability to meet their own needs) is primarily a problem of whether the resources (including capital) needed to fuel a development process were foreseeably available. Daly thought that Solow was too sanguine about the future availability of natural capital, and that GDP was quite misleading about what to count, in the first place. Still, sustainability is primarily about ensuring that the resources needed to produce (or reproduce) key goods continue to be available. Daly and Solow both construe sustainability largely in terms of maintaining the inflow for whatever stock (capital, biodiversity) has been chosen as the target for sustainable management practice. People who share a commitment to this overall understanding of sustainability as such may nonetheless disagree sharply about *which* stocks really matter: is it economic activity, or is it biodiversity? What they agree on is that the complex systems models that are used to calculate stocks and flows are tools for getting a better grip on how much of the key stock (however they conceptualize *that*) will be available.

2.6 The Resilience Paradigm

But systems thinking can also yield an entirely different kind of understanding. The same year as the Stockholm conference alluded to above, *The Limits to Growth* (1972) was released with lead authorship attributed to Donella Meadows. To many mainstream economists, *Limits to Growth* looked rather much like an update for Barnett and Morse. It provided a complex systems model linking sectors of the industrial economy to various processes of resource exhaustion and damage to ecosystem services. Even its title suggested a similar orientation: identifying how increasing scarcity of inputs to various production processes would eventually constrain the quarter to quarter growth in GDP that had come to symbolize a healthy economy. This was in fact the approach taken by William Nordhaus in a critique of the *Limits to Growth* model that was written just as the weak and strong sustainability debate was heating up (Nordhaus 1992).

But as Meadows made clear in later publications, the significance of the systems modeling done for this project lay not in the absolute numbers generated for future economic activity (or for stocks of natural resources such as fossil fuels or biodiversity). Rather, the point was to illustrate how the feedback relationships that were governing stock and flow relationships were themselves unstable. The point of modeling was not to predict a shortage of resources that would constrain the growth process; it was to identify vulnerabilities in system organization. Whether changes in system functioning took the form of declines in top predators or distortion of economic incentives, the stocks and flows crucial to economic growth would inevitably lead to unstable oscillations, to overshoot (as rates of consumption or population growth are stimulated by temporary feedbacks) and finally to total system collapse. While it might not be clear *when* this would happen, it was the overall structure of the system that was at the root of humanity's predicament. The system itself was unsustainable, (Meadows 2008).

The Limits to Growth relied upon an exceedingly complicated systems model, but elsewhere ecologists were building on the modeling techniques being promulgated by Patten to develop a repertoire of systemic interactions, some of which produce surprising patterns of behavior, while others lead to catastrophic failure and system collapse. Ecological models exploring the relationship between stocking rates for livestock, grasses and woody plants showed that for some configurations of these variables, a savanna ecosystem could potentially return to previous levels of biological productivity even after periods of drought or flooding (e.g. resilience), while other configurations would lead to stable ecological states with lower overall productivity (Walker et al. 1981; Ludwig et al. 1997). Other models including human decision making showed that decision makers using highly simplified mental models could outperform decision makers attempting to optimize performance, even when equipped with total knowledge of ecosystem functions (Grant and Thompson 1997). In this kind of system modeling, the point was to identify structural features that affect system function, especially when they can lead to collapse.

It may be easier to illustrate the significance of system integrity in engineering applications or the built environment. A robotic control assembly, an information processor or the infrastructure for electrical power distribution need to function under the actual environmental conditions in which they operate. A mechanical assembly that vibrates as it moves can generate oscillating feedback that eventually shakes the entire apparatus apart. People of a certain age may recall information processors that crashed whenever they encountered improperly formatted data. An electrical grid that can be taken down by a simple thunderstorm is too vulnerable to external perturbations. In the first two cases, the system is susceptible to collapse from an internally generated feedback, while in the latter case it is exposed to externally imposed disruption. In either instance, the system in question is *fragile,* and an engineer must take such weaknesses into account in the design process. In no case is a depletion of inputs or resources the underlying source of system failure.

The contrasting notion of sustainability that Meadows and her coauthors were trying to get across is analogous to this source of fragility in designed or engineered systems. In short, engineered systems must be *robust:* they must resist both internal and external sources of failure. But robustness is not everything. Contemporary information processing systems have been designed to recover from minor errors—an instruction to search for some missing bit of code, for example—and electric grids are designed to reset and restore power quickly when a lightning bolt or a falling limb disables a power line or substation. Such systems are designed to be *resilient.* They recover essential functions after momentary or episodic perturbation. Adaptive capacity (important for Norton) is a further feature that allows for continuous modifications of system structure in response to changes in the system's operating environment. For engineered systems, adaptive capacity is generally the responsibility of human managers, while the beauty of many natural systems is their inherent capacity for evolutionary change.

It is useful to relate robustness, resilience and adaptive capacity to engineered systems because it is easy to see how these system-level goals are distinct from whatever goods the systems were intended to deliver in the first place: a manufactured good in the case of a robotic assembly system, a control procedure in the case of an information processor, or electric power in the case of the grid. The goods produced or delivered by such engineered systems are the desiderata that led engineers to design and build them in the first place, but once in place a new set of systemic objectives emerge reflecting the integrity with which system components have been assembled and interact with the system's external environment. A hospital with a gasoline-powered generator may be able to sustain patient services even if the power grid goes down and stays down (e.g. is not resilient). The hospital is sustainable in a resource sufficiency sense because it can secure the flow of power resources it needs to deliver services through switching to an alternative source. But the grid itself has clearly failed to be sustainable in a functional integrity sense.

But what if we were to think of the hospital itself in systems terms? Here, we would include both its connections to the power grid and its back-up generator as system components. We might then say that as a system, the hospital has been made more sustainable (more resilient) because it has the capacity to restore functions even when one system component (the power it typically gets from the utility company) is disrupted. Notice that in thinking of the hospital this way, we are no longer talking about the underlying capacity of the hospital to deliver health services. It is not as if switching to back-up power gives the hospital more resources to produce healthcare (it is, in fact, probably less capable). But including a back-up generator in the hospital's overall system design makes it more sustainable in the sense that it is more robust, resilient and has greater adaptive capacity. When viewed as a system, the hospital has greater ability to maintain its functional integrity. The turn toward thinking of buildings and infrastructure in such terms has been characterized as "regenerative thinking," (Svec et al. 2012).

The hospital example shows how functional integrity becomes apparent through a perspectival shift—a shift of paradigm. It is often possible to take a systems perspective on an object or set of practices that might more typically be conceptualized purely in terms of the particular goods and services that the object or practices produce. In making this shift, one can articulate features of the system that are intrinsically valuable for the system's integrity and ability to function, but that may be less apparent when one is viewing the object or practice strictly in terms of its instrumental value for producing certain goods. This perspectival dimension is characteristic of a paradigm shift. As the Copernican astronomer looks out at sundown and sees the horizon moving up, a systems perspective shifts the attention from specific goods being delivered (healthcare, in the case of our hospital) to those systemic features that confer sustainability in the sense of resistance to internal or external threats that can cause the system itself to collapse.

2.7 Conclusion

Given Norton's longstanding commitment to adaptive management, one might think that his approach to sustainability would align with the resilience paradigm. Yet in sticking to the debate over fungibility of capitals even through his most recent work, he has implicitly committed to an approach that is grounded in the accounting mentality of economic development. One might more charitably read Norton's commitment to adaptive management as a strategy for adding adaptive capabilities to socio-ecological systems in order to enhance their ability to resist threats (their robustness) and their ability to bounce back after perturbation (resilience). If only Norton himself could provide more support for such a reading! Indeed, in a longer treatment of these issues, I would stress how adaptive capacity takes the resilience paradigm into a set of values-based issues that it sorely needs to confront, and I hope that Norton would be an ally. Today, authors are suggesting that it is now time to "get beyond" sustainability by focusing on resilience (Benson and Craig 2014; Hoffman 2014; Manyena and Gordon 2015). The notion of Kuhnian paradigms is relevant to these suggestions, and I believe it would be more helpful for theoretically grounded scholars to view resilience as part of a more general approach to sustainability, rather than as an alternative. It is not possible for me to provide a detailed argument for this view in this context, but I will close with a few of the reasons why I think so.

First, historically oriented reviews of the discourse around sustainability support the claim that functional integrity has indeed been at the forefront of sustainability thinking since the beginning. My own work on sustainable agriculture documents the way that soil scientists and sociologists each focused on the fragility of food systems that that were incorporating industrial inputs, as opposed to questioning whether a continuous flow of these inputs could be sustained (Thompson 2010). Debates over sustained yield in fisheries managements also exhibited critiques that emphasize system integrity over future flows of edible protein, (Ludwig et al. 1993; Costanza and Patten 1995). Grober's study of timber resources shows that at least some foresters also saw sustainable yield primarily as an approach to maintaining the forest's ability to reproduce itself. The availability and demand for wood was a secondary consideration—though often foremost in the minds of those who *employed* foresters (Grober 2012). It is, in short, simply *inaccurate* to associate the concept of sustainability as such so strongly with the resource sufficiency orientation that became dominant after the publication of the Brundtland report.

It is nonetheless true that someone whose focus is on maintaining a foreseeable flow of resources will take an interest in the systems that generate flows of timber or fish. It is not as if those who conceptualize sustainability in terms of resource sufficiency are ignorant of the need to maintain the systems that produce these stocks. There is, then, a connection between the functional integrity of natural systems that produce key resources and natural capital. Someone who is thinking about sustainability in terms of resource flows will thus *translate* the concerns about functional integrity into the concepts that drive a resource sufficiency approach. But if there are truly paradigms at work here, something will be lost in the translation. What could that be? Arguably, it is a normative commitment to the integrity, stability and beauty of the system itself. The resonance to Leopoldian language here is suggestive, and what it suggests is that probing the intransitivity of translation in this particular case might have something to do with environmental ethics.

But there are other and more practical reasons to hold that we should not simply cast sustainability aside in favor of resilience. One problem with the new calls for resilience [something seen even in the approach of the Resilience Institute itself (Rockström et al. 2009)] is a tendency toward highly technical approaches to the concept of resiliency. Mathematical models play a large role in revealing the complexity of resilient or adaptive natural systems. It is easy to dissociate the activity of modeling these systems from the value commitments that come to the fore in a resource sufficiency approach to sustainability. The system models and the limits derived from them are too often reminiscent of the positivist conceptualization of scientific knowledge that Norton has so ably critiqued in the Appendix to *Sustainability*. The Brundtland Commission's commitment to distributive justice both between developed and less developed nations and between present and future generations is an element of the sustainability discourse that is not well captured by our fascination with integrity and stability of beautifully complex natural systems. An emphasis on making systems adaptive might also correct the tendency to argue as if resilience were good in and of itself, as if many of our most resilient social institutions were not deeply problematic. Structural racism would be an example.

Norton's larger environmental philosophy provides many resources for addressing this list of problems, yet when Norton turns directly to a discussion of sustainability as such, his approach seems to operate *within* the Brundtland paradigm. Although adaptive management is focused on political and managerial processes in developed industrial states, it is consistent with an approach to sustainability that simply adds distributive justice criteria to those strategies that try to extend the availability of resources by increasing the efficiency with which they are utilized. We see this in Norton's response to key issues faced by the poor. While people in extreme poverty may consume fewer resources than those who enjoy a higher standard of living, they also have far less ability to adjust their activity under changed conditions. This holds whether change is broadly positive or negative. When technological advances do become available, the poor have a limited ability to take advantage of them, hence damaging practices persist. When conditions take a turn for the worse, as is currently taking place due to climate change, the poor shoulder a disproportionate share of the harm. Thus the philosophical position Norton advocates condones temporary use of natural capital (such as fossil fuels) in the hope that it will limit present-day overfishing, scavenging for food in forests, and the destruction of ecosystems. Yet at the same time, Norton recognizes that continued sustainability will require the development of less destructive forms of total consumption (including energy) in the future (Norton 2005). In short, although Norton's contribution has been to argue for a procedural understanding of sustainability as adaptive management, he does not contest the underlying paradigm put forward by the Brundtland Commission in 1987.

I hasten to add that in calling for an appreciation of the values being celebrated by a resilience perspective I am not arguing against distributive justice. The Brundtland commission's ethical commitment to equalizing opportunity for those in the global South is one of its strengths. Yet at the same time debates over the tenor of Western efforts to promote economic development have only become more controversial since 1987. Some critiques of Western style development stress systemic linkages between traditional cultures and the integrity of local ecosystems (Rozzi 2012). An exploration of the connection (or lack thereof) between such critiques of the development ideal and the resilience paradigm is beyond the scope of this chapter. Yet the possibility that there may be limitations to the conceptualization of justice that Norton shares with the Brundtland commission is all the more reason to consider the possibility that paradigmatic contrasts in theoretical approaches to sustainability are worth pursuing.

Finally, in calling for environmental philosophers to consider how competing paradigms have shaped the discourse of sustainability, I am not also proposing that one of these paradigms must necessarily win out over the other. Kuhn's use of the debate over geocentric and heliocentric models of the solar system (see Kuhn 1957) was not intended to imply that when paradigms compete one of them must be "right" while the other is "wrong." As my comments on the importance of maintaining the Brundtland perspective on distributive justice suggest, I believe that there are strengths and weaknesses to each of these paradigms. Rather than one of them displacing the other, a more positive development would be for our discourse on sustainability to develop the capacity to accommodate both resource sufficiency *and* functional integrity within the ways that we conceptualize sustainability. For all that Norton has done for us, his reluctance to acknowledge and challenge the way that both weak and strong sustainability have been framed within the accounting mentality of GDP and economic development is a lacuna that limits the contributions environmental philosophy might make to a more sustainable future.

To conclude, in reinforcing the dominance of the weak versus strong sustainability debate, Norton has failed to bring the underlying conceptual commitments of this accounting mentality to light. Although Kuznet's brilliant conceptualization of national accounts involves thinking about the national economy as a system of interlinked stocks and flows, there is an important sense in which the legacy of GDP has been to reinforce economic growth and a budgeting mind-set when scholars think about sustainability. For a philosopher who challenges environmentalists and environmental scientists to be responsive to a broader range of value perspectives, Norton's failure in this respect is doubly problematic. Ironically, it is the development mindset that encourages the laser-beam focus on efficiencies and optimization that Norton critiques so ably in *Sustainable Values, Sustainable Change.* If sustainability is primarily about ensuring that the "stuff" we need to produce even *more* stuff is available, then anything that improves the efficiency of the processes we use to transform stuff into more stuff is going to be a potential advance in sustainability. Whether or not various kinds of stuff are fungible for the larger processes of economic growth is not really material to the need broadening of this underlying mindset. We can, I think, see our tendency to think and talk in this way as a vulnerability—a source of fragility—in that system of ordinary language that is our environment.

References

Barnett, H.J., and C. Morse. 1962. *Scarcity and growth: The economics of natural resource availability*. Baltimore, MD: Johns Hopkins Press.

Benson, M.H., and R.K. Craig. 2014. The end of sustainability. *Society and Natural Resources* 27: 777–782.

Biermann, F.K., A.S. Andresen, K. Bäckstrand, S. Bernstein, M.M. Betsill, H. Bulkeley, B. Cashore, J. Clapp, C. Folke, A. Gupta, J. Gupta, P.N. Haas, A. Jordan, N. Kanie, T. Kluvánková-Oravská, L. Lebel, D. Liverman, J. Meadowcroft, R.B. Mitchell, P. Newell, S. Oberthür, L. Olsson, P. Pattberg, R. Sánchez-Rodríguez, H. Schroeder, A. Underdal, S. Camargo Vieira, C. Vogel, O.R. Young, A. Brock, and R. Zondervan. 2012. Navigating the anthropocene: Improving Earth system governance. *Science* 335: 1306–1307.

Costanza, R., and B.C. Patten. 1995. Defining and predicting sustainability. *Ecological Economics* 15: 193–196.

Daly, H.E. 2005. Sustainable development—definitions, principles, policies. In *The future of sustainability*, ed. M. Keiner, 39–53. Dordrecht, NL: Springer.

Daly, H.E., and J.B. Cobb Jr. 1989. *For the common good: Redirecting the economy toward community, the environment, and a sustainable future*. Boston: Beacon Press.

Grant, W.E., and P.B. Thompson. 1997. Integrated ecological models: Simulation of socio-cultural constraints on ecological dynamics. *Ecological Modeling* 100: 43–59.

Grober, U. 2012. *Sustainability: A cultural history*. Trans. R. Cunningham. Dover, UK: Green Publishing.

Hoffman, L.M. 2014. From sustainability to resilience: Why locality matters. *Research in Urban Sustainability* 14: 341–357.

Kapuria-Foreman, V., and M. Perlman. 1995. An economic historian's economist: Remembering Simon Kuznets. *The Economic Journal* 105: 1524–1547.

Kuhn, T.S. 1957. *The Copernican revolution; Planetary astronomy in the development of Western thought*. Cambridge, MA: Harvard University Press.

Kuhn, T.S. 1962. *The structure of scientific revolutions*. Chicago, IL: University of Chicago Press.

Ludwig, D., B. Walker, and C.S. Holling. 1997. Sustainability, stability, and resilience. *Conservation Ecology* 1: 7.

Ludwig, D., R. Hilborn, and C. Walters. 1993. Uncertainty, resource exploitation, and conservation: Lessons from history. *Ecological Applications* 3 (4): 547–549.

Manyena, S.B., and S. Gordon. 2015. Bridging the concepts of resilience, fragility and stabilisation. *Disaster Prevention and Management* 24: 38–52.

Meadows, D.H. 2008. *Thinking in systems: A primer*, ed. D. Wright. White River Junction, VT: Chelsea Green Publishing.

Misak, C. 2016. *Cambridge pragmatism: From Peirce and James to Ramsey and Wittgenstein*. New York, NY: Oxford University Press.

Nordhaus, W.D. 1992. Lethal model 2: Limits to growth revisited. *Brookings Papers on Economic Activity* 2: 1–59.

Norton, B.G. 1987. *Why preserve natural variety?* Princeton, NJ: Princeton University Press.

Norton, B.G. 1992. *Toward unity among environmentalists*. New York: Oxford University Press.

Norton, B.G. 1995. Evaluating ecosystem states: Two competing paradigms. *Ecological Economics* 14: 113–127.

Norton, B.G. 2003. *Searching for sustainability: Interdisciplinary essays in the philosophy of conservation biology*. New York: Cambridge University Press.

Norton, B.G. 2005. *Sustainability: A philosophy of adaptive ecosystem management*. Chicago, IL: University of Chicago Press.

Norton, B.G. 2007. A reply to my critics. *Journal of Agricultural and Environmental Ethics* 20: 387–405.

Norton, B.G. 2009. Convergence and Contextualism: Some clarifications and a reply to Steverson. In *Nature in common? Environmental ethics and the contested foundations of environmental policy*, ed. B. Minteer, 36–48. Philadelphia: Temple University Press.

Norton, B.G. 2015. *Sustainable values, sustainable change*. Chicago, IL: University of Chicago Press.

Patten, B.C. 1971. *Systems analysis and simulation in ecology*. New York: Academic Press.

Rockström, J., W. Steffen, K. Noone, Å. Persson, F.S. Chapin III, E.F. Lambin, T.M. Lenton, M. Scheffer, C. Folke, H.J. Schellnhuber, B. Nykvist, C.A. de Wit, T. Hughes, S. van der Leeuw, H. Rodhe, S. Sörlin, P.K. Snyder, R. Costanza, U. Svedin, M. Falkenmark, L. Karlberg, R.W. Corell, V.J. Fabry, J. Hansen, B.H. Walker, D. Liverman, K. Richardson, P. Crutzen, and J.A. Foley. 2009. A safe operating space for humanity. *Nature* 461: 472–475.

Rozzi, R. 2012. Biocultural ethics: From biocultural homogenization toward biocultural conservation. In *Linking ecology and ethics for a changing World*, ed. R. Rozzi, S.T.A. Pickett, C. Plamer, J.J. Armesto, and J.B. Callicott, 9–32. Dordrecht, NL: Springer.

Solow, R.M. 1992. *An almost practical step toward sustainability*. Washington, DC: Resources for the future.

Svec, P., R. Berkebile, and J.A. Todd. 2012. REGEN: Toward a tool for regenerative thinking. *Building Research & Information* 40 (1): 81–94.

Thompson, P.B. 2007. Norton's sustainability: some comments on risk and sustainability. *Journal of Agricultural and Environmental Ethics* 20: 375–386.

Thompson, P.B. 2010. *The agrarian vision: Sustainability and environmental ethics*. Lexington: University Press of Kentucky.

Walker, B.H., D. Ludwig, C.S. Holling, and R.M. Peterman. 1981. Stability of semi-arid savanna grazing systems. *The Journal of Ecology* 69: 473–498.

Paul B. Thompson holds the W. K. Kellogg Chair in Agricultural, Food and Community Ethics. His book *The Spirit of the Soil: Agriculture and Environmental Ethics* was released in a revised 2nd edition from Routledge in 2017. Thompson's research on sustainability has been supported by the U.S. National Institute for Food and Agriculture (NIFA) Project # MICL02324: Theory Development for Sustainability in Agriculture and Natural Resource Management.

Chapter 3
Ecological Sustainability

J. Baird Callicott

Abstract **Sustainability** is simply a property of an activity or complex system of activities capable of going on indefinitely. The concept of sustainability, however, is most often implicitly associated with human economic activities or complex systems of human economic activities that are constrained by *environmental* limits. From the perspective of current environmental concerns, the essentially economic concepts of **sustainable yield** and **sustainable development** are flawed. The former is strongly anthropocentric and has historically ignored ecological (and mathematical) complexity. The latter, while based on weak anthropocentrism, is also based on **weak sustainability**—artificial capital may be substituted for depleted natural capital; and thus the only things the present generation need bequeath to future generations are wealth and an entrepreneurial culture. Norton rejects weak sustainability, but also **strong sustainability**—the view that artificial capital cannot substitute for natural capital—championed by ecological economists and conservation biologists. Strong sustainability, according to Norton, is about "stuff," while sustainable societies commit themselves to "to perpetuate place-based values and project them into the future." Place-based values, however, are not always environmentally sound—think, for example, of coal-country values or desert-casino-country values. Alternatively, **ecological sustainability** conceives the human economy to be (as in fact it is) a subset of the economy of nature. The economy of nature is powered by solar energy and efficiently re-cycles materials. A sustainable human economy would thus pattern itself as microcosm to macrocosm on the economy of nature, lest growth of the former seriously disrupt the latter.

Keywords Sustainability · Sustainable yield · Sustainable development
Weak sustainability · Strong sustainability · Ecological sustainability
Economy of nature

J. B. Callicott (✉)
University Distinguished Research Professor Emeritus,
University of North Texas, Denton 76203-5017, USA
e-mail: johnbaird.callicott@unt.edu

S. Sarkar and B. A. Minteer (eds.), *A Sustainable Philosophy—The Work of Bryan Norton*, The International Library of Environmental, Agricultural and Food Ethics 26, https://doi.org/10.1007/978-3-319-92597-4_3

3.1 A Personal Introduction

Bryan Norton and I have been acquainted since the early-to-mid 1980s, just as environmental ethics was coalescing as a field in philosophy. We first met at a conference, the date and venue of which I do not recall. Soon thereafter, Bryan graciously invited me to participate in a project on which he was embarking that was focused on the preservation of species. That project too included a conference, the papers from which were published in a volume that he edited (Norton 1986). So Bryan and I go way back. Over the years our philosophical disagreements have become legend in the field—he an anthropocentrist, albeit a "weak" one, I an ecocentrist—and while it was never publicly stated until now, I confess that my ecocentrism is "weak" as well.

On at least two occasions our philosophical arguments became arguments in the other sense of the word—red-faced and shouting arguments. Ben Minteer, as I recall, was present on one such occasion—a dispute over the intrinsic versus the instrumental value of nature—in a restaurant in Burlington, Vermont. (As a last resort, Norton invoked Kant in defense of anthropocentrism.) The other was in the living room of Gene Hargrove's house in Athens, Georgia. That one was a dispute over whether Aldo Leopold's brief quotation from A. T. Hadley, who had jumped on the Pragmatist bandwagon, was ironic (as anyone with an ear for irony could tell) or approving (Leopold 1979).

Our weak–centrisms have pretty much converged (a little inside-baseball joke here) theoretically as well as pragmatically, but concerning that Hadley quote, neither of us has budged an inch. Somewhere Norton even wrote that he believed that Hadley, the President of Yale during Leopold's Yale-Forest-School student days, had given Leopold his book, *Some Influences in Modern Philosophical Thought*, in manuscript (Hadley 1913). "Yeah," I said to myself upon reading that, "I guess he emailed it to him. Oh wait, no; there was no email in 1909 (the year Leopold graduated). So I guess Hadley must have dropped off a Xerox copy for Aldo in his dorm room." (Can you hear the irony in the previous sentence?). But surely the fact that the titles of both Hadley's book and the paper in which Leopold quotes Hadley begin with the word "Some" should settle that dispute in Norton's favor (Ditto).

But through it all, Bryan and I have remained the best of friends. I love Bryan. He has a big heart and a generous disposition. Now when I see him—I stayed over night with Bryan in Atlanta last year on my way from Memphis to Charleston to visit my son and grandson—it's like family gatherings at Thanksgiving. You steer clear of politics and religion. Me and Bryan, we steer clear of Leopold, Hadley and value theory in environmental ethics. Thus no arguments in either sense of the word ensue (or erupt). We confine our discourse on the bouillabaisse, the pinot grigio, and the Atlanta Braves, just like at Thanksgiving you confine your comments to the turkey, the dressing, and the football.

Here I critique Norton's understanding of *sustainability*, a concept to which he has devoted much recent work. I contextualize my critique by first examining two classic conceptual permutations of the naked concept of sustainability: *sustainable yield* and *sustainable development*. *Sustainability* is most often associated implicitly with human economic activities or complex systems of human economic activities that are constrained by *environmental* limits—sustainable agriculture for example. From the perspective of current environmental concerns, both of these familiar—and essentially economic—ways of understanding sustainability are fatally flawed. Beyond my critique of Norton's understanding of sustainability, I introduce and commend a novel way of understanding sustainability that I dub *ecological sustainability*.

3.2 Sustainable Yield and Sustainable Development

The concept of *sustainable yield* itself has two permutations: *maximum* sustainable yield and *optimum* (sustainable) yield, the former being the more prominent and having the longer history of application. Classic natural resource management is focused on single species—this kind of game animal, that kind of commercially harvested fish—the goal of management being to attain a *sustainable yield* of the target "resource."

The principal application of the concept of *maximum sustainable yield* has been in marine fishery management (Schaefer 1954). The formulae for calculating maximum sustainable yield are appropriately sophisticated mathematically—involving variables for the species' fertility, growth rate, age of reproductive maturity, and so on (Bousquet et al. 2008). They are not, however, at all sophisticated ecologically. The logistic equations for representing the dynamics of species populations are assumed to actually represent the dynamics of species populations—and that assumption is questionable, to say the least (Botkin 2012). Little attention is paid to the feedback loops between those species that strongly interact with the target species and even less attention is paid to those other species that strongly interact with species that strongly interact with the target species (Larkin 1977). As a result, the calculation of population growth, stabilization, harvest, and rebound bear little resemblance to the dynamics of actual populations of target species on the ground or in the oceans (Botkin 1990). In general, the mismatch between classic population models and actual population facts is attributable to the impact of artificially fluctuating populations of the target species on other species in the biotic communities of which the target species is a member. Thus the impact of the fluctuating populations of those species on that of the target species is unaccounted for and may even be chaotic (in the mathematical sense of the word) and thus be unpredictable in principle (Walters and McGuire 1996). This classical way of understanding sustainability is further vitiated ethically by the assumption in economics that the natural world is nothing but a "pool" resources existing to service human wants as well as human needs—strong anthropocentrism, in other words.

The concept of optimum (sustainable) yield—the word "sustainable" is usually omitted—is more inclusive of desiderata beyond the "resource" and the industry (mainly commercial fishing) exploiting it. Its locus classicus is the US Magnuson-Stevens Fishery Conservation Act (Public Law 96–265) of 1976 (and amended as recently as 1996) which defines "optimum yield" thusly: "The term 'optimum,' with respect to the yield from a fishery, means the amount of fish which (A) will provide the greatest *overall benefit* to the Nation, particularly with respect to food production and recreational opportunities, and taking into account the protection of marine *ecosystems*; (B) is prescribed as such on the basis of maximum sustainable yield of the fishery, *as reduced* by any relevant economic, social, or *ecological factor*; and (C) in the case of an overfished fishery, provides for rebuilding to a level consistent with producing the maximum sustainable yield in such fishery." To generalize the concept of optimum sustainable yield from this legislative definition, three aspects stand out: (1) benefits to humans collectively other than narrow commercial benefits are taken into account—weak anthropocentrism, in other words; (2) by definition the optimum sustainable yield will be lower than the maximum sustainable yield; (3) in calculating the optimum sustainable yield, ecological factors are taken into account as is the protection of ecosystems. The optimum-sustainable-yield concept has not been operationalized with the same degree of mathematical precision as has been attempted in the case of the maximum-sustainable-yield concept.

In response to the politically sensitive tension between economic development for impoverished human populations and the destruction of natural capital that, historically, accompanies human economic development, the concept of *sustainable development* was offered as a resolution. Perhaps it would be possible to eat our environmental cake and have it too—by means of *sustainable* development. Many skeptical environmentalists of a Malthusian bent regard "sustainable development" to be an oxymoron (Bartlett 2012). If poverty is defined as a low standard of living and a low standard of living is equated with low levels of consumption of goods and services, then economic development for impoverished human populations entails increased consumption of goods and services, which in turn entails increased conversion (destruction) of natural capital on a finite, small planet—which is not sustainable—or so the neo-Malthusians grumble (McKee 2012).

Personally, I am not convinced that the concept of sustainable yield or that of sustainable development cannot be successfully operationalized. Calculating the former is a technical problem—a daunting one to be sure, but a technical problem nevertheless. Increased mathematical sophistication by, for example, the use of nonlinear equations; increased modeling sophistication by, for example, the use of cellular-automata and multi-agent-based models; and the ever increasing computational power of evolving computer hardware and software—all may make it possible for modelers to approximate the behavior of target species populations in a complex nexus of coupled natural and human systems (Filatova et al. 2013). And taking the concept of sustainable development seriously might require challenging the rather conventional assumptions that its critics make (Roudsepp-Hearne et al. 2010). Might not an impoverished people's standard of living be increased without a corresponding and proportionate increase in its levels of consumption of goods and services? Possibly, I would think. But if not, Can increases in the consumption of goods and services be attained without a corresponding and proportionate increase in the conversion of natural capital? Again, possibly, I would think. And here too, the problems are partly technical, but they are also partly psychological, cultural, and political. Better agricultural techniques, for example, might increase food production without converting more land to crops; alternative techniques of generating energy might raise standards of living without adversely affecting the environment. Re-envisioning the good life—from consumerist values to those of association, environmental aesthetics, education, vegetarianism, human and ecological health—might increase standards of living without an inevitable adverse impact on the natural environment (Ybarra 2016).

The very famous definition of "sustainable development" in the Brundtland Report—"development that meets the needs of the present without compromising the ability of future generations to meet their own needs"—reveals a deeper, essentially philosophical flaw in the concept of sustainable development, if so defined (United Nations World Commission on Environment and Development 1987). To be sure, the Brundtland Report's definition of "sustainable development" is laudable to the extent that it implicitly distinguishes between *needs* and *wants*. But, as things presently stand in our consumerist culture, satisfying wants is thought to be a need. So, even the Brundtland Report's laudable focus on needs is less than satisfactory, in my opinion, without explicitly divorcing needs from wants. That it makes no reference whatever to environmental constraints on development is what is really striking about this definition. Consider a possible alternative definition: development that meets the needs of the present and those of future generations without compromising environmental quality—without increasing atmospheric, oceanic, or freshwater pollution; without intensifying species extinction and the erosion of biodiversity; without accelerating deforestation and global climate change…

Doesn't the Brundtland Report's definition of sustainable development at least imply sustainably harvesting renewable natural resources (sustainable yield) and preserving ecological services, without which future generations could not meet their own needs? The Report may not mention environmental conservation explicitly, but it implies the necessity of sustaining environmental quality as a condition for future generations meeting their own needs, doesn't it? No, it doesn't. In neo-classical economics there is the "gross substitution axiom" (Fisher 1972). As a heavily exploited natural resource becomes scarce its price increases, making investment in finding a substitute become increasingly attractive. Thus, there is no need to conserve any particular natural resource. For example, when we begin to run short of copper for making telephone wires, someone will (as someone did) invent fiber optics and then wireless cell phones. Such accumulated anecdotal evidence suggests that market forces will always stimulate the discovery or invention of substitutes for any natural resource—from petroleum to Madagascar periwinkles. As one species of marine finfish is harvested to commercial extinction, the fishing industry just exploits another; when all marine finfish stocks are depleted, someone will figure out how to make sushi with a fish-flavored and fish-textured soy product. According to this prevailing way of thinking in economics, the present generation can, therefore, meet its own needs—including the presently prevailing need to satisfy wants—by rapidly exploiting existing organic natural resources to commercial if not to biological extinction and bequeathing a legacy of wealth and technology and a culture of business and inventiveness to future generations, by means of which they can meet their own needs. That in nutshell is weak sustainability (Cabeza-Gutez 1996).

3.3 Sustainability Plain and Simple

As noted in beginning, sustainability is most often implicitly associated with human economic activities or complex systems of human economic activities that are constrained by *environmental* limits. But as the foregoing considerations demonstrate, the two classic conceptual permutations of the naked concept of sustainability—*sustainable yield* and *sustainable development*—are embedded in the larger conceptual domain of neo-classical economics and share its assumptions. The former regards the natural environment as a pool of resources and the latter, even more radically, regards the natural environment not at all. To arrive at an environmentally sensitive concept of sustainability, we might make a fresh start by examining the naked concept of *sustainability*.

Sustainability is a property of an activity or complex system of activities capable of going on indefinitely. Any human activity, whether economic or otherwise, may or may not be sustainable; moreover, sustainability does not necessarily reference environmental constraints. Gambling losses of a thousand dollars a week, week after week, are not sustainable by a person who earns fifty thousand dollars a year. A national health-care system the cost of which increases by 10% per year is not sustainable.

Because the property of sustainability is a property of activities—processes—the concept of sustainability implicitly, but obviously, references time. Further, although the temporal scale of sustainability is rarely specified, we unconsciously and often very vaguely scale the temporal parameters of sustainability relative to the human activity or complex system of human activities in question. Continuous running is a human activity that may or may not be sustainable, but the temporal scale of sustained continuous running is calibrated in minutes and hours. I am unable to sustain running for much more than 30 min. One might ask of a passing marathon runner, on pace for a three-hour finish, if that pace is sustainable. I would suppose that a national health-care system that is designed to function well for a century would be deemed sustainable. But I would also suppose that an agricultural system would be deemed unsustainable if it too were designed to function well for only a century, after which period it would collapse. One would probably not, however, withhold the sustainability descriptor from an agricultural system that functioned well for a million years followed by collapse. Where should the temporal parameter of sustainability be drawn for an agricultural system? At a thousand years, two thousand, ten thousand…? I don't know. My point simply is that the temporal scale of sustainability is not infinity and, though often both implicit and vague—as the word "indefinitely" in the first sentence of the previous paragraph is meant to suggest—the temporal scale for assessing sustainability is relative to the human activity or complex system of human activities in question.

3.4 Sustainability à la Norton

And now we come to the definition of "sustainability" proffered by Norton (2005). Norton rejects weak sustainability, the economistic understanding of sustainability that the famous Brundtland-Report definition of *sustainable development* invites. He also rejects strong sustainability—the view that artificial capital cannot substitute for natural capital—championed by ecological economists and conservation biologists (Common and Perrings 1992). What they regard as a foundational natural legacy for future generations—biodiversity, for example—Norton demeaningly refers to as "stuff." Advocates of strong sustainability think it necessary to bequeath some natural "stuff" to future generations. Norton (2005: 336) goes on to define "sustainability" in terms of a less tangible legacy: "*sustainable* and *sustainable development* are not themselves general *descriptors* of states of societies or cultures but rather refer to many specific sets of commitments on the part of specific societies, communities, and cultures to perpetuate place-based values and project them into the future."

What those place-based values are is up to specific societies, communities, and cultures to determine democratically. While I certainly agree with Churchill (2013: 574) that "democracy is the worst form of government, except for all the other forms," I am less sanguine than is Norton that place-based values will always project environmental quality into the future. That may happen in some specific societies, communities, and cultures—say Ashland, Oregon and Boulder, Colorado. But it may not happen in others—say Talladega, Alabama and Odessa, Texas, where the place-based values likely to be projected into the future center, respectively, on NASCAR racing and wildcat drilling for the petroleum industry's gold standard—light, sweet Texas crude. Thus even the concept of sustainability proffered by Norton, makes no provision for environmental constraints on human economic activities—biodiversity being just a bunch of stuff.

Norton (2015) would argue, perhaps, that his framing emphasizes the constraints created by the duty to preserve ecological options in the future tied to core values of communities—calling it "normative sustainability." But that, in my opinion falls short of an adequately robust notion of sustainability, although, it does seem to me to at least make a provision for some environmental constraints. Such constraints, for Norton, remain to be defined both contextually and democratically. Given the direction that democracy has taken in the US—what with climate deniers, as I write, completely controlling the government—I think that Norton is still proffering a notion of sustainability that inclines way too far in the direction of weak sustainability.

Further, Norton's implicit bioregionalism—a geographical ontology of "place" and a sociological ontology of "*specific* societies, communities, and cultures"—is quaint. The twenty-first century is characterized by globality—for better or worse. Our most daunting and urgent environmental challenge is indeed *global* climate change. Specific places, societies, communities, and cultures can no longer be considered in isolation from one another and can no longer plausibly make independent commitments to sustainability. Without international cooperation to abate carbon emissions, no place will look like the same place fifty years from now, no matter what commitments to place-based values its denizens make and project into the future.

Twenty-first century ecologists have also abandoned the concept of closed, self-regulating, ontologically robust ecosystems (Pickett and Ostfeld 1995). Ecosystems are partly artifacts of ecological hypotheses as Tansley (1935) long ago noted. Once bounded as such, they are porously open to all sorts of comings and goings, from invasive organisms to minerals from afar blown in on the wind and washed down by the rain. For example, Korin et al. (2006) argue that the scant fertility of Amazonian soils depends on African dust blown across the Atlantic Ocean on the prevailing easterlies, about half of it coming from one specific location. If global climate change affects wind patterns across the tropical Atlantic Ocean, the Amazon rainforest could be starved for nutrients, with a huge loss of biodiversity. For another example, the drought cycle, constraining the ecosystems in the American Southwest that Leopold (1979) long ago noticed and wondered about, is traceable to the Pacific Decadal Oscillation cycle (McCabe et al. 2004). Sustainability has an implicit spatial scale, as well as an implicit temporal scale, and it is becoming increasingly clear that that scale is necessarily global, not local. Sustainability is thus not a matter of "place-based" independent commitments on the part of "specific societies, communities, and cultures" as per Norton's definition.

Although problematic in the ways just indicated, there is a kernel of insight in Norton's understanding of sustainability in terms of commitments to values projected into the future. Dedichotomized—by rejecting Norton's zero-sum stuff versus values dichotomy—and scaled up from place to planet, Norton has something valuable to contribute to an understanding of sustainability in the third millennium and its inescapably global dimensions. To that something I return at the conclusion of this chapter.

3.5 Ecological Sustainability

To fully capture the implicit sense of sustainability ambient in contemporary discourse—human economic activities or complex systems of human economic activities that are constrained by *environmental* limits—I would characterize sustainability ecologically as follows. The human economy is a subset of the economy of nature. Need it be recalled that the words "ecology" and "economy" were coined from the same Greek root, *oikos*, meaning "home"? Less well known, the notion of an "economy of nature" is our oldest proto-ecological metaphor, coined by none other than Carl Linnaeus in the mid-18th century (Linnaeo 1749). Sustainable human economic activity would not disrupt the globally integrated ecological processes and functions of our global home, planet Earth. Obversely put, sustainable development consists in devising artificial human economic systems—call them econo/ecosystems—that are modeled on and symbiotically adapted to the economy of nature (the global ecosystem, the living biosphere). The macro-economy of nature is the model for a sustainable human economic microcosm.

How does the economy of nature work? In other words, what are the fundamental principles of the economy of nature?

That it is solar-powered is its first principle. Green plants convert radiant energy from the sun into potential chemical energy by stripping the carbon atoms from atmospheric carbon dioxide (CO_2) and photosynthesize them with the hydrogen and oxygen atoms in water (H_2O) to form carbohydrates—for example, glucose ($C_6H_{10}O_6$), among the simplest such compounds. Respiring herbivorous animals oxidize (burn as it were) the hydrocarbons composing plants, converting the chemical energy therein into thermal and kinetic energy to power their metabolisms and locomotion, scattering the thermally degraded energy to their surroundings.

That it is cyclical is the second principle of the economy of nature. While energy flows through the economy of nature on a one-way trip from solar source to the sink of outer space, the materials of nature's economy cycle. In the simplified example of the previous paragraph, green plants dump the excess oxygen atoms—generated as a waste product (O_2) of their photosynthesis of carbohydrates—into the atmosphere. When respiring animals use atmospheric O_2 to burn carbohydrates they, in turn, dump the waste products (CO_2 and H_2O) of oxidization back into the atmosphere, completing a cycle. But the carbon cycle is just one of many materials cycles in the biosphere—others involve nitrogen, phosphorus, calcium, potassium, iron, and other plant "nutrients."

In the simplest terms, therefore, an ecologically sustainable human economy—an econo/ecosystem—will be powered by solar energy and will thoroughly recycle its material components.

A good, explicit example of devising artificial econo/ecosystems that are adapted to and modeled on natural ecosystems is the perennial polyculture envisioned by Jackson (1980) and in process of development at the Land Institute in Salina, Kansas. Jackson's model is the prairie, which, under various names (grassland, savannah, steppe) on various continents (North and South America, Asia, Africa) is a globally distributed biome. According to Jackson (1980), four kinds of plants must constitute a perennial polyculture: cool-season grasses, warm-season grasses, legumes, and sunflowers—all perennialized. Fossil-fuel input is reduced by eliminating the need for annual plowing; and artificial nitrogen input is eliminated by mixing grains with nitrogen-fixing legumes. If successful (or rather when successful, as progress has been notable), Jackson's would (will) represent a second agricultural revolution, the first, based on annual monocultural grasses, having occurred worldwide about ten or eleven thousand years ago (Barker 2009). Another example is the suite of artificial econo/ecosystems going under the rubric of "industrial ecology"—the general idea being that the waste product of one industry is the resource of another (Ehrenfeld 2004). Spent fry oil, to take but one instance, from fast-food restaurants can be (and is in some places) the feedstock for the manufacture of biodiesel fuel for the engines of cars, trucks, and buses (Kemp 2006).

3.6 Sustainability Ethics

This way of conceptualizing sustainability—as *ecological* sustainability—points to the emergence of a new field in philosophy: sustainability ethics. Sustainability ethics emerges at the interface of environmental ethics and business ethics and is currently being developed at several institutions.[1] Ecological sustainability revolves around the concept of economy—the planetary economy of nature and the globalized human economy. Ecological sustainability is a matter of adapting human economic systems to and modeling them on the economy of nature in which the globalized human economy is embedded and in relation to which it should stand as microcosm to macrocosm. This ecological way of *understanding* sustainability devolves from environmental ethics. But to *actually achieve* sustainability, so understood, would seem to me to be a matter of business practice, constrained by environmental ethics wed to business ethics.

If the concept of economy is what ecological sustainability revolves around, why would actually achieving sustainability not be a matter of applied economics, instead of being a matter of a hybrid environmental-business ethics constraining business practices? Because economics is a descriptive social science, with predictive ambitions based on a number of controversial assumptions, among them the invidious gross substitution axiom. That all values are preferences and that human welfare consists of maximizing "preference satisfaction" is also among those controversial assumptions (Randall 1988). To treat all values as preferences enables economists to quantify the way humans value various things in a monetary metric for purposes of comparison—thus making economics a totalizing discipline, beyond the reach of which there is nothing. But all values are not preferences. One may prefer strawberry ice cream to chocolate, but one does not just prefer human autonomy to slavery (I hope). One might actually prefer to have a few slaves at one's beck and call. However, no moral human being would consent to own slaves, even if given the opportunity to do so, because slavery is a gross violation of human dignity and human dignity is something that moral human beings *value*, not

[1]For example, The Markkula Center of Applied Ethics at Santa Clara University http://www.scu.edu/ethics/practicing/focusareas/environmental_ethics/lesson4.html (Retrieved 12 Feb 2015) and The Rock Ethics Institute of Penn State University http://rockethics.psu.edu/climate/sustainability-ethics (Retrieved 12 Feb 2015).

something that we merely prefer. Such "transcendent values" (as we might call them) constrain preferences and things so valued should not be subjected to shadow pricing and benefit-cost analysis (Callicott 2017). Nor does human welfare consist of maximizing preference satisfaction; to think that it does is to think like a two-year old.

Business ethics consists precisely in exploring various transcendent values, and recommending that such values negatively constrain and positively inspire and guide human economic activities (Goodpaster et al. 2006). At a minimum, business ethics demands that human economic activities be constrained by the aforementioned value of human dignity, which would, among other things, prohibit cost-cutting by employing child or prison labor in workplaces that are a hazard to life, lung, liver, or limb. Business ethics should also inspire and guide human economic activities by recommending that businessmen and -women set a goal of enhancing the commonweal as well making a private profit.

At the conjunction of environmental and business ethics, sustainability ethics would demand that transcendent *environmental* values also serve as constraints on and inspirational guides for human economic activity. Such values would range from a concern for animal welfare to clean air and water to biodiversity conservation. Sustainability ethics, as I conceive it, is business ethics informed not only by the extra-economic values of traditional social ethics, but also by those of contemporary environmental ethics, all in service of the goal of achieving ecological sustainability—that is, atuning the human economic microcosm to the macrocosmic economy of nature.

3.7 Comparative Philosophy

We live in a global environment culturally as well as environmentally. Thus any viable 21st-century understanding of sustainability must include representation from non-Western traditions of thought. Thus, comparative philosophy can make a crucial contribution to sustainability ethics in a variety of ways. At the metaphysical level or the level of "first philosophy" in Aristotle's sense of the term, many of the great philosophical traditions of Asia locate human being, human society, and, yes, the human economy in a cosmic setting and urge a harmonization of things human with cosmic goings on. Indeed, the first (in the ordinary ordinal sense of the word) philosophy in the Western tradition of thought—that of the Presocratics—also did just that (Callicott et al. 2018). Such a harmonization of "society" with "heaven" (cosmically not religiously understood) is also at the heart of Confucianism (Tucker 2014). Daoism, of course, leaps to mind as well, especially with its concepts of the *dao*, *wu wei*, and *feng shui* (Lai 2014). At the moral level, certainly Buddhism draws a sharp distinction between needs and wants and between preferences and values proper, and offers as searing a critique as one could hope to find of a consumerist lifestyle (Kaza 2014).

Comparative environmental philosophy and ethics has by now been very well explored and developed in theory and in the abstract (Callicott 1994; Callicott and Ames 1989; Callicott and McRae 2014, 2017). Thus I can add little of substance here—which has not already been worked out by scholars more able and more expert than I—to that body of scholarship. As it seems to me, less well explored territories for working by comparative environmental philosophers and ethicists are the ecologically sustainable economies that remain in place in various parts of the world. The dream of Japanese conservationists to restore the iconic Crested Ibis (*Nipponia nippon*), first on Sado Island and eventually on Honshu and other Japanese islands, depends directly on ecologically sustainable paddy-rice cultivation (Nishimiya and Hiyashi 2010). The Toki, as the bird is affectionately known in Japan, was driven to near global extinction by hunting, pollution, and habitat loss. It nests in pine trees on wooded uplands and feeds on amphibians, small fish, and small mammals that inhabit nearby wetlands. In the traditional paddy-rice system of farming, uplands are kept forested to protect the watershed and stream flow from which water is diverted to the small rice fields, in which the Crested Ibis feeds symbiotically with the rice cultivation (Totman 1989). The paddies are, as it were, artificial wetland habitat for the bird. And the Toki performs pest-and-vermin control services for the Japanese rice farmer. One might even imagine that the Toki and the Japanese rice paddy mutually co-evolved, they are so well adapted to one another.

But what of distinctly philosophical substance does one find in this example of ecological restoration and ecologically sustainable economics? The future of philosophy—mainstream, as well as comparative and environmental—lies in a turning outward to other intellectual disciplines for the purpose of interdisciplinary problem solving. The inward-looking, over-specialized preoccupations of twentieth-century philosophy are no longer viable as the twenty-first century unfolds. Ornithologists work with the bird. Ecologists work with the bird's habitat. Agronomists work with the rice cultivation. Anthropologists work with the local folklore. Political economists work with Japanese government policymakers to keep rice cultivation small-scale and traditional. (Japan prohibits the importation of rice, thus helping to preserve its traditional rural landscape and thus helping to restore its emblematic avifauna.) And what can philosophers do? We can synthesize all the knowledge gathered by the other disciplinarians into a coherent whole and locate it in a fitting metaphysical superstructure—in the case of Sado Island, into a syncretic Buddhist-Shinto metaphysical superstructure (Grapard 1984; Toyoda 2013). And, most importantly, we can try to understand how such local efforts to achieve ecological sustainability and ecological restoration can be integrated into a global network of resonant efforts around the world.

3.8 What Value Does Philosophy Add to Ecological Sustainability

What unique contributions can the more general discipline of philosophy make toward enhancing our understanding of what ecological sustainability is and how ecologically sustainable goals can be accomplished? The foregoing is a discussion based in the discipline of philosophy primarily aimed at enhancing our understanding of what sustainability is and, secondarily, aimed towards how sustainable goals can be accomplished. So let's stand back and look at what is going on.

First, the discipline of philosophy is concerned with conceptual clarification and precision of expression. Sometimes philosophical conceptual clarification and precision of expression can be annoying because it may seem too fastidious and (indeed be as well as seem) trivial. But clarifying the difference between the concepts of sustained yield, sustainable development, and ecological sustainability goes to the substantive heart of the matter at hand.

Second, the discipline of philosophy is concerned with critical thinking. Here I subjected the concept of maximum sustained yield and the definitions of "sustainable development" in the Brundtland Report and "sustainability" in Norton's work to extensive critical examination and clearly exposed the inadequacies of each. I also subjected the assumptions of mainstream economics to a more cursory regimen of critical thought.

Third, the discipline of philosophy is concerned with values. In other disciplines, especially those with scientific pretenses, values are often treated as purely subjective, personal, arbitrary, and irrational. Preferences may be all these things, but, as noted, not all values are preferences. And values are of the utmost importance. Whether we admit it or not, values drive all human activity and all public policy. Recognizing values for what they are, exposing them to people who have been taught to marginalize and trivialize them, and stressing their ultimate importance is a central and unique contribution that the discipline of philosophy can make to any human endeavor, including enhancing our understanding of what sustainability is and how ecological sustainability can be achieved.

Fourth, the discipline of philosophy is concerned with ontology, with what exists and what does not. I criticized Norton's ontological assumptions about geography and society, his hypostatization of place and specific communities. To make more explicit another ontological issue lurking in the foregoing discussion, let me ask: Do ecosystems exist robustly—as robustly as say a snake or a monkey? I hinted that they do not. They exist, but their existence is less robust than that of snake or a monkey. That ecosystems have fuzzy boundaries is the least ontologically problematic thing about them (Pickett et al. 1992). More problematic ontologically, their (often fuzzy) boundaries are determined by the particular interests of the ecologists who investigate them (Tansley 1935). In the instance mentioned in the foregoing discussion, the Amazon rainforest is taken to be an ecosystem sustained by dust blown in from Africa. But by an ecologist interested in energy flows in Amazonian food webs, the boundaries of the ecological object of study would be drawn very much more narrowly. Ecosystems are thus partly scientific artifacts. They come into being, partly, only when interrogated by ecologists.

Fifthly and finally, the discipline of philosophy is concerned with epistemology—with what we know, how we know it, and the limits of human knowledge. Scientific knowledge is often privileged epistemologically in comparison with other knowledge claims. At the same time, uncertainty is becoming an ever more prominent epistemological issue in science itself, an epistemological issue that is also ever more prominent at the interface of science and policy and politics (Jamieson 2014). Epistemological issues hardly appear at all in the foregoing discussion, but perhaps they should have taken more of a center-stage position. Who gets to declare what is and what is not a sustainable human activity? Perhaps unconscionably, I disparaged the epistemic credentials of Talladega rednecks and Odessa roughnecks. Perhaps arrogantly, I dismissed without hesitation their various claims to know what is and what is not sustainable. A signal and laudable virtue of Norton's definition of "sustainability" is that it is not so dismissive as is my account of ecological sustainability. But should every claim to knowledge be treated with respect and given an equal hearing with every other? What about the knowledge claims of Scientologists who believe that they arrived on a spaceship from another planet or evangelical Christian fundamentalists who claim to know that the Earth has only been around for about six thousand years? Aren't some extra-scientific knowledge claims just as worthy of contempt as others are of respect? How do we determine which knowledge claims warrant appropriate respect and which warrant appropriate contempt? That is an epistemological problem; and epistemology is a subdiscipline of philosophy.

3.9 Important Topics for Future Inquiry

Looking toward the future, we might ask, what are the most important topics of future inquiry that ecological sustainability theorists need to investigate?

Assuming that ecological sustainability requires a multidisciplinary approach and that ecological sustainability theorists inhabit a wide variety of disciplines, then the topics of future inquiry might be sorted by discipline. In the foregoing discussion I mention two disciplines in which topics of future inquiry are quite technical: perennial polyculture agronomy and biofuels engineering. In the case of other disciplines, economics, for example, the concept of ecological sustainability challenges some very deep assumptions: substitutability, for one example; the reduction of all values to preferences, for another. More basically, the concept of ecological sustainability challenges the economistic understanding (in terms of preference satisfaction) of what it means for human beings to fare well. In addition, the concept of ecological sustainability challenges the practice by economists of discounting future benefits at the current rate of interest. Ecological sustainability is, as noted, an inherently temporal concept, and implicitly references an indefinite future. As long as future benefits are discounted at a significant rate—or, really, at any discount rate at all—economics as it is now practiced will be an impediment to an ecologically sustainable economy, not an analytic aid toward achieving it.

As these reflections indicate, the topics of future inquiry by ecological-sustainability theorists are myriad. I will limit my remaining reflections on topics of further inquiry for ecological-sustainability theorists in the discipline of philosophy.

One topic for future philosophical inquiry is the temporal scale of ecological sustainability—as I have characterized it in terms of atuning the human economy to the economy of nature. The spatial scale, as noted, must be global because of two primary considerations. The first relates to the indistinct ontology of the Earth's many ecosystems in comparison with the robust ontology of the living Earth itself considered as a systemic unit, the biosphere. The second relates to the globalization of the human economy, rendering its local and regional subdivisions no more ontologically independent than are its local and regional ecosystems. So what *temporal* scale is a fitting correlative to the global scale of a sustainable human economy? It lies somewhere between a scale calibrated in decades and one calibrated in millions of years, but narrowing that vast range to the proper temporal scale is indeed a topic of future philosophical inquiry.

A second topic for future philosophical inquiry into sustainability is articulating a fitting moral ontology. To whom—or perhaps better, to what—is the present generation obligated by way of our widely recognized obligation to "future generations"? Resolving this problem does not depend on a definitive resolution to the problem of determining a fitting temporal scale for ecological sustainability; nevertheless that topic for future inquiry puts this one in temporal perspective. Whatever the fitting temporal scale of ecological sustainability, we can be sure that it extends beyond what we might think of as the *extended personal future*.

In the very paper of his, "Some Fundamentals of Conservation in the Southwest," the interpretation of which Norton and I have so hotly disputed, Leopold (1979) distinguishes between "immediate posterity" and the "Unknown Future" (thus capitalized). By "extended personal future," I mean something much the same as Leopold meant by "immediate posterity." For members of my generation, future generations already exist; and my age mates and I are personally acquainted with some members of those future generations. Being a bit north of seventy-five years of age, my own personal future is very limited. My son is nearing fifty years old and may well live into his nineties, as did my father and his namesake. That forty-plus year temporal horizon is thus part of my *extended* personal future—because I am personally very concerned about the world that my son will live in as he approaches my present age. His son, my grandson, is coming up on twenty years old. He may well live another seventy-plus years, enlarging my extended personal future—members of the future generations with whom I am personally acquainted and for whose welfare I am personally concerned—almost throughout the whole twenty-first century. If I live to see the birth of my grandson's son or daughter, my extended personal future—as here characterized, my window of personal concern for the future persons with whom I am acquainted and for whom I care—will grow well into the twenty-second century.

I want to help insure that the world that my son, grandson, and great grandchild will live in remains habitable and pleasant. For that to happen, I and others with similar concerns—that would be pretty much everyone who has an extended personal future—must join together and effect radical changes in the human economy. Creating a sustainable econo/ecosystem (as here characterized) will be necessary to mitigate and to adapt to global climate change. But radical changes in the human economy will entrain radical changes in human lifestyles, which will affect the reproductive chances and choices of the members of my great grandchildren's generation. That leads to the "Parfit paradox," as I call it, or the non-identity problem as it is more widely known (Parfit 1984). If business goes on as usual, my great grandchildren and other members of their generation will meet, mate, and have one set of children. But if radical changes in the human economy are initiated, with attendant lifestyle changes, my great grandchildren and other members of their generation will meet, mate, and have a set of children different from the set of children they would have had had business gone on as usual. So it is impossible for me to be concerned *now* about the individual welfare of my unborn great grandson or -daughter, *qua individual*, and that of his or her unborn cohort, *considered individually*. For if I succeed in helping make the world of the twenty-second century habitable and pleasant, those individuals who would have existed, had business gone on as usual, would not exist at all; rather, different individuals would exist in their stead. That's what I take Leopold to have meant by the "Unknown Future" beyond "immediate posterity." But, of course, Norton will probably not agree.

Assuming that the temporal scale of ecological sustainability, as I understand it, extends beyond one century into the future, the ontology of future-generations ethics must also be scaled up from one of individual persons to something more proportional to the temporal scale of ecological sustainability, whatever that turns out to be (Callicott 2013). As just noted, the appropriate *objects* of moral considerability for intergenerational ethics cannot be yet unborn future *individual* human persons—because their very existence or nonexistence, as individual persons, will depend on what we presently do or leave undone. Well then, is the appropriate object of moral considerability for intergenerational ethics *Homo sapiens*—the human species? I think not. The natural lifespan of a large mammalian species is about a million years—which, I would think, exceeds the temporal scale of ecological sustainability. Further, our highly adaptive species may well survive an environmental apocalypse and hang on with a much-reduced population living in a scarcely habitable and very unpleasant world—if we can trust the vision of post-apocalypse fiction writers, film makers, and Gaia theorists (Lovelock 2006; McCarthy 2006; Robertson 2013).

No, I think that the appropriate object(s) of moral considerability for intergenerational ethics is (are) neither individual human persons nor the human species, but global human civilization and biodiversity. In addition to biodiversity and more generally environmental quality, what I most care about projecting into the future are things like the visual arts, music, poetry, literature, science, philosophy, architecture, law, and government. One aspect of what I am getting at here is "stuff" sensu Norton—the cultural achievements of the past—Plato's *Republic*, the *Bhagavad Gita*, *Hamlet*, the "Mona Lisa," the Taj Mahal, ... The other aspect is the visual arts, music, poetry, literature, science, philosophy, architecture, law, and government of the future. That's essentially a commitment to perpetuate cultural values and project them into the future sensu Norton. But aren't such future achievements of global human civilization equally subject to the Parfit paradox? Yes, they are—if we think of the future achievements of global human civilization in terms of particular artifacts. Radical changes in the human economy will result in radical changes in the visual arts, music, poetry, literature, science, philosophy, architecture—and, indeed in law and government (because the latter two must needs be fully globalized). Making the radical lifestyle, economic, and political changes necessary to avert catastrophic global climate change will surely affect the cultural productions of the future; but if we go forward with business as usual cultural productions worthy of bing called the visual arts, music, poetry, literature, science, and philosophy may not be forthcoming at all in a world of failed states and remnant bands of naked apes led by sociopathic warlords. So it's not a matter of what specific visual arts, music, poetry, literature, science, philosophy, and architecture will be forthcoming than a matter of perpetuating these human *activities* themselves. Thus moral concern for the future achievements of human civilization, though subject to, is not vitiated by the non-identity problem.

3.10 Conclusion

Therefore in some ways, this view of ecological sustainability as adjusting the human economy to the economy of nature aligns with that of Bryan Norton, but in other ways it does not.

This way of understanding sustainability, like Norton's way, projects values into the future (those of global human civilization) but it also projects stuff (the precious past achievements of human civilization). Indeed, without projecting the stuff as exemplars of the values of global civilization it would be virtually impossible to project into the future the values that they embody and exemplify. Thus to understand sustainability in terms of projecting global human civilization into the future is to project both the accumulated artifacts (stuff) of human civilization—the extant literature, art, music, architecture, science, philosophy that is our precious human heritage—and the culture, the values, and skills which fostered the creation of those artifacts.

And what of sustaining biodiversity? Well, we have long realized that the species concept in biology is an abstraction—and one that remains contested (Agapow et al. 2004). Organisms exist concretely and by a process of abstraction we organize them into Linnaean categories. Biodiversity, conceptually comprising not just species diversity, but genetic diversity and various kinds of ecological diversity—for example, β and γ, as well α diversity (sensu Whittaker 1960)—is an even more abstract idea. So to project biodiversity into the future by actively conserving it (whatever "it" may be) is not exactly to project "stuff" into the future. However, without projecting the actual organisms that embody biodiversity along with their ecological inter-relationships we cannot project that non-thing that we call biodiversity into the future.

The further difference I have with Norton's understanding of sustainability is partly a matter of scale and partly a matter of governance. He thinks locally; I am suggesting we think globally about sustainability; nor should we view it through a narrow window of time, but in the open vista of thousands of years past and thousands of years future. Moreover, the sustainability of biodiversity and global human civilization should not depend on the vagaries of democratic governance. Sustaining global human civilization and biodiversity should be more like an article of the implicit constitution of the social and natural contracts (sensu Serres 1992); it should be beyond the vagaries of parliamentary legislation and town-hall decision making.

That's just my opening gambit concerning the moral ontology of intergenerational ethics. The question has scarcely been asked, and so a definitive answer is impossible to even broach. This along with many others is an important topic of future inquiry in the nascent field of ecological-sustainability philosophy.

We all owe Bryan Norton a hardy thank you for putting the concept of sustainability up for sustained (no pun intended) philosophical analysis and development (again, no pun intended). More generally, we all owe Bryan a hardy thank you for his sustained (ok, this time the pun is intended) commitment to environmental philosophy—he was there at its birth and, although institutionally retired, he remains an active and prolific contributor to its advancement. And I, in particular, owe Bryan a hardy thank you for our career-long adversarial professional relationship. His criticism of my work has led me to think more deeply and clearly about the issues in contention between us. And most of all, I owe Bryan a hardy thank you for a friendship lasting now for well over a quarter of a century.

References

Agapow, P.-M., G.M. Mace, O.R. Biinda-Emons, and K.A. Crandall. 2004. The impact of species concept on biodiversity. *Quarterly Review of Biology* 79: 161–179.

Barker, G. 2009. *The agricultural revolution in prehistory: Why did foragers become farmers?*. New York: Oxford University Press.

Bartlett, A. 2012. Reflections on sustainability and population growth. In *On the brink: Environmentalists confront overpopulation*, ed. P. Cafaro and E. Christ, 29–40. Athens: University of Georgia Press.

Botkin, D. 1990. *Discordant harmonies: A new ecology for the twenty-first century*. New York: Oxford University Press.

Botkin, D. 2012. *The moon in the nautilus shell: Discordant harmonies reconsidered*. New York: Oxford University Press.

Bousquet, N., T. Duchesne, and L.P. Rivest. 2008. Redefining the maximum sustainable yield for the Schaefer population model including multiplicative environmental noise. *Journal of Theoretical Biology* 254: 65–75.

Cabeza-Gutez, M. 1996. The concept of weak sustainability. *Ecological Economics* 17: 147–156.

Callicott, J.B. 1994. *Earth's insights: A multicultural survey of ecological ethics from the mediterranean basin to Australian outback*. Berkeley: University of California Press.

Callicott, J.B. 2013. *Thinking like a planet: The land ethic and the earth ethic*. New York: Oxford University Press.

Callicott, J.B. 2017. 12. What good is it, anyway?" In *The routledge handbook of philosophy of biodiversity*, ed. Justin Garson, Anya Plutynski and Sahotra Sarkar, 168–182. London: Routledge.

Callicott, J.B., and R.T. Ames (eds.). 1989. *Nature in Asian traditions of thought: Essays in environmental philosophy*. Albany: State University of New York Press.

Callicott, J.B. and J. McRae (eds.). 2014. *Environmental philosophy in Asian traditions of thought*. Albany: State University of New York Press.

Callicott, J.B., and J. McRae. 2017. *Japanese environmental philosophy*. New York: Oxford University Press.

Callicott, J.B., J. van Buren, and K.W. Brown. 2018. *Greek natural philosophy: The presocratics and their importance for current environmental philosophy*. San Diego: Cognella Academic Publishing.

Churchill, W. 2013. *Churchill by himself: The definitive collection of quotations*. London: Ebury Press.

Common, M., and C. Perrings. 1992. Toward an ecological economics of sustainability. *Ecological Economics* 6: 7–34.

Ehrenfeld, J. 2004. Can industrial ecology be the "science of sustainability"? *Journal of Industrial Ecology* 8: 1–3.

Fisher, F.M. 1972. Gross substitutes and the utility function. *Journal of Economic Theory* 4: 82–87.

Filatova, T., P.H. Vurburg, D.C. Parker, and C.A. Stanard. 2013. Spatial agenr-based models for socio-ecological systems. *Environmental Modeling and Software* 45: 1–7.

Goodpaster, K.E., L.L. Nash, and H.C. deBettignies. 2006. *Business ethics: Policies and persons*, 4th ed. New York: McGraw Hill Irwin.

Grapard, A. 1984. Japan's ignored cultural revolution: The separation of Shinto and Buddhist in Meiji: A case study. *History of Religions* 23: 240–265.

Hadley, A.T. 1913. *Some influences in modern philosophical thought*. New Haven, Conn.: Yale University Press.

Jackson, W. 1980. *New roots for agriculture*. Lincoln: University of Nebraska Press.

Jamieson, D. 2014. *Reason in dark time: Why the struggle against climate change failed—and what it means for our future*. New York: Oxford University Press.

Kaza, Stephanie. 2014. Acting with compassion: Buddhism, feminism, and the environmental crisis. In *Environmental philosophy in Asian traditions of thought*, ed. J.B. Callicott, and J. McRae, 71–98. Albany: State University of New York Press.

Kemp, W. 2006. *Biodiesel: Basics and beyond*. Tamworth, Ontario: Aztext.

Korin, I., Y.J. Kaufman, R. Washington, M.C. Todd, J.V. Matins, and D. Rosenfeld. 2006. The Bodélé depression: A single spot in the Sahara that provides most of the mineral dust to the Amazon forest. *Environmental Research Letters* 1: 014005.

Lai, K.L. 2014. Conceptual foundations for environmental ethics: A Daoist perspective. In *Environmental philosophy in Asian traditions of thought*, ed. J.B. Callicott, and J. McRae, 173–195. Albany: State University of New York Press.

Larkin, P.A. 1977. An epitaph for the concept of maximum sustained yield. *Transactions of the American Fisheries Society* 106: 1–11.

Leopold, A. 1979. Some fundamentals of conservation in the southwest. *Environmental Ethics* 1: 131–141.

Linnaeo, Carolo. 1749. *Specimen Academicum de Oeconomi Naturae*. Stockholm: Isacus J. Bieberg.

Lovelock, J. 2006. *The revenge of Gaia*. New York: Basic Books.

McCabe, G.J., M.A. Palecki, and J.L. Betancourt. 2004. Pacific and Atlantic Ocean influences on multidecadal drought frequency in the United States. *Proceedings of the National Academy of Sciences* 101: 4136–4141.
McCarthy, C. 2006. *The road*. New York: Alfred A. Knopf.
McKee, L. 2012. The human population footprint on global biodiversity. In *On the brink: Environmentalists confront overpopulation*, ed. p. Cafaro and E. Christ, 91–97. Athens: University of Georgia Press.
Nishimiya, H and K. Hiyashi. 2010. *Reintroducing the Japanese crested Ibis in Sado*. Japan: TEEBweb (Retrieved 30 Jan 2015).
Norton, B.G. (ed.). 1986. *The preservation of species*. Princeton, N. J.: Princeton University Press.
Norton, B.G. 2005. *Sustainability: A philosophy of adaptive ecosystem management*. Chicago: University of Chicago Press.
Norton, B.G. 2015. *Sustainable values, sustainable change*. Chicago: University of Chicago Press.
Parfit, D. 1984. *Reasons and persons*. Oxford: The Clarendon Press of Oxford University.
Pickett, S.T.A., V.T. Parker, and P.L. Fielder. 1992. The new paradigm in ecology: implications for conservation biology above the species level. In *Conservation biology*, ed. P.L. Fielder, and S.K. Jain, 50–65. New York: Chapman and Hall.
Pickett, S.T.A., and R.S. Ostfeld. 1995. The shifting paradigm in ecology. In *A new century for natural resource management*, ed. R.L. Knight, and S.F. Bates, 261–279. Washington: Island Press.
Randall, A. 1988. What mainstream economists have to say about biodiversity. In *biodiversity*, ed. E.O. Wilson, 217–223. Washington: National Academy Press.
Robertson, M. 2013. *Crash: A post-apocalyptic Tale*. Kindle Edition.
Roudsepp-Hearne, C., G.D. Peterson, M. Tengö, E.M. Bennett, T. Holland, K. Benessaiah, G.K. McDonald, and L. Pfeifer. 2010. Untangling the environmentalist's dilemma: Why is human well-being increasing while ecosystem services degrade? *BioScience* 60: 576–589.
Schaefer, M.B. 1954. Some aspects of the dynamics of populations important to the management of the commercial marine fisheries. *Bulletin of the Inter-American Tropical Tuna Commission* 1: 25–56.
Serres, M. 1992. *Le contrat naturel*. Paris: Éditions Flammarion.
Tansley, A.G. 1935. The use and abuse of vegetational concepts and terms. *Ecology* 16: 284–307.
Totman, C. 1989. *The green archipelago: Forestry in preindustrial Japan*. Berkeley: University of California Press.
Toyoda, M. 2013. Revitalizing local commons. *Environmental Ethics* 35: 279–293.
Tucker, M.E. 2014. The relevance of Chinese neo-Confucianism for the reverence for nature. In *Environmental philosophy in Asian traditions of thought*, ed. J.B. Callicott, and J. McRae, 133–148. Albany: State University of New York Press.
United Nations World Commission on Environment and Development. 1987. *Our common future*. New York: United Nations.
Walters, C., and J. McGuire. 1996. Lessons for stock assessment from the northern cod collapse. *Reviews in Fish Biology and Fisheries* 6: 125–137.
Whittaker, R.H. 1960. Vegetation of the Siskiyou Mountains, Oregon and California. *Ecological Monographs* 30: 279–338.
Ybarra, P.S. 2016. *Writing the goodlife: Mexican American literature and the environment*. Tucson: University of Arizona Press.

J. Baird Callicott is University Distinguished Research Professor Emeritus at the University of North Texas. He is co-Editor-in-Chief of the *Encyclopedia of Environmental Ethics and Philosophy* and author or editor of a score of books and author of dozens of journal articles, encyclopedia articles, and book chapters in environmental philosophy and ethics. Callicott has served the International Society for Environmental Ethics as President and Yale University as Bioethicist-in-Residence. He is perhaps best known as the leading contemporary exponent of Aldo Leopold's land ethic and has elaborated an Earth ethic, *Thinking Like a Planet* (OUP 2013), in response to climate change.

Chapter 4
Sustainability and the Currency of Intergenerational Obligations: Norton, Solow, Rawls, Mill, and Sen on Problems of Intergenerational Allocation

Clark Wolf

Abstract Concepts of sustainability guide policy and environmental management decisions. But when goals are articulated badly, they provide poor decision guides, and may lead to serious mistakes. This paper reviews and critically evaluates a series of popular conceptions of 'sustainability,' with special focus on a conception advocated by Bryan Norton. While no conception of sustainability (not even Norton's) is problem-free, we gain by understanding the limitations of each. Adaptive management, as I understand it here, is not a *conception* of sustainability, but a view about how conceptions of sustainability might be used in practical reasoning about policy or management.

Keywords Adaptive management · Capability theory · Environmental management · Sustainability · Intergenerational justice · Sufficientarianism
Resource management

4.1 Sustainability: Shifting Standards, Shifting Problems

In 1934 an American forester named Nelson Courtland Brown evaluated the state of forest harvest in the United States. Brown was concerned about the communities supported by the lumber industry, and the ability of that industry to maintain thriving permanent towns and settlements. He was disappointed to find that much U.S. forestry was, in effect, following a "slash and burn" model. The virgin timber was so valuable that logging camps would follow the stands of old-growth trees as they gave way before a steadily receding frontier. Brown looked forward to a time

---

C. Wolf (✉)
Departments of Philosophy, Political Science, and Sustainable Agriculture, Iowa State University, 435 Catt Hall, Ames, IA 50011, USA
e-mail: jwcwolf@iastate.edu

S. Sarkar and B. A. Minteer (eds.), *A Sustainable Philosophy—The Work of Bryan Norton*, The International Library of Environmental, Agricultural and Food Ethics 26, https://doi.org/10.1007/978-3-319-92597-4_4

in the future when forestry would provide a more stable and organized way of life for those who pursued it. But before that could happen, he argued, the virgin timber needed to be removed from the land so that a sustainable and stable second-growth timber industry could develop:

> The nomadic character of the [timber] industry has resulted in the rapid building of camps and forest communities, the concentration of labor in temporary quarters, and the subsequent abandonment of woods settlements and sawmill towns. Until the virgin forests are removed, the stabilization of forest communities as found throughout Europe is not to be expected. With the increasing importance of second-growth forests, however, stable forest communities will gradually develop and continue. (Brown 1934 pp. 54–55)

Brown's views on these issues are not simple productionism—the view that the rate of production is the only standard we should use to measure success. He was concerned about the economic and social sustainability of the communities that developed around the business of logging production. His words embody a principle of environmental management that led him to recommend the cultivation of what he believed would be a more effectively productive and sustainable ecosystem—an ecosystem of second-growth lumber—that would, he hoped, be capable of sustaining the economic and social needs of the communities of people who depended on the logging industry for their livelihoods. But, recommended Brown, before we can have sustainable logging communities, we first need to cut down all the old growth timber.

In retrospect we see the flaw in Brown's theory of sustainability: the careless logging of old-growth U.S. and Canadian forests constitutes one of the worst environmental disasters of the last two centuries. For decades, the forests of North America had seemed so vast that some thought we could never deplete them. But deplete we did. The indigenous forest ecosystem of North America was almost entirely *removed.* Was it worth it? Defenders note that logging made possible economic development and settlement, as European immigrants and their children moved across the continent, changing it forever. To recognize the logging era as an environmental disaster is not to overlook the benefits that have resulted from economic and industrial development, or to deny those aspects of our present well-being that essentially depend on that development. Perhaps it is unreasonable to expect that we might have done differently, perhaps the development and settlement of the western United States was inevitable, perhaps our present well-being depends, in part, on processes that involved removal of the old-growth timber. But to think that present economic prosperity depended on the wholesale destruction of North American old-growth forests is just a mistake.

Oftentimes the ancient trees were cut for trivial purposes: on the campus of the University of California at Santa Cruz, there are ruined furnaces where old-growth trees were burned to fire-harden clay bricks. The bricks can be seen in the older buildings of downtown Santa Cruz, but they are crumbling away after a century. In retrospect, it seems clear that removing valuable ancient trees to make 100-year bricks made us poorer, not richer. Taking down these trees was a serious mistake, and this mistake resulted, in part, from a flawed view of forest management.

It is worthwhile to count the environmental and human cost of deforestation and to think about different management decisions that might have been made. By coming to an understanding of the values and objectives that informed the decisions of Brown and others whose vision shaped the nation, we might hope to learn how to make better decisions in the future.

Environmental ethicists have often focused on problems of *value*, and on which objects and/or processes may have it. Such issues are important. But we cannot settle questions about environmental decision-making simply by showing that ecosystems or trees have (or do not have) intrinsic value. The foresters of Santa Cruz valued the bricks, and to them the value of the bricks seemed greater than the value—intrinsic or no—of the trees they burned. In a world where values clash, we often face trade-offs among values we regard as intrinsically, or non-instrumentally valuable. Perhaps the most productive way to proceed in environmental ethics would be to examine the process of environmental decision making and the various values and goals that inform that process.

The environmental philosopher who has most consistently focused on environmental decision-making is Bryan Norton, whose work is honored in the present volume. Norton has developed an articulate theory of adaptive environmental management, a flexible decision-making process that requires collection of data and vigilant responsiveness to changing circumstances (esp. Norton 2005, 2015). He has also emphasized, in his work, the different, plural objectives we have when we make management decisions, and the difficulties involved when we must make tradeoffs among competing aims. This paper will evaluate one important element of Norton's account of environmental decision-making. While it is in one sense unfair to extract an element of Norton's complex view and subject it to independent analysis and criticism, the aim of this paper is constructive: by extracting an articulate statement of Norton's criterion of sustainability and critically comparing it with relevant alternatives, we may hope to expand and improve Norton's project. This process also, I believe, helps to draw out implications of that project for development theory, for economics, and for public policy.

4.2 Adaptive Management and Obligations to Future Generations

Environmental management involves present decisions that will shape the opportunities and even the circumstances of life for later generations. Our present choices may preserve or create opportunities for future generations, as they do when we take steps to conserve our accessible resources so that they will still be available for use in the future, or when we develop new technologies that enable future people to accomplish what would have been impossible or difficult otherwise. But our choices may also impose constraints: we can significantly alter, use up, or destroy resources that could have been used by future generations, and our present choices may, in a wide variety of different ways, impose limits on what they can do with their own lives. Norton's account of sustainability is designed in response to our power to constrain or enable people in the future: "A set of behaviors is… understood as sustainable" writes Norton, "if and only if its practice in generation *m* will not reduce the ratio of opportunities to constraints that will be encountered by individuals in generations *n,o,p*." (Norton 2009, p. 41).

Unfortunately, the process of environmental management involves daunting challenges: environmental systems are complex, and the effects of our interventions are often unpredictable. For this reason, among others, Norton recommends that we adopt a process of *adaptive management* for making environmental decisions. Adaptive managers adopt an experimental process, taking actions that are reversible (when possible) and studying outcomes of prior decisions to inform the process of subsequent decision-making. Adaptive managers model environmental problems as multi-scalar systems, and must be cognizant of the way management decisions influence environmental systems at different scales. Finally, adaptive management requires sensitivity to *place*. Management decisions involve interaction between human communities and the environmental systems in which those communities exist. Just as human communities are substantially unique, environmental systems are unique to the region and the place where they exist. The requirement that management must be sensitive to local human communities means that management decisions should have a democratic component. And the requirement that they be sensitive to local natural systems means that management decisions must be informed by the unique peculiarities of the locale where they are implemented.

As Norton argues, the problem of environmental management is inherently value-laden. We bring values to the process in a variety of different ways: some values are built into the analytic tools we use to evaluate policies and actions, as they are, for example, when it is assumed that goods are fungible, or that human well-being can accurately be measured by the rate at which people consume market goods. Values also drive our interest in the more purely scientific aspects of environmental management: while our effort to understand environmental systems may sometimes reflect our interest in finding undiscovered truths, our interest in these systems reflects a wide range of underlying attitudes that are linked to human interests and human concerns. The very scientific concepts we employ are inextricably entangled with underlying values in a way that makes it impossible to draw a sharp boundary between facts, as the realm of science, and values as a realm of human interests. For this reason, argues Norton, it is a mistake to think that physical or social scientists can engage management or policy problems by resolving the objective facts, and that they can then depart to leave the final choice to the political process. Ideally, adaptive management will constitute a kind of 'mission oriented science,' a process of information gathering that leads managers and communities to alter their objectives as they gain information and improve their understanding.

One important way that management is value-laden is that it is forward-looking, since present decisions may decisively shape future circumstances. At the intergenerational scale, management choices made in the present may shape the circumstances and the lives of people who don't yet exist. The concept of sustainability embodies a conception of our obligations to these future people: an obligation not to constrain the circumstances of their lives as compared with our own. It will be illuminating to compare this conception of sustainability with relevant alternatives from economics, from development theory, and from political philosophy.

4.3 Sustainability as a Ratio of Opportunities and Constraints

Recall Norton's criterion of 'sustainability' (Norton 2009, p. 41), stated in a slightly expanded version below:

> **Norton's sustainability criterion:** A set of behaviors (an institution, a policy, or a management practice) is sustainable if and only if its practice in generation *m* will not reduce the ratio of opportunities to constraints that will be encountered by individuals in generations *n, o,p…*

To take this principle out of context, one might define it as a *goal*, and shape environmental practices around the instrumental achievement of this goal. One could, for example, define a way to *optimize* the ratio of opportunities and constraints left as a bequest for future generations, subject to constraints imposed by budget, resource availability, and protections for later generations. To do so, however, would be at odds with the spirit and form of Norton's project. Norton *contrasts* his view with others that recommend optimization, and recommends that we should instead adopt an *adaptionist* model that takes goals (like the criterion above) and constraints to be provisional (Norton 2015). Adaptive management involves incremental decision-making and continued revision of goals in light of new information that becomes available. Even within an adaptionist model, however, decision-making will involve prioritization of plural goals, and consideration of value trade-offs. Such tools, however, will be used in the context of a decision structure that involves continued openness to information, and reconsideration and reformulation of decision frames.

Norton applies this criterion to 'sets of behaviors,' but here it has been extended to institutions, policies, and management practices. I believe this to be in the spirit of Norton's proposal, and it is a relevant extension in any case, since we need to evaluate sustainability in each of these domains. As Norton defines it, this criterion applies only to a few subsequent generations *n, o,* and *p,* but one might assume ellipses (…) after those three listed generations to extend the criterion into the indefinite future. It would seem that we have good reason to effect such an extension, even though our ability to apply such a criterion will be limited by our understanding of the way present actions and institutions will influence the lives of people who are increasingly distant from us in time. It is sometimes assumed (wrongly, I think) that our knowledge and ability to predict will decrease in a regular geometric sequence as events are projected further and further into the future. Sometimes this assumption is used to justify the practice of *discounting* future events. But even if our confidence about the future diminished in this way, it would not imply that we should regard institutions or practices that impoverish the opportunities of people in the more distant future to be *sustainable.* Rather it would mean that our ability to ensure the long-term sustainability of these practices will be limited by constraints in our knowledge and understanding of their predicament.

Norton's sustainability criterion is *anthropocentric,* since it focuses our attention exclusively on human beings (future human beings). But to say that Norton's criterion is anthropocentric is not a criticism of the criterion, but a comment about it. Norton advocates a weakly anthropocentric view that recognizes valuation as a human activity, but does not assume that it is irrational or inappropriate for people to place non-instrumental value on environmental preservation. By contrast, we could articulate a non-anthropocentric conception of sustainability that does not assume that nature (or natural objects or natural systems) has *intrinsic* value. For example, consider the following criterion:

> **Non-Anthropocentric Sustainability:** Institutions, policies, or management practices are sustainable if and only if their operation leaves environmental systems no more damaged in subsequent generations than they were in earlier generations.

This principle is non-anthropocentric because it focuses not on preservation of some good for people, but on environmental preservation as such. But a person's reasons for advocating non-anthropocentric sustainability as a value (as some do) might themselves be anthropocentric reasons. It is possible to read human interests into the concept of 'damage' to environmental systems, though there may be ways to avoid this. Some advocates find non-anthropocentric sustainability appealing because they regard environmental systems to have intrinsic or non-instrumental value. Others might favor the same ideal for an anthropocentric reason: because they believe that we have an obligation to preserve environmental systems for future generations.

Norton's criterion defines sustainability in terms of the opportunities and constraints afforded to later generations. It is satisfied if the *ratio* of opportunities to constraints is not reduced for later generations as compared with earlier ones. Constraints might be thought to include restrictions, burdens, and deprivations. If the existence of thriving ocean fisheries constitutes an opportunity (or more likely, a large set of varied opportunities), then the destruction of ocean fisheries will constitute a constraint, since those opportunities can no longer be exercised if the fishery is destroyed. The oil reserves in the North Sea, similarly, constitute an *opportunity*, but if we use that oil we may impose different kinds of constraints, since the oil won't be there for later generations to use, and since our present use of oil will create pollution and release CO_2.

In applying Norton's criterion we may need a way to model opportunities and constraints as quantities that can be represented in a ratio. Since the criterion is satisfied if that ratio is nondecreasing over time, we can represent the value of that ratio as a constant S, representing the present ratio between opportunities and constraints faced by the extant generation. Then Norton's criterion for sustainability can be represented as a simple equation:

$$\frac{\text{Opportunities}}{\text{Constraints}} = \text{S}$$

Several questions might be posed about this equation: Note that the ratio S can stay constant even if opportunities decline, as long as constraints decline proportionally. If there is some minimum level of opportunity that is necessary if people are to live reasonable lives, then we might reasonably be concerned not simply to keep the ratio stable, but to ensure that the numerator, expressing the quantity or value of opportunities available in subsequent generations, should not fall below a certain level. Perhaps this problem will not arise, if we assume that opportunities and constraints are causally related (as they sometimes are) so that whenever opportunities diminish, constraints increase. Once again, however, this will importantly depend on how opportunities and constraints are defined and represented, and how the values representing them are determined. In what follows, I will identify Norton's account with the view that sustainability requires maintenance of nondecreasing opportunity for later generations. It should be understood, by this, that I mean to refer to the nondecreasing ratio of opportunities to constraints, as Norton more specifically requires.

The idea that sustainability might be represented as a ratio of opportunities and constraints is intuitively plausible, if what we hope to sustain is a set of life circumstances that will leave future persons as well-off as present persons, in the specified respect. Even so, operationalizing a model that incorporates such a value is a more daunting task than it might seem. To effect such a formalization, we would need a method to quantify the values involved. For example, we might try to apply Norton's criterion by simply counting opportunities (and constraints) and listing the numbers in the numerator and denominator of this ratio. Norton does not recommend such a method, and there are good reasons to avoid such simple enumeration. First, as Norton certainly recognizes, opportunities are not all created equal. To quantify by counting would be to give equal significance to life-changing excellent opportunities, mediocre opportunities, and decidedly bad opportunities. But second, the sheer *number* of opportunities we face, at any given moment, may be infinite. At this moment, I have the opportunity to hold my arm straight, to hold my arm at a 90° right angle, or to hold my arm at any of the infinite number of angles that lie between these two alternatives. But these opportunities are insignificant. The fact that I have an infinite number of opportunities available to me, in this sense, does not mean that I am very well-off. Similar difficulties apply, *mutatis mutandis,* to the quantification of constraints. Infinite values in the numerator and denominator of the ratio would render Norton's criterion mathematically useless. So we will need a method for setting such values that avoids infinite lists of opportunities, and which appropriately evaluates opportunities of different kinds and assigns them appropriate quantificational weight. This may be very difficult to do.

Norton's response to this problem is to specify the meaning of "opportunity" in a way that is specifically designed to incorporate natural values in human decision-making processes. He distinguishes an 'option,' which he identifies as "a natural resource available for human use," from an 'opportunity,' which is "a situation in which all the conditions are right to allow the choice to use that resource at that time," including "the power to act to use the resource at a particular time" (Norton 1999, p. 135). In an early discussion of this distinction, Norton writes

> For a resource option to truly present an opportunity, the resource must continue to exist in a non-degraded and usable form at the time the person faces the choice, and the person must have the power, both physical and political, to gain access to that option. (Norton 1999, p. 135)

Even with this explanation, however, Norton's sustainability criterion will be difficult to apply in practice. There are myriad conditions necessary for a person to have physical and political power to access and use an option, but specifying them all may be impossible. And to specify these conditions for future people and future times and future conditions of the world is sure to be impossible. Application difficulties like these, however, are in no way a "knock down argument" against Norton's account of sustainability. First, the fact that a technical problem is difficult does not mean that it is impossible. In fact, as I will argue later in this paper, the development of capabilities theory by Amartya Sen and others might be interpreted as an attempt to provide a way to operationalize and quantify the key elements of a conception of sustainability like Norton's. Second, Norton's criterion might be the right one, in the sense that it captures what *matters* when we discuss sustainability, even if it were *impossible* to operationalize because of limitations in our knowledge or modeling abilities. It would be disappointing to discover that other human limitations prevent us from effectively pursuing what we care about, but if that were the case the problem would be with our knowledge, not with the criterion. But finally, and importantly, in order to use a criterion like Norton's as a guide for environmental decision making, it will not always be necessary to provide a strictly formal interpretation or to specify the impossible range of powers and conditions needed by each future person. Even for large-scale decisions, adaptive managers can guide decisions by considering the opportunities and constraints that might be implied by alternative courses of action, even if they cannot specifically quantify all of the opportunities and constraints.

It is worth noting that similar problems arise for other suggested sustainability criteria, and arguably for most practical decision-making criteria we might use. As we will see, Norton's criterion is not worse than relevant alternatives, and is better in several important ways. And there is good reason to think that we need a criterion like this one, to gauge the weight and content of our obligations to future generations, and to evaluate the effects of present actions on future persons. If the best criterion we have turns out to be a criterion that is difficult to apply, this will be a reason to work to make it more articulate and applicable. If the effects of our present choices on future persons *matter,* then we can't just throw up our hands and give up.

Difficulties associated with formal quantification of Norton's criterion do not render it useless. In some cases, we might be fairly confident in evaluations based on such an intuitive projection of future constraints and opportunities. For example, consider the case mentioned earlier, concerning the decision to cut down the ancient redwood forest around Santa Cruz California in order to make cheap bricks with a 100 year lifespan, or the decision to cut old-growth trees to make hot-tub decking. The present value of living old-growth trees is enormous, even if we restrict our estimate of that value to their economic existence-value: the amount people are presently willing to pay just to preserve them. Of course, that value is partly contingent on scarcity, which resulted from decisions to develop, like the decision under consideration here. If old-growth trees are to be cut and "developed," there are many more value-preserving ways to develop them than burning them for fuel or cutting them into deck boards. And these environmentally destructive alternatives must be evaluated against the option not to develop such a resource at all, but to leave the trees standing. In this case, as in many environmental management decisions, there are clearly-better (and clearly-worse) development paths available and we can be reasonably confident that the path selected by the brick makers of Santa Cruz reflects a net loss of opportunity. In contexts where such a comparison is difficult and judgment is uncertain, at least Norton's criterion (or its close relatives) point us in the right direction, directing our attention to the effects that our choices might have to diminish the opportunity or freedom of later generations. In a world in which processes are underway that threaten resources that might provide options and opportunities for future generations, it is not difficult to predict that our present actions—the status quo—are compromising opportunities that might otherwise have been enjoyed by future persons.

Is it a concern that Norton's criterion represents the value of environmental resources in terms of their ability to provide opportunities for human beings? In some contexts of environmental choice, this might be a serious concern: Nelson Courtland Brown and Gifford Pinchot famously saw environmental systems as *resources*, and often they judged that the best way to realize the value of these resources was to harvest and destroy. In the context of present-day environmental decision-making, the concern is more minimal. We live in an era of swift environmental change and, in many instances, destruction that threatens the most serious imposition of constraints, and deprivation of opportunities for later generations. While anthropocentric and biocentric theories of value might have had important implications for environmental policy in an earlier age, we now live in an age where advocates of different value theories are likely to converge in their judgments about most of the pressing environmental policy and environmental management decisions we presently face. This *convergence hypothesis* is another important claim that has been associated with Norton's work. If the convergence hypothesis is true, then we don't need to resolve underlying theories of environmental value before we can figure out what to do.

Norton's sustainability criterion is a principle of intergenerational distribution. In fact, it is an egalitarian principle that identifies sustainability as intergenerational equality. As such, it can be usefully compared with other views of intergenerational allocation, some of which are expressly identified as principles of intergenerational justice. Alternative conceptions of intergenerational allocation differ on at least two dimensions: they differ in the *currency* that is to be allocated among members of different generations, and they differ in the shape of the function that allocates whatever currency is chosen. The question of currency has been an important one in other areas of political theory and policy. If equality is valuable, what is it that should be equal? Candidate theories have proposed to equalize welfare, opportunities, resources, primary goods, capabilities, and freedoms, among other candidates.

Conceptions of intergenerational equity (and sustainability) differ on another important dimension as well: While some (Norton, Solow, Hartwick, Barry, Dasgupta) identify sustainability with non-decreasing availability of some good thing (opportunity, welfare, capital, productive opportunities, renewable resources), others (Mill, Brundtland, and Rawls) identify key sustainability concepts with need satisfaction, or with the provision of a sufficient minimum. Still others (Sen) urge that we should measure sustainability in a currency of freedom or capability. In what follows, I will compare Norton's criterion with some of these relevant alternatives.

4.4 Sustainability as Intergenerational Equality

The model of sustainability as intergenerational equality has become standard in most economic theory and social science literature. This model is often associated with the work of Robert Solow (1974, 1993a, b), who argues that we should equalize the *capacity to achieve welfare* across generations. According to Solow, sustainability implies "an obligation to conduct ourselves so that we leave to the future the option or the capacity to be as well-off as we are" (Solow 1993b, p. 181).

> **Sustainable Welfare (Solow Sustainability-1):** Institutions, policies, or management practices are sustainable if and only if their operation in the present generation leaves later generations the capacity to be as well off as (no worse off than) earlier generations.

Note that Solow's view is one-step removed from the view that we should equalize intergenerational welfare. It is not welfare itself, but the *capacity* for welfare that is to be equalized. If we were to interpret 'capacity' as 'opportunity,' then we might be tempted to see Solow's view as a version of Norton's, and to interpret both as recommending a principle of *equal opportunity for welfare.* Since the principle of equal opportunity for welfare has articulate contemporary philosophical defenders (Arneson 1989), this interpretation would be interesting and potentially promising.

There are several reasons why this promising interpretation must be ruled out: First, Solow makes it clear that his model equates welfare with consumption—a view that might reasonably be rejected, since there are many things that contribute to make good lives good. And as Solow explains the principle that we should equalize 'the capacity for welfare' it soon becomes clear that what he has in mind is equalization of 'access to consumable resources.' So from the view that we should equalize intergenerational welfare, he much too quickly moves to the view that our obligation to future generations is to ensure that the available stock of consumable goods will not decrease from one generation to the next. This takes Solow one step from the view that sustainability requires equalization of intergenerational *consumption*, but advocates of that view (Heap et al. 2000) can and do see Solow as their inspiration.

Sustainable Consumption (Solow Sustainability-2): Institutions, policies, or management practices are sustainable if and only if their operation in the present generation leaves later generations with consumption opportunities that are no worse than the consumption opportunities available to earlier generations.

These views are problematical. But if we were to try to follow Solow's criterion, what would we do? Even if we assume, with Solow, that the welfare of future generations depends only on their consumption opportunities, we don't know what future people will want to consume. In response to this concern, Solow argues that we cannot save *everything*. It is impossible for us to leave the world just as we found it, and Solow insists that we cannot be responsible to do what is impossible. In response to this impossibility, Solow flies to the opposite extreme: Since we cannot reasonably be required to save *everything*, Solow argues that we therefore (therefore?) have no obligation to save anything in particular. Instead, we should use resources that are at our disposal, but should set aside a pot of fungible resources that future generations can use in place of the resources we use up. Norton calls this peculiar move the "Grand Simplification," since Solow simplifies the saving problem by reducing 'saving' to 'capital.' Following Solow's earlier (1974) work on this problem, John Hartwick (1977) famously worked out the rate of investment necessary of offset declining resource stocks, so that consumption will not fall from present to future generations. The rule that we should invest at an appropriate rate so that we precisely offset the value of declining nonrenewable resources as we use them up is often called "Hartwick's Rule."

Non-Depletionary Use (Hartwick/Solow Sustainability): Institutions, policies, or management practices are sustainable if and only if their operation leaves later generations *capital resources* that are value-equivalent to the capital resources available to earlier generations.

The most common interpretation of Solow's suggestion is that we should set aside a portion of the funds generated from the use of irreplaceable resources to generate a growing investment that will be at the disposal of later generations. A more accurate account of the Solow/Hartwick view would note that they recommend investment in durable or renewable capital resources, including intellectual property and infrastructure that will that have long-term value for future generations. Hartwick proposed that we invest in "reproducible capital." He intended that we should gradually replace our reliance on natural resources with reliance on "machines" (Hartwick 1977, p. 972). The goal, however, was to guarantee nondecreasing intergenerational "consumption" under the assumption that population size remains constant from one generation to the next.

While he is at pains to distinguish his view from Solow's, Norton's suggestion that we equalize intergenerational opportunity is similar to the Solow/Hartwick view: while Solow and Hartwick focus on the preservation of a non-decreasing stock of capital resources, others (Heap and Kent 2000) focus on intergenerationally equitable *consumption.* By contrast, Norton requires nondecreasing opportunities, so that the actions of the present generation will not unduly constrain later generations. In his critique of Solow, Norton has focused on what he calls "the grand simplification:" the claim that goods are fungible, so that we need not save anything in particular since we cannot save everything. Norton urges, among other things, that environmental resources are heterogeneous on a variety of different dimensions. He argues that our choices should not be measured not merely in terms of the welfare costs our resource-use might impose. We need to consider, in addition, the extent to which the environmental damage we cause is *reversible*, and the magnitude of environmental impact as independent decision factors. I will not emphasize these issues here not because I regard them to be unimportant, but because I find Norton's arguments for these claims to be conclusive. As Norton insists, treating environmental systems and resources as an undifferentiated and fungible mass of goods is a fatal error that will lead (and has led) to disastrous decisions (Norton 1995).

Aside from the "grand simplification," the main difference at play between Norton and the economists he addresses is one of *currency*: While Solow and Hartwick focus on welfare and resource availability as the value to be maintained at a nondecreasing level from generation to generation, Norton measures sustainability as nondecreasing *opportunity.* As others have noted, there are definite advantages to the focus on opportunity instead of focusing on goods or welfare (Dworkin 1981a, b). Ideals that focus on opportunity recognize future persons as *agents*, not as mere recipients of what we pass on to them. To urge that we should protect future opportunities instead of goods or welfare is to recognize that future people must be substantially responsible for their own lives.

A more serious problem arises for these conceptions when we add the possibility of changes in population size. Solow and Hartwick both expressly assume that population will remain the same from one generation to the next, but this assumption is counterfactual, and severely limits the applicability of their account. It is easy to see why they impose this condition as a modeling requirement: a sustainability criterion that requires a nondecreasing supply of consumable resources or nondecreasing aggregate welfare will imply increasing want and deprivation, or decreasing per-capita well-being from one generation to the next, if population size is increasing rapidly from one generation to the next. Paradoxically, these criteria are fully satisfied, so under plausible population predictions, increasing deprivation or decreasing per capita welfare is to be expected on the optimal development path if these measures are used. This would seem sufficient reason to reject them both. The most obvious adjustment to make would be to move to per capita provision, or average welfare measures. But these adjustments bring their own characteristic problems (Parfit 1981; Wolf 1996, 1997a, b, 2004, 2009a).

Norton's criterion of intergenerationally nondecreasing opportunity may also be ambiguous when applied to contexts where population is changing. Norton's account of options and opportunities would seem to imply that we must not only save resources to equalize access across generations, but we must also ensure that future persons enjoy conditions that facilitate the just (permissible) use of those resources. This seems right, but less specific than one might wish. And Norton does not specify how opportunities and constraints are to be evaluated in circumstances of changing population size. Perhaps Norton's criterion should be read to require that no future person should lack adequate opportunities. This interpretation would bring Norton's view in line with the sufficientarian views to be discussed in the next section. But such a view might require *more* than merely preservation of intergenerationally equal opportunity. Since there are people in the current generation who lack adequate opportunities, such a criterion would require more than simply maintaining the status quo. Perhaps this is a reason to think that such an interpretation diverges from Norton's intent.

I have argued that criteria like those recommended by Solow and Hartwick, and even the criterion recommended by Norton will not adequately account for what we owe to future generations, this does not mean that they are entirely off-track. When we use renewable resources, it is crucial to understand how much and at what rate these resources can be used without *depleting* them. Solow's sustainability criteria have an important role to play in such calculations, since they are, in essence, principles that forbid us from using resources at depletionary rates. For an account of intergenerational justice or intergenerational obligations, however, we need to look further.

4.5 Sustainability as Sufficiency

Egalitarian conceptions of 'sustainability' like those of Solow and Norton focus on the maintenance of an intergenerationally equal, or at least nondecreasing stock of some good thing that must be preserved for the future. Such conceptions can be contrasted with another conception that has roots in the work of John Stuart Mill.

Mill's most famous discussion of sustainability appears in his *Principles of Political Economy,* Book IV, Chapter VI, "The Stationary State" (Mill 1879, p. 452). In that brief but brilliant chapter, Mill describes economic circumstances that are 'stationary' in the sense that there is no further growth, but human needs are satisfied and people are free to pursue whatever projects or whatever way of life seems best to them. Mill's 'stationary state' embodies a conception of what we would call 'sustainability,' but one that is focused on sufficiency, not on stable non-decreasing availability of resources or opportunities or any other good. This ideal also finds expression in Mill's *Principles.* Discussing the decision whether to accept a system of collectively held property and liberty, or an alternate system of private property and liberty, Mill writes:

> If a conjecture may be hazarded, the decision will probably depend mainly on one consideration, viz. which of the two systems is consistent with the greatest amount of human liberty and spontaneity. After the means of subsistence are assured, the next in strength of the personal wants of human beings is liberty; and (unlike the physical wants, which as civilization advances become more moderate and more amenable to control) it increases instead of diminishing in intensity, as the intelligence and the moral faculties are more developed. The perfection both of social arrangements and of practical morality would be, to secure to all persons complete independence and freedom of action, subject to no restriction but that of not doing injury to others. (Mill 1879, p. 129)

Mill regarded need satisfaction—universal access to the "means of subsistence"—to be an urgent social priority. This makes Mill a 'sufficientarian' in an important sense: according to Mill, satisfaction of fundamental needs (subsistence needs) is necessary so that people will be able to enjoy the benefits of "liberty and spontaneity." Once subsistence needs have been met, Mill argued, then after that point public institutions can then address other most important requirements, which are, in Mill's view, to keep people from harming one another, and to avoid interfering with their liberty of choice and freedom of action. Mill's statement here is not expressly intergenerational, but his later discussion of stationary-state economics makes it clear that he has the same goal over time and across generations. Like most utilitarians, Mill advocated intertemporal neutrality: it doesn't matter when or where needs and unhappiness exist, our obligation is to address them if it is in our power to do so. Extrapolating Mill's commitments, we might formulate this into a sustainability criterion of its own, though we might be cautious about attributing the result too directly to Mill:

> **Sufficientarian Sustainability:** Institutions are sustainable just in case they minimize unmet needs, regardless of the time or generation in which the needy persons exist.

Mill is not usually interpreted to be a sufficientarian, and one might wonder whether sufficientarianism is consistent with Mill's utilitarianism. If we read Mill to be a sufficientarian, his still turns out to be 'utilitarian' in an important sense, since he regards it as paramount to alleviate deprivation and suffering and to promote well-being. But Mill's view differs dramatically from most contemporary utilitarian ethical theories, which focus on maximizing happiness not on minimizing unmet needs or misery. Mill's view differs even more sharply from the utilitarianism of most contemporary economic theory, which does not usually distinguish between deprivation with respect to needs, and deprivation with respect to non-need-wants. The fundamentals of economic utility theory *prevent* one from making such a distinction. This has in turn led many economists to deny its relevance. But if Mill's empirical claim is true—the claim that people require only liberty once their subsistence needs are met—then institutions built on modern utility theory will not lead toward truly "utilitarian" policy recommendations. For an economist using von Neumann-Morgenstern utility theory, it may be difficult to decide whether it is better to provide food for people who are hungry or to provide theater tickets to people who are bored. The choice involves interpersonal comparisons, which cannot be made using standard economic tools. But the choice was easy for Mill, and should be easy for us: hungry people must have their subsistence needs met first. Bored people should have liberty to pursue other amusements, but they do not have unmet needs that constitute a public priority, and boredom does not give bored people a claim to public assistance.

What reasons favor a sufficientarian conception? Mill's sufficientarianism is in part based on or constituted by an empirical claim about the relationship between sufficiency and happiness. According to Mill, beyond a certain point, material goods are no longer necessary or even particularly relevant to happiness. Interestingly this thought has been confirmed by contemporary work in economics and psychology. In 2010, Kahneman and Deaton (2010) found that people who earned more than ~$75,000 per year were as well-off, in terms of their emotional well-being, as people who earned much more. Their finding supports Mill's view that subsistence needs should be met, but once they are met, we don't increase utility by increasing income or goods beyond that point. "After the means of subsistence are assured," writes Mill, "the next in strength of the personal wants of human beings is liberty" (Mill 1879, p. 129).

Other considerations are commonly cited in support of sufficientarianism, at least as a high-priority principle among others. Needs are a high-priority category, often distinguished from mere 'wants,' or from other goods that are less urgently required. A sufficientarian conception recognizes this priority of needs in the strongest terms. Needs do not always provide a justificatory foundation for rights—the fact that a person needs something is not always recognized as sufficient reason to think that she has a right to it. But on a plausible theory of rights (Hohfeld 1917; Feinberg 1992; Rainbolt 2006), a person has a right to something if she has an undefeated *prima facie* claims to it. Needs constitute a strong *prima facie* reason in favor of a person's claim. Unless that reason is in conflict with the prior or stronger claim of someone else, that reason will support a right in this minimal sense. Accounts of sustainability (and of justice) that recognize such rights will have good reason to favor sufficientarianism.

If the function of a theory of sustainability is to promote long-term human interests, sufficientarian views may do better than conceptions of sustainability that instead require non-depletionary use of resources. In the context of increasing population size, sufficientarian principles direct attention to the needs of those who require resources, not to the resources themselves. If the value of managed resources *as resources* is their ability to contribute to the lives of those who have the opportunity to use them, then it is better to adopt a theory that addresses the human need for resources first, and which treats the resources themselves as instrumentally valuable.

Mill's sufficientarian account of public priorities is naturally allied with a popular contemporary conception of sustainability, one that is usually associated with the "Brundtland Report"—the World Commission report titled "Our Common Future" (WCED 1987). While the Brundtland Report specifically offered an account of 'sustainable development,' the expression of the Brundtland criterion given below has been adapted for comparison with the other views described above:

> **Brundtland Sustainability:** Institutions, policies, and management practices are sustainable if and only if they "meet the needs of the present generation without compromising the ability of future generations to meet their needs." (Adapted from WCED 1987, p. 43)

Note that in a wide range of possible cases, this conception of sustainability, focused as it is on minimizing unmet needs, has quite different implications from the conceptions offered by Solow and Hartwick. Note, for example, that the Brundtland condition can be satisfied where there is no saving for the future at all, as long as there will be no future people to be deprived. But the Brundtland criterion does not tell us what to do in cases of tragic choice, when the two identified goals conflict with each other: What if we find ourselves in circumstances in which it is impossible to meet the needs of the present generation without compromising the needs of future people? Brundtland's criterion gives no unambiguous guidance, but can be read as giving nominal priority to the needs of the present generation—they are listed first. Otherwise, without this implied priority, the Brundtland criterion will be equivalent to Mill's more general and generation-neutral directive to 'Minimize deprivation," or "Minimize unmet needs."

Where population size is increasing, sufficientarian conceptions of sustainability will sharply diverge from intergenerational egalitarian conceptions like those of Solow and Norton. For this reason, it would also be a mistake to think that Brundtland sustainability, or sufficientarian conceptions in general, are either weaker or stronger than intergenerational egalitarian conceptions. Where population size is increasing, and future needs will be greater than present needs, it will not be enough simply to preserve non-decreasing stocks of resources or goods or to preserve an equal set of opportunities over constraints, to be distributed among members of the increasingly larger group. Sufficientarianism requires more than preservation of nondecreasing resources or opportunities in such circumstances. But where needs are met and populations size is decreasing, sufficientarianism may require less: as long as depleted future resource stocks don't cause deprivation, sufficientarianism does not require that they be preserved. Sufficientarian principles also have direct implications for fertility decisions: sufficientarianism recommends against bringing children into the world if their needs cannot be met.

Sufficientarian conceptions of sustainability might seem to mesh well with the social priorities of justice. John Rawls recommends liberal principles of justice among members of a given generation, but recommends a mixed sufficientarian principle in his discussion of justice between generations. Like Solow and Hartwick, Rawls frames this principle as an obligation to *save* for future generations:

> The purpose of a just (real) saving principle is to establish (reasonably) just basic institutions for a free constitutional democratic society (or any well-ordered society) and to secure a social world that makes possible a worthwhile life for all its citizens. Accordingly, savings may stop once just (or decent) basic institutions have been established. At this point real saving (that is, net additions to real capital of all kinds) may fall to zero; and existing stock

> only needs to be maintained, or replaced, and nonrenewable resources carefully husbanded for future use as appropriate. (Rawls 1999, p. 107)

Rawls's account here describes a two-stage process: In the first stage, before just institutions have been established, the purpose of saving is to ensure that later generations will have it better than earlier ones. Earlier generations save so that just institutions can be put in place, where the required institutions must meet needs and protect fundamental rights and liberties. As Rawls recognizes, this requires stable access to resources, but resources are not an end in themselves. Rawls does not specify exactly what should be saved, but he is usually assumed to have meant investment in long-lived capital resources. Rawls assumes without argument that population will cease to grow in the second stage, since his view will otherwise be subject to increasing scarcities.

4.6 Sustainability as Opportunity, Capability, and Freedom

Norton urges that we should pass on an undiminished ratio of opportunities to constraints. But how should we understand 'opportunity?' Norton conceives of an opportunity as a certain kind of access to a material resource, and recommends that we should pass on a bequest of undiminshed opportunities to constraints. Interestingly, this makes Norton's account surprisingly similar to Solow's intergenerational resource egalitarianism. It is not identical to Solow's criterion, since Norton denies that resources are fungible, insisting that we should leave a complex and structured resource bequest to later generations. On Norton's view, we should take careful account of the extent to which our present management decisions put at risk resources that are irreplaceable, and the effects on ecosystems are not local but ecosystem-wide (Norton 2005, p. 353). In spite of these important differences, Norton's association of 'opportunities' and 'options' with specific resources that can be used, and his insistence that his conception of sustainability is based on "stuff" not "welfare" make his view quite similar to Solow's notion of sustainability as non-depletionary use (Norton 2005, p. 306).

But we might instead read Norton's criterion in a sufficientarian mode. Note that this is not the obvious reading, and may not be what Norton has in mind. But a sufficientarian reading might be better for several reasons. A moderate sufficientarian reading would recommend that present actions should not reduce the average opportunity range available to people at present. A somewhat stronger sufficientarian reading would hold that people who lack an adequate range of opportunities are relevantly deprived, and would urge that we ensure that later generations should at least be no more deprived than the current generation with respect to the opportunity/constraint ratio we face. A fully sufficientarian reading would hold that we should minimize deprivation with respect to opportunity, between and across generations.

On Norton's view, a resource option does not constitute a proper opportunity unless it is available for use. This restriction makes his view both more complicated and more plausible than many relevant alternatives. Note that subsequent generations will have the same obligation to preserve opportunities for later generations that we in the present generation have to preserve opportunities for them. So *depletionary* use of resources would seem to be off the table: if a resource can be used only by violating one's obligations to future generations, then the conditions of one's life are not such as to make that resource "available for use" in Norton's sense. Norton insists that opportunities include resource options, combined with social and economic circumstances that render it permissible for people to *use* those options. This, in turn, means that a full account of the view will require that we specify those circumstances more fully. One way to do this would be to link the project to the development economics of Amartya Sen.

Instead of focusing on 'opportunity,' Sen argues that appropriate sustainable development involves preservation and promotion of substantive freedoms and capabilities, where capabilities provide people with the ability to do or be what they choose, with minimal constraints. Of course Sen recognizes that the freedoms of different people need to coincide, so we cannot increase the freedom of some people in ways that inappropriately encroach on the freedom of others. Sen insists that social circumstances that enable the exercise of freedom include protections for political liberties, economic facilities, and social opportunities. Sen de-emphasizes goods and resources in his discussion of substantive liberties. Although he does identify poverty as "capability deprivation" (Sen 1999, p. 87), his primary focus is on capabilities, not on the substantive resources people need to survive. In his more recent work, Sen expressly urges that sustainability is "*the preservation, and when possible expansion, of the substantive freedoms and capabilities of people today without compromising the capability of future generations to have similar - or more - freedom*" (Sen 2009, pp. 251–2). This enables us to articulate a version of Sen's view that can be compared with others we have considered above:

> **Sustainable Freedom and Capability (Sen):** Institutions, policies, or management practices are sustainable if and only if their present operation preserves or expands the substantive freedoms and capabilities of people in the present generation without compromising the ability of later generations to have similar or greater freedom.

Sen's criterion is obviously designed to mirror the structure of the Brundtland criterion. But Sen's version is ambiguous, and departs from the Brundtland approach in important ways. The requirement to "preserve substantive freedoms" and to avoid "compromising the ability of later generations to have similar or greater freedom" appears to be an injunction to maintain intergenerational equality of substantive freedoms. The objective to "Satisfy needs" is satiable—once people's needs are satisfied, the objective is completely fulfilled. But Sen's substantive freedoms do not work that way: unlike needs, freedoms are not satiable. A person can always expand one's capability set—there is no endpoint, on Sen's view, where one is maximally free. Further, while sufficientarian proposals keep us focused on those who are worst off, and recommend that we satisfy the needs of those who are badly off as a first priority, Sen's criterion is ambiguous about tradeoffs between the freedoms of those who are better off, and the freedom of those who are worse off. Sen's long career, focusing on issues of poverty and need, might lead one to think that he would favor an interpretation that would give priority to those who are worse off. But unlike the sufficientarian principles considered in the previous section, his criterion does not express this priority.

Sen's criterion also diverges from sufficientarian formulations in its implications for development paths that involve changes in population size. But when the same level of freedom is preserved for later generations, we can once again ask whether the capabilities associated with later freedoms are more widely dispersed in cases of population increase. This may seem an odd interpretation of Sen's dictum, but Sen's text makes it clear that he sees his account as a descendant of Solow, and only indirectly a descendant of Brundtland. In fact, Sen appears not to notice the importantly different implications of Brundtland's and Solow's very different accounts—he presents Solow's view as a refinement and an 'elegant extension' of the Brundtland conception, (Sen 2009, p. 250) and apparently regards his own proposal as an extension of the Brundtland-Solow view. It should be evident that the views of Brundtland and Solow—intertemporal sufficientarianism and intertemporal egalitarianism—have very different implications in a variety of different circumstances. Sen's account of sustainability, as presently expressed, does not resolve these differences. In fact, it incorporates them both. In this, it introduces serious and perhaps unmanageable ambiguities.

4.7 A Return to Adaptive Management: Paradox or Reconciliation?

All of the authors discussed in this paper—all including Nelson Courtland Brown–agree that we owe *something* to future generations. All are concerned to find a precise way to specify the value of sustainability so that we can use it to make practical decisions about environmental policy and management. Brown's example shows that the way we articulate our values can have important practical implications. The cost can be high if we get things wrong. I have raised objections to each of the formulations discussed, and none of them would seem to define an unambiguously pursuit-worthy goal. As should be clear from my remarks above, and for reasons I have explained elsewhere, in cases concerning our obligations to future generations I favor sufficientarian principles (Wolf 1995, 1996, 1997a, b, 2004, 2009a, b, 2010, 2012, 2013). Sufficientarianism focuses our attention on prevention of misery and deprivation. Equalizing opportunities between generations won't be enough to provide sufficient opportunities for everyone in contexts where population is increasing, and might not be necessary for the achievement of that goal if population levels were to decrease. But 'sustainability' is a broad concept, and it should be equally clear that we should not always use the same sustainability criterion in different practical contexts. Different sustainability criteria will be appropriate in different circumstances, for managers and policymakers who face different concrete problems.

For example, effective environmental management often means that managers will compare damage/consumption and growth rates, and will use Solow's formula to figure out what management options will avoid depletion. It is Solow's conception of sustainability that is implicit in the annual calculation of "Earth Overshoot Day" (www.overshootday.org), the day when the human population of the earth has used more resources than can be produced in the course of a year. But that conception is not, I have argued, the one that should be primary in our determination of what we owe to future generations. When we use these different criteria to guide our choices, it will be crucial to keep in mind their limitations, and to be aware of objections like those discussed above. But the 'upshot' of a paper like this one should not be to select a single winner. There is no single criterion for sustainability that can fill all the various functions we need that concept to fill.

This brings us back to consideration of the concrete circumstances of environmental decision making and policy choice, and to Bryan Norton's advocacy of adaptive management. Norton often writes as if optimization reasoning were an alternative to adaptive management. In his recent book *Sustainable Values, Sustainable Change* (2015) the two are expressly presented as alternative modes of thinking. But perhaps it is better to treat optimization reasoning as a procedure that will often take place within the context of adaptive management. Pragmatists will recognize that optimization exercises may sometimes be useful, but will recognize that such analyses are always exercises in partial reasoning, limited by the conceptual frame in which they take place. Because of these limitations, the results of an optimization exercise is never a full exercise of pragmatic practical reasoning.

The discussion above considers alternative criteria of sustainability—alternative ways to *analyze* the concept of sustainability. But adaptive management is not a criterion of sustainability, nor is it a conception of sustainability. It is, instead, an approach to sustainable decision-making that will *use* conceptions like those discussed above as tools, but will not do so in a way that is rigid or insensitive to new data, or to the changing context of choice. As I hope the discussion above shows, it matters how we specify the content of our obligation, and will be important to find unambiguous ways to express our objectives. But there is no single simple sustainability criterion that will be appropriate for use in all circumstances. Several of the principles identified above have appropriate domains where they capture important intergenerational values. The best management strategy may, as Norton suggests, involve careful and ongoing re-evaluation not only of the effects of prior management decisions, but of the way objectives and goals have been articulated and re-articulated over the course of time. Environmental management problems are complex, and the constraints of our reasoning make it impossible adequately to account for all variables and uncertainties. We cannot simply settle on a strategy as optimal, and then pursue it to the grim conclusion, we need to stop and take stock as we go. This is one of the most important lessons of Norton's pragmatist theory of management.

What happens when this simple, intelligent, pragmatist advice is ignored? If we allow our free capacity to reason about policy and management to be coopted by a simplistic maximization exercise, then we, like Nelson Courtland Brown, may find ourselves overlooking important costs as we make the decisions and evaluate the tradeoffs involved in environmental management. The values Brown failed to recognize are, in fact, values Norton has articulated with great clarity: the old-growth forests were irreplaceable, and the effects of deforestation were ecosystem-wide. Appropriate environmental management would require care to ensure that these resources would be protected and valued appropriately. Norton recommends the limited use of a precautionary approach, or of safe minimum standard principles, in contexts where management decisions have irreversible and ecosystem wide implications. Would Brown have done better if he had selected a different objective? Recall that Brown hoped to describe a system of forest management that would encourage the rise of thriving sustainable human communities. Brown envisioned communities that would exist in harmony with forest ecosystems, with workers harvesting forest resources at nondepletionary rates. It is a good vision, one that is shared by many contemporary advocates of sustainable forestry and sustainable agriculture. But it is also a seriously flawed and incomplete vision. It led to serious mismanagement of irreplaceable resources. Perhaps if Brown had reflected on the receding lines of old-growth forests, he would have realized their irreplaceable value. Perhaps if he had reflected on the continent-wide implications of deforestation, he would have come to a richer understanding of what was being lost as the great ancient trees came down.

References

Arneson, Richard. 1989. Equality and equal opportunity for welfare. *Philosophical Studies* 56: 77–93.

Brown, Nelson Courtland. 1934. *Logging: Principles and practices in the United States and Canada*. New York: John Wiley & Sons Inc.

Dworkin, Ronald. 1981a. What is equality? Part 1: Equality of welfare. *Philosophy & Public Affairs* 10(3): 185–246.

Dworkin, Ronald. 1981b. What is equality? *Philosophy & Public Affairs* 10(4): 283–345.

Feinberg, J. 1992. *Freedom and fulfillment*. Princeton: Princeton University Press.

Hartwick, J.M. 1977. Intergenerational equity and investing rents from exhaustible resources. *American Economic Review* 66: 972–974.

Heap, B., and J. Kent. 2000. *Towards sustainable consumption: A European perspective*. London: The Royal Society.

Hohfeld, W.N. 1917. Fundamental legal conceptions as applied in judicial reasoning. *Yale Faculty Scholarship Series*. Paper 4378. http://digitalcommons.law.yale.edu/fss_papers/4378.

Kahneman, Daniel and Deaton Angus. 2010. High income improves evaluation of life but not emotional well-being. *Proceedings of the National Academy of Sciences* 107(48). https://doi.org/10.1073/pnas.1011492107.

Mill, J.S. 1879. *Principles of political economy*. New York: Lee, Shepard, and Dillingham.

Norton, Bryan. 1995. Evaluating ecosystem states: Two competing paradigms. *Ecological Economics* 14: 113–127.

Norton, Bryan. 1999. Ecology and opportunity: Intergenerational equity and sustainable options. In *Fairness and futurity*, ed. Andrew Dobson. New York: Oxford University Press.
Norton, Bryan. 2005. *Sustainability: A philosophy of adaptive ecosystem management*. Chicago: University of Chicago Press.
Norton, Bryan. 2009. Ethics and sustainable development: An adaptive approach to environmental choice. In *Handbook of sustainable development*, ed. Giles Atkinson, T. Tietenberg, et al., 27–44. Northhampton: Edward Elgar Publishers.
Norton, Bryan. 2015. *Sustainable values, sustainable change: A guide to environmental decision making*. Chicago: University of Chicago Press.
Parfit, Derek. 1981. *Reasons and persons*. New York: Oxford University Press.
Rawls, John. 1999. *The law of peoples*. Cambridge: Harvard University Press.
Rainbolt, G. 2006. *The concept of rights*. Dordrecht: Springer.
Sen, Amartya. 1999. *Development as freedom*. New York: Alfred A. Knopf.
Sen, Amartya. 2009. *The idea of justice*. Cambridge: Harvard University Press.
Solow, R.M. 1974. The economics of resources or the resources of economics. *American Economic Review: Papers and Proceedings* 64: 1–14.
Solow, R.M. 1993a. An almost practical step toward sustainability. *Resources Policy* (Sept 1993): 162–172.
Solow, R.M. 1993b. Sustainability: An economist's perspective. In *Economics of the environment: Selected readings*, ed. J. Gaus, R. Dorfman, and N. Dorfman, 179–187. New York: W.W. Norton.
WCED (World Commission on Environment and Development). 1987. *Our common future*. New York: Oxford University Press.
Wolf, Clark. 1995. Contemporary property rights, Lockean Provisos, and the interests of future generations. *Ethics* 105(4): 791–818.
Wolf, Clark. 1996. Social choice and normative population theory: A person-affecting solution to Parfit's Mere-addition paradox. *Philosophical Studies* 81(2–3): 263–282.
Wolf, Clark. 1997a. Person affecting utilitarianism and population; or, Sissy Jupe's theory of social choice. In *Contingent future persons*, ed. J. Heller and N. Fotion, 99–122. Dordrecht: Kluwer.
Wolf, Clark. 1997b. Property rights, economic inequalities, and international obligations. In *Ethics in practice*, ed. H. LaFollette, 559–570. London: Blackwell Publishers.
Wolf, Clark. 2004. O repugnance, where is thy sting? On the neutral level of existence and Parfit's repugnant conclusion. In *The repugnant conclusion*, ed. J. Ryberg and T. Tännsjö, 61–80. Dordrecht: Kluwer.
Wolf, Clark. 2009a. Do future persons presently have alternative possible identities? In *Harming future persons*, ed. M. Roberts and D. Wasserman. Dordrecht: Springer.
Wolf, Clark. 2009b. Intergenerational justice, human needs, and climate policy. In *Justice between generations*, ed. A. Gosseries and L. Meyer, 347–376. Oxford: Oxford University Press.
Wolf, Clark. 2010. Intergenerational Justice and Saving. In *Values, justice, and economics*, ed. J. LaMont, K. Favor, and J. Gaus, 269–290. Stanford: Stanford University Press.
Wolf, Clark. 2012. Environmental ethics, future generations, and environmental law. In *Routledge companion to the philosophy of law*, ed. A. Marmour, 397–414. London: Routledge.
Wolf, Clark. 2013. Intergenerational distributive justice. In *Routledge companion to social and political philosophy*, ed. J. Gaus and F. D'Agostino, 467–479. London: Routledge.

Clark Wolf is Professor of Philosophy and Professor of Political Science at Iowa State University, where he also serves as Director of the Bioethics Program. He is also a faculty member in the Graduate Program in Sustainable Agriculture, and regularly a seminar titled "Foundations of Sustainable Agriculture" for students in that program. He is author, most recently, of *This Is Political Philosophy*, an introduction to political philosophy co-written with Alex Tuckness.

Chapter 5
Norton and the Search for Sustainability in Hawai'i

Jennifer M. Chirico

Ua Mau ke Ea o ka 'Āina i ka Pono
The life of the land is perpetuated in righteousness
Hawaiian State Motto

Abstract Hawai'i is the most isolated landmass in the world. It is a unique island chain where its beauty and isolation form the ideal vision of paradise. Its isolation, however, creates an environment of opportunity and challenge. Prior to European contact, Hawai'i maintained a population of a million people or more. They developed a land management system that adhered to the ecological limits of the island, and they learned to live healthy and sustainably within those limits. Modern-day Hawai'i, however, has increased anthropocentric impacts that strain current resources and deplete local ecosystems. Today, the people of Hawai'i are dependent upon other areas of the world for food, energy, and other materials. In this chapter I draw upon Leopold's essay, "Thinking Like a Mountain," and Bryan Norton's important work on how multi-scalar relationships can be significantly affected by human decisions and the importance of valuing the entire island system. Today, Hawai'i is experiencing a surge of new strategies that make living in the islands more ecologically, economically, and socially resilient. This chapter establishes Hawai'i as a model from which other communities can draw examples for setting sustainability goals, adaptive management techniques, and planning for the future.

Keywords Sustainability · Hawai'i · Adaptive management · Systems thinking

5.1 Introduction

Hawai'i is the most remote group of islands in the world, located thousands of miles from the nearest landmass. Its remoteness is often seen as a unique benefit—a tropical paradise where visitors and residents can seemingly forget the woes of the world. However, its isolation, combined with westernized technological conveniences of the

J. M. Chirico (✉)
Sustainable Pacific Consulting, PO Box 790726, Paia, HI 96779, USA
e-mail: jennifer@sustypacific.com

S. Sarkar and B. A. Minteer (eds.), *A Sustainable Philosophy—The Work of Bryan Norton*, The International Library of Environmental, Agricultural and Food Ethics 26, https://doi.org/10.1007/978-3-319-92597-4_5

twenty-first century, has led to economic challenges that manifest as dependence on the rest of the world for basic resources. Hawai'i depends on oil shipped from other parts of the world for almost 80% of its energy needs (HCEI 2017); food-supply is dependent on long-distance transportation from global shipments; available land to cultivate more food is scarce and expensive; and Hawai'i's fragile ecosystem suffers from substantial anthropogenic impacts from tourists—who, at the same time, play a central role in the state's economy. Hawai'i's governmental leaders recognize these issues and the limitations of being a remote island chain, and have actively engaged stakeholders in a facilitated, iterative process of defining and pursuing sustainability initiatives that combine knowledge from the past with technologies of the present. The stakeholder processes identified history and culture as key variables for future planning, and defining sustainability in Hawai'i has extended beyond the scope of neoclassical economics.

The process of defining and measuring sustainability in Hawai'i has been developed over the last decade. Sustainability in this context is not defined by economic welfare, but by a more comprehensive approach that takes into account community values and the unique culture history of Hawai'i. Bryan Norton's work (2005, 2015) provides a foundational understanding of how sustainability is approached in Hawai'i. He states that sustainability can be a measure of economic welfare, referred to as weak sustainability based on Solow's work (1993), or welfare and "stuff" (i.e. other variables that are important to communities but not included in traditional CBA models), which Norton refers to as strong sustainability. I argue that Norton's definition of strong sustainability is the most resilient path forward for Hawai'i (and other isolated communities), cultivated today by a strong native history, a fragile young island ecosystem, and accelerated technological advances.

This chapter provides an overview of Norton's work and how it serves as a foundational understanding of the process that Hawai'i is taking to define and measure its sustainability progress over time. It provides background on the various ways that sustainability can be defined, and how it is currently being defined in Hawai'i. A case study of the organization, Hawai'i Green Growth, shows how adaptive management is being utilized to define, measure, and track sustainability performance through its Aloha + Challenge dashboard. The chapter concludes with how Norton's framework is being applied within the Hawaiian context of sustainability, and how this process can be used in other communities around the world to advance sustainability efforts.

5.2 Defining Sustainability in Hawai'i

How we choose to define sustainability is important because it determines how it can be measured over time since there is not a standard scientific framework. Definitions range across scientific disciplines, organizations, government, environmentalists, and anyone seeking to define it for their particular community or set

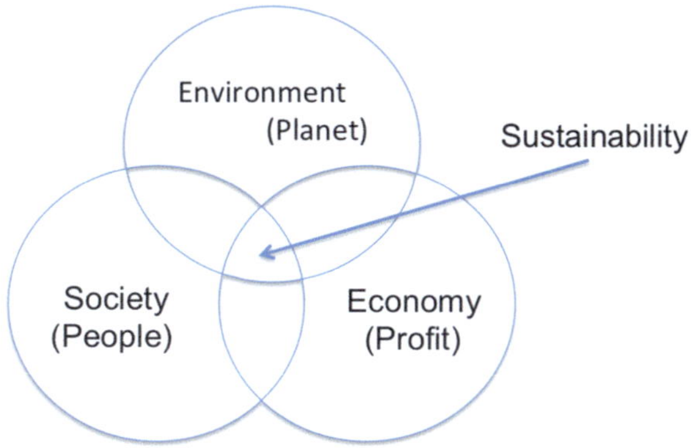

Fig. 5.1 The triple bottom line in sustainability

of circumstances. Most organizations refer to the "Triple Bottom Line" (TBL)—the integration of economic, social, and environmental balance (Fig. 5.1)—when defining sustainability. Many business and organizational models utilize this definition to provide a foundation for setting sustainability goals and communicating their efforts to the public. Sustainability reporting, and more specifically the Global Reporting Initiative (GRI), the most utilized and universal sustainability reporting framework in the world, uses the TBL to create a framework for companies to publicly report on sustainability indicators.[1] Using this definition provides an organized approach to track and report sustainability performance.

What the TBL fails to do is provide a model that illustrates the importance of community and cultural values that also informs and contributes to sustainability decisions. An alternative approach is taken in the book, *Thinking Like an Island: Navigating a Sustainable Future in Hawai'i* (Chirico and Farley, 2015). In their contributions to the volume sustainability experts from across Hawai'i identified culture as one of the important components of understanding sustainability from a place-based perspective. Many concluded that a project's success often depended upon recognizing and including culture in its sustainability efforts. Companies initiating new sustainability projects in Hawai'i, such as in renewable energy, watershed maintenance, and regenerative agriculture—and especially projects originating from the mainland United States—have found it impossible to ignore cultural values. Project managers often hire consultants to facilitate a community stakeholder process for the public, and particularly Native Hawaiians, to provide input (Chirico and Farley 2015).

While culture is often nested within the "social" component of the TBL, the contributions to *Thinking Like an Island* emphasized the importance of including and acknowledging culture as another key pillar of defining sustainability. This led the authors to advance the TBL one step further: culture was added as a fourth

---

[1]https://www.globalreporting.org/.

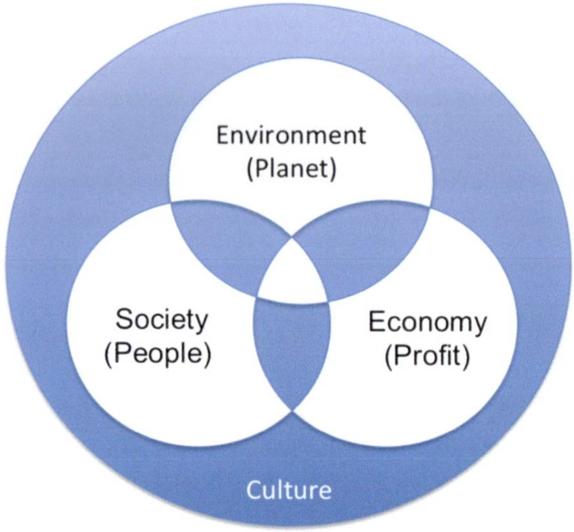

Fig. 5.2 An emerging model of sustainability representing a Quadruple-Bottom-Line (QBL): four concentric circles with culture encompassing all other components. It is possible for inner rings to be arranged differently (Chirico and Farley 2015)

dimension to the Venn diagram, which is referred to as the Quadruple Bottom Line (QBL) (Fig. 5.2).

The TBL and QBL provide helpful frameworks for communities and organizations to define and organize their sustainability priorities. Norton (2005), however, reviews various definitions of sustainability and concludes that a core meaning exists across all of them: "sustainable living is forward-looking living…it is about the future, our concern toward it and our acceptance of responsibility for our actions that affect future people" (p. 304). People living in democratic societies generally believe they have an obligation to the future and prefer to live in a society that cares for the future, although disagreement exists about *what* to care for, and about *how* to care for it. In the past, such as in Hawai'i, tribal leaders and generational knowledge was used to guide actions for preserving and conserving resources. Today scientific modeling is the predominant method for predicting future outcomes, yet accelerating technological change and the human ability to control nature create conflicts with traditional place-based knowledge and values. Different community and cultural values exist that delineate agreement about the best process for identifying and implementing sustainability measures. This value dilemma is witnessed within recent environmental policy debates in the U.S. Some people believe that there is too much spent on environmental protection and conservation, and some believe that there is not enough. The reasons on both sides can be rooted in either cultural or scientific values. These differing values are illustrated in Norton's spectrum of weak-to-strong sustainability (Fig. 5.3).

Weak sustainability is based on neoclassical economics and counts only welfare. It asserts that sustainability is assured as long as generational capital savings are non-declining (Norton 2015). "It is governed by the comparatively weak constraint that total capital available must be nondeclining from generation to generation…

	Weak Sustainability	Strong Economic Sustainability	Strong Sustainability	Normative Sustainability
Home Discipline	Mainstream Economics	Ecological Economics	Systems ecology	Policy science/ Environmental ethics
Paradigm	Welfare economics	Welfare econ + natural captital	Complex dynamic systems theory/AM	Complex dynamic systems theory/AM
Definition	Maintenance of undifferentiated capital	Weak sustainability + maintenance of natural capital	Weak sustainability + maintenance of resilience	Weak sustainability + maintenance of options
Key Concepts	Nondeclining wealth	Maintaining natural capital	Maintaining resilient ecosystems	Integrity of place
Key Advocates	Solow	Pearce & Barbier	Holling, Lee	Leopold, Norton

Weak Sustainability – Welfare Counters ⟷ Strong Sustainability – "Stuff" Counters

Fig. 5.3 Weak-to-strong sustainability spectrum (Norton 2005)

[and] is achieved when total societal wealth is maintained across generations" (Norton 2005, p. 307). One of its most important distinctions is that it only focuses on welfare changes, and it allows for unlimited substitution of natural resources, claiming that fungibility is possible.

Strong Economic Sustainability represents a more integrated approach between economists and ecologists. Sustainability is economically evaluated, but it is acknowledged that some aspects of ecological change are damaging to the future. It claims that using "natural capital" can protect aspects of natural systems. Since advocates have not been able to define natural capital in noneconomic terms, it is more of a modified version of weak sustainability that accentuates natural substitutions when pursuing economic sustainability (Norton 2015).

Strong Sustainability focuses on systems theory and embodies adaptive management for the resilience of ecosystems on multiple levels using empirical approaches. It counts both welfare and "stuff." Norton defines "stuff" as welfare and anything else that matters to a community that is not always easily measurable: ecosystems, clean water, parks, clean air, history, and much more. "Stuff" is based on what is important to a community—what they value and what is most important to protect for future generations. Strong sustainability places limits on substitutions, protocols for the replacement of natural assets with human-built ones, and more constraints on destroying "stuff" (Norton 2005).

Normative Sustainability expands from strong sustainability to include community values. Community values guide sustainability and the management and resiliency of ecosystems. Given that some communities might have short-term desires for consumption that curtail global sustainability, Norton (2015) claims that normative sustainability is about values that emerge during processes focused on protecting natural systems. He defines normative sustainability as "[sustaining] the values a community would espouse as the outcome of an appropriate process of deliberation and social learning" (Norton 2015, p. 78).

Strong and normative sustainability lean more toward viewing the world and all of its processes from a systems perspective. Aldo Leopold's (1949) "Thinking Like a Mountain," an essay in *A Sand County Almanac*, illustrates the importance of understanding sustainability from this perspective. In his essay, Leopold shows the value of the whole ecosystem, and how one poor decision that does not take into account a holistic perspective can affect systemic natural relationships. "Thinking like a mountain is thinking about human values as time-sensitive and as produced by specific processes and dynamics that unfold on identifiable scales" (Norton 2005, p. 230).

History and cultural knowledge of systemic natural relationships has moved to the forefront in defining sustainability for the State of Hawai'i. In 2008, stakeholders from across the state participated in facilitated working sessions to develop a Hawai'i 2050 Sustainability Plan (Higa 2008). The report defined sustainability as: "a Hawai'i that:

- Respects the culture, character, beauty and history of our state's island communities;
- Strikes a balance between economic, social and community, and environmental priorities; and
- Meets the needs of the present without compromising the ability of future generations to meet their own needs" (p. 13).

This definition is 3-fold: 1) it provides a uniquely crafted definition of sustainability that highlights the importance of Hawaiian culture and history—the other "stuff" that the community values as important to the future, illustrating greater alignment with Norton's strong sustainability; 2) it captures the TBL, recognizing the balance of economic, social, and environmental priorities; and 3) it recognizes the most universally used-definition of sustainability, articulated by the United Nations in the Brundtland Report, *Our Common Future*: "Meeting the needs of the present without compromising the ability of future generations to meet their needs" (United Nations 1987).

5.3 The History of Sustainability in Hawai'i

Culture and history were identified as important aspects of sustainability in the Hawai'i 2050 Sustainability Plan partly because they represent the community values and, partly, because only several hundred years ago, Hawai'i fostered a highly sustainable community and lifestyle. Prior to European contact in 1778, Hawai'i's resources nurtured a population of about one million people (Dye 1994).

> In precontact Hawai‘i, the community sustained itself through the establishment of reciprocal and familial relationships between humans and the natural world, as well as among the various human communities. In this regard, Hawaiians asserted their right to appropriately harvest those items critical to human welfare while also acknowledging an obligation to ensure the long- term health of the ecosystems upon which all life depended. (Fisher 2015, p. 9)

Hawaiians cultivated the natural environment and were able to feed the entire population without dependencies on foreign resources.

The heart of this strategy was the *ahupua‘a,* a mountain-to-sea land and ecosystem management system (see Fig. 5.4). The islands were divided into wedge-shaped parcels that contained, and were sometimes defined by, riverine watersheds. "The *ahupua‘a* comprised the most basic unit of land within which an individual had the right to access resources" (Fisher 2015, p. 11). Every Hawaiian community within each *ahupua‘a* had fresh water that flowed from the mountains to the sea, nurturing the growing food supply and providing clean water for human and animal needs. The center of each *ahupua‘a* system included thriving community centers for farming taro[2] and building shelter. The oceanside areas were abundant with fresh fish and vegetation that provided healthy protein diets. Banana and coconut trees provided housing, household items, and additional food. The *ahupua‘a* system represented a cyclical process that was aligned with the biorhythms of nature: all of the materials within the system were biodegradable and returned to the land after use to build healthy soil for growing food for the next season (Kane 1997; Chirico and Farley 2015). The system was interconnected and interdependent.

> …access to cool, clean water made the difference between wealth (waiwai) and impoverishment (hune), and ultimately life and death. While the ease of available water allowed for the community to flourish, a reliance on a readily abundant quantity of water was not seen as an absolute barrier to human habitation. Rather, through the ingenuity that comes from a culture that evolved for millennia on islands, an intimate knowledge of and familial connection to the environment and the natural world that shaped the land, and an interdependent community, the Hawaiian people flourished as stewards of a sustainable island community (Fisher 2015, p. 14).

Similar to Leopold's "Thinking Like a Mountain," it created a balanced ecosystem of systemic natural relationships among people, animals, plants, and the earth. This system provided a sustainable foundation for the Native Hawaiian culture for hundreds of years.

Although demographers often disagree on the Hawaiian population in 1778 when Captain Cook arrived, Hawaii's isolation suggests a high level of interdependent sustainability. The arrival of foreigners in the eighteen-century led to profound changes that ultimately resulted in the loss of political and economic independence. The changes were a result of the introduction of foreign weapons (e.g., guns, canons) and foreign diseases, which decimated a large portion of the indigenous population. "The arrival of foreigners in general, and congregational

---

[2] *Colocasia esculenta,* a staple food for Native Hawaiians that is also features in Hawaiian creation stories and has vast cultural importance.

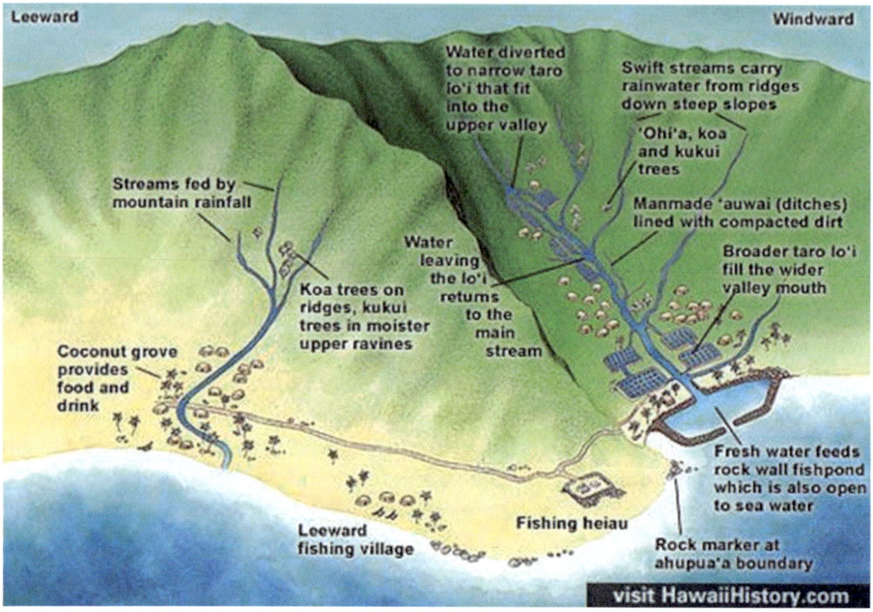

Fig. 5.4 Hawai'i historical Ahupua'a system (www.HawaiiHistory.com)

missionaries in particular, inaugurated a rapid transition from an island worldview to a European or western worldview" (Fisher 2015, p. 15).

Missionaries began introducing westernized cultural approaches to the people. Prior to this time, land ownership did not exist—the Hawaiians believed the land belonged to the people and to the earth. Missionaries introduced the concept of land ownership and began trading materials for land. They introduced westernized land management practices related to agriculture and husbandry, and new foreign materials. Slowly, the islands started departing from traditional practices and adopting more westernized approaches for land management, education, and ways of living, as well as outside resources and materials. Eventually, these practices, coupled with technological advances, brought new challenges to Hawai'i that resulted in dependence on resources from other places.

Energy dependency has become one of the biggest challenges. Today, Hawai'i is 80% dependent on oil from other parts of the world to meet its electricity and transportation needs (HCEI 2017). This dependency makes the state vulnerable to sharp price fluctuations in oil, which have occurred regularly over the last several decades. Residents use the fourth-lowest amount of energy per capita in the U.S., yet they pay the highest prices for energy in the nation (EIA 2017). Fossil fuels provide electricity for most homes and businesses, and power most equipment, such as computers and phones, which have become essential tools for conducting business and participating in the current economy. Food supply in Hawai'i is

85–90% dependent on imports, and in the event of an emergency where oil is not available, it is estimated that there are only five to seven days of food supply for the islands (DEBDT & Hawai‘i Department of Agriculture 2012). In addition, energy is required for pumping fresh water to homes, businesses, and farms, as well as for pumping used water from those same places to waste water facilities for processing. Imported consumer goods create additional economic, environmental, and societal costs in the form of solid waste management. As more packaged goods are imported to the islands that cannot be disposed of naturally, more land and air quality are sacrificed to manage the waste through landfilling and incineration. Many landfills in Hawai‘i are nearing or at capacity, and finding space for new landfills is difficult and expensive. Alternatives to landfilling, such as recycling, are more challenging due to the low economies of scale. The population on each island in Hawaii is small, and most recycling is collected individually and shipped to Asian countries rather than processed in local markets.

As these issues are addressed, maintaining the integrity of, and respect for, Hawaiian culture has been a key component of working toward greater sustainability in Hawai‘i. Many traditional Hawaiian resource-use practices help model modern solutions to resource management issues. A resurgence of historical efforts to restore Native Hawaiian sustainability using the resources from the past to create a more sustainable future is in effect. In *Thinking Like an Island: Navigating a Sustainable Future in Hawai‘i,* the authors discuss examples of various sustainability projects that have been successfully implemented, as well as the accompanying challenges. While the contributing authors were not asked to include culture within their chapters, culture emerged as a strong theme in every area. The authors illustrated how people came together to grow food, generate clean energy, build homes, and educate students, while preserving the Hawaiian Islands’ unique native culture (Chirico and Farley 2015). Hawaiian culture is also being revitalized and integrated into local business models, government, and scholarly research on sustainability. The consistent, central role for culture is a clear indicator that in Hawai‘i, successful sustainability approaches are guided by cultural and community values. These approaches are consistent with Norton’s strong sustainability perspective.

5.4 Adaptive Management and Measuring Sustainability in Hawai‘i

Recognizing the unique cultural and historical values in Hawai‘i requires sustainability models that are beyond those upon which neoclassical economic models are based. Local and national groups that commit to protecting the environment in Hawai‘i recognize this need, and many of them embrace commitments to protect places and create institutions that will continue to save them. For example, the Hawaiian Island Land Trust (HILT), and other similar non-profits in Hawaii, are using place-based values to plan and work toward common goals for the future. HILT preserves sacred Hawaiian lands and provides free education to the public about the history of Hawaiian sustainability and preserving it into the future through its Waihe'e Coastal Dunes and Wetlands Refuge project.[3] Such commitments draw on continuity between the past and the future and attempt to create models that "are founded on the commitments that a community makes to continuity with its past, to its natural and cultural histories, and to a future in which its roots in nature are revered, protected, learned from, and cared for" (Norton 2015, p. 334). Norton recognized this type of effort as vital to sustainability efforts:

> …The key terms sustainable and sustainable development are not themselves general descriptors of states of society or cultures but rather refer to many specific sets of commitments on the part of specific societies, communities, and cultures to perpetuate place-based values and project them into the future. This projection must include a strong sense of community and a respect for the natural history of the place in question. The problem of how to measure success and failure in attempts at living sustainably is not the problem, for each community, of choosing a fair natural legacy for the future and then operationalizing its commitments as concrete goals to be measured by democratically agreed-upon indicators. (P. 336)

Creating a unique process for choosing "a fair natural legacy for the future" helps set goals that, in turn, create the foundation for developing measurable indictors for sustainability performance.

The State of Hawai'i has embraced this community-based process. Hawai'i Green Growth (HGG) is an organization tasked by the Hawaiian government to following through on the goals of the Hawai'i 2050 Sustainability Plan. One of the goals of the Plan is to measure sustainability performance. To reach that goal, HGG started implementing the Aloha + Challenge in 2014, a dashboard for sustainability indicators that could be measured over time (HGG 2017). The unique aspect of the Challenge was the process used to determine what to measure and how to measure it. HGG facilitators used an extensive stakeholder engagement process for identifying the "stuff" that matters to the Hawaiian community and determining appropriate indicators to measure it over time. Stakeholders from across the four counties within the state of Hawai'i (Oahu, Maui, Kauai, and Hawai'i) were invited to participate in facilitated working groups to discuss common community values and determine the best indicators for measuring sustainability performance. The working group meetings were two-day events that were held on each island. The public was invited to attend the first day of meetings, and people who wished to engage more were invited to participate on the second day to help determine the most appropriate indicators. The engagement process was primarily funded by the governor's office, and associated travel costs were provided to fly stakeholders to other island counties to participate.

[3]http://www.hilt.org/protected-lands/maui/waihee-refuge/.

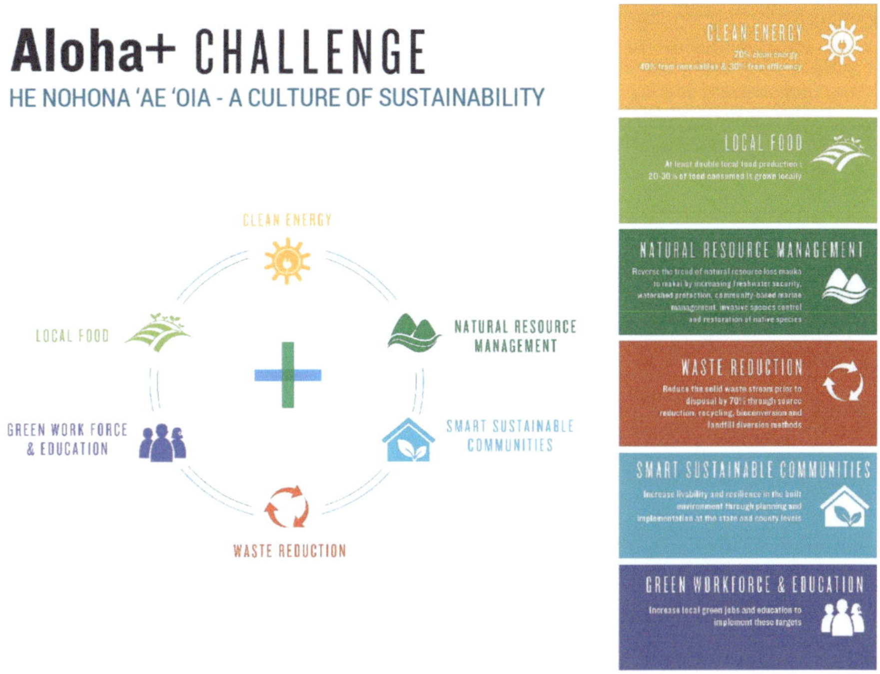

Fig. 5.5 Hawai'i Aloha + Challenge sustainability targets (Hawai'i Green Growth 2017)

It was a lengthy, iterative process that reflects the significance that the Hawai'i government places on engaging the public and identifying the values that matter most to the community. The process engaged stakeholders from across the state in an on-going series of meetings. The most important sustainability issues were identified, reviewed by the committee and outside experts, and then revised and strengthened to better reflect the values and scientific knowledge available. The results of the process was the Aloha + Challenge. The dashboard[4] provides measures of "stuff" that are important to community and cultural values—and corresponds to goals that are manageable and measurable. The dashboard is referred to as "he nohona 'ae 'oia: a culture of sustainability" and focuses on six target areas of sustainability (Fig. 5.5).

The process embraced culture and the ahupua'a as a central theme for community-driven data to guide decisions about what to measure and how to measure them. Figure 5.6 provides a diagram of the goals connected to the Hawaiian ahupua'a. Goals for the six sustainability targets—clean energy, local food, waste reduction, natural resource management, green workforce and education, and smart sustainable communities—are identified within the ahupua'a.

[4]https://dashboard.hawaii.gov/aloha-challenge.

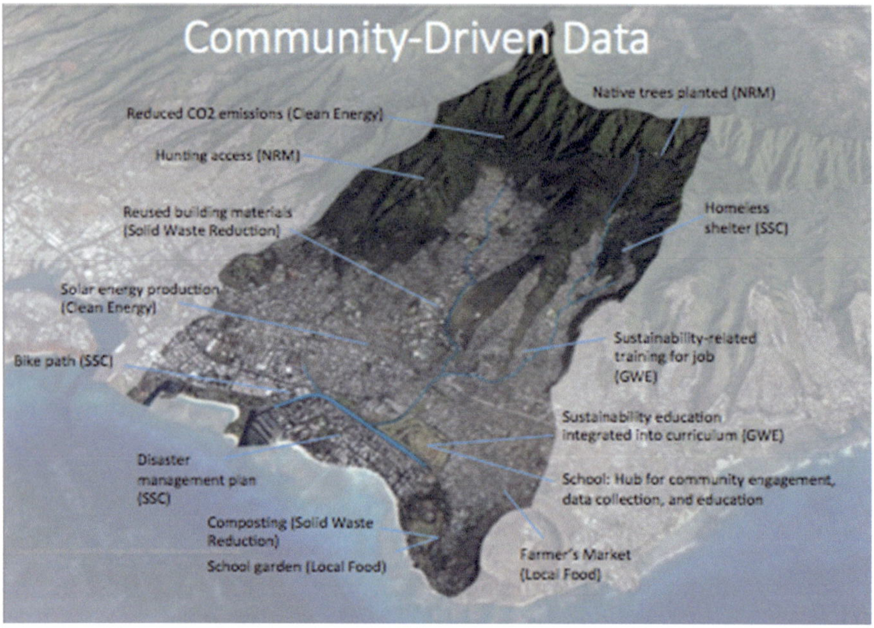

Fig. 5.6 Hawai'i Green Growth adaption of Hawaiian ahupua'a with community driven data (Hawai'i Green Growth, 2017)

Similar to Norton's adaptive management process, this process included real world scenarios and environmental managers working in the field. The Challenge process embraced multiple feedback loops from the community and used adaptive management techniques to identify measurable indicators. According to Norton (2005),

> Choosing measureable indicators is logically subsequent to a commitment to moral goals, but the tasks of choosing measurable indicators can, and must, proceed simultaneously with the articulation of long-term environmental goals. It cannot be otherwise because the choices that are made by real communities regarding which indicators are relevant to their moral commitments represent, in effect, an operationalization of moral commitments… [and], the specification of a legacy or bequest for the future must ultimately be a political question, to be determined in political arenas. The best way to achieve consensus in such arenas is to involve real communities in an articulation of values, in a search for common management goals, and to include in that process a publicly accountable search for accurate indicators—measures of the stuff that the community cares about—to correspond to management goals. (p. 335).

For the Aloha + Challenge, once the indicators were determined through a community-based process, the working group discussions continued through continuous feedback loops to improve upon and provide greater clarity about the goals, indicators, and measurements over time. With input from the community, government, business, and scientific experts, the process allowed for indicators to be identified for each target area that were realistic to measure over time but also viewed as important based on economic, environmental, or social priorities.

5.5 Norton and the Search for Sustainability in Hawai'i

Hawai'i's search for sustainability will continue in the years to come. In the interim, Hawai'i is forging forward with bold sustainability goals. For instance, Hawai'i recently became the first U.S. state to set a goal of using 100% clean energy by 2045. The Aloha + Challenge continues to gain more momentum, and stakeholder engagement remains at the forefront for maintaining an open, transparent process where all community members are invited to participate in public meetings.

The Challenge provided a model for using adaptive management techniques that involved an experimental approach with the local community and key stakeholders in value discussions, and decision-making to determine measurable indicators. Norton (2005) discussed three things that were important factors to Aldo Leopold that also apply to the sustainability efforts in Hawai'i. The first principle was relying on experience for learning. To Leopold this was the "measure of all truth." His understanding was that learning often takes place through experience, and it is difficult to know what is best for the land and people without actually being engaged or actively and carefully observing it. This requires a more bottom-up approach of engaging people who are working in the field and have real world experience with the subject matter. Science can be used in conjunction with experience to provide validation and protection from observation bias. Most of the stakeholders engaged in the Challenge process were actively involved in professions, including science, that were identified as important to sustainability in Hawai'i, such as in natural resources, energy, and waste management.

The second important factor to Leopold was the use of group adaptation and that anthropocentric activities must be "embedded in institutions and practices that leave the land alive and undamaged" (p. 122). History shows that the Native Hawaiians were able to sustainably manage the ecosystem without damaging the environment, while also meeting their basic needs of food and shelter. The introduction of western approaches to land and consumerism affected Hawai'i, and in some cases, left it damaged, such as with coral reef degradation (Szmant 2002). However, the recent Hawai'i Sustainability Plan clearly articulates that sustainability means respect for the "culture, character, beauty and history." While there is still progress to be made, the Plan illustrates that a respect for the land, and approaches that leave the land alive and undamaged are recognized by Hawai'i government institutions, and often actively sought after to provide guidance for the future. Leopold's (1949) trio of integrity, stability, and beauty, all understood within an expansive community concept, is similar. Indeed, in *A Sand County Almanac*, Leopold's essay *The Land Ethic* provides a vision in which the traditional (human) community boundaries are enlarged to "include soils, waters, plants, and animals, or collectively: the land (p. 204).

The third important factor identified by Leopold is that "people's values have consequences for the land" (Norton 2005, p. 90). Values are real, and they have impacts. They are equally subject to the regimens of observation as scientific hypotheses. Norton indicated that practitioners see problems, including scientific ones, through the lens of values, and that even a scientific hypothesis is invoked by someone's belief that something "ought" to be done. Norton discusses how Leopold's thinking is similar to the American philosopher John Dewey, who argued that social learning is the key to expanding our experiences. This is witnessed in the HGG process, where community values were identified prior to setting goals and making decisions about the most appropriate indicators and measurements. Hawai'i leaders recognize that sustainability projects benefit from, and require, stakeholder engagement in order to succeed. Government Request for Proposals often identify stakeholder engagement as essential to any planning process for new projects, and the HGG process assessed the community values while engaging the public and key stakeholders in its efforts to map a sustainable future.

One of the most vital results of the HGG process was recognizing that Hawai'i stands as a leader in navigating community processes that place community values at the forefront of sustainability discussions. This places Hawai'i firmly within the strong and normative sustainability areas of Norton's weak-to-strong spectrum. While neoclassical economics is still used to inform decision-making, sustainability is not defined based solely on welfare economics. Instead, current processes utilize a systems approach for understanding the island ecosystem and for facilitating an adaptive management process. Historically and cultural values matter and have moved to the forefront of sustainability conversations. Hawai'i has also begun designating more limits on substitution. For instance, in the goals of the Aloha + Challenge, there is more emphases on planting native plants and trees that have been destroyed, rather than non-native ones. Natural waste cycle alternatives to reducing waste, such as composting and recycling, rather than building new landfills, are explored first. This is also illustrated by the rapid replacement of fossil fuels with renewable energy from earth elements—primarily energy from the sun (solar panels), wind (wind turbines), earth (geothermal), and ocean (wave energy, which is currently being tested in Hawai'i as a viable option). Non-profit organizations, largely funded by government organizations, have begun actively removing invasive species and protecting the islands from new invasive species. The number of non-profits with missions to protect fragile watersheds has increased on all of the islands, and watershed collaborations have been established to develop more holistic approaches for ecosystem balance. Cultural education and preservation has become a cornerstone for many environmental non-profits. They collaborate with schools to teach students across the islands about the rich cultural history of Hawai'i and began engaging them in hands-on activities at early ages.

These examples show that the integrity of place and community involvement, identified as key aspects of the strong and normative sides of Norton's sustainability spectrum, are evidence that "stuff" matters in Hawai'i. Adaptive management, combined with a rich cultural history and new technologies that create more systemic natural processes, provide unique learning experiences. These efforts illustrate that historical and cultural values are important variables in planning for the future, and that strong sustainability is required for creating a sustainable, and more importantly, resilient and regenerative economy, ecosystem, and society in Hawai'i.

Norton's work provides a foundation for other communities around the world to address sustainability issues through a community-based process that is built upon the understanding that people and the land are intertwined and interdependent. Adaptive management and the concept of "thinking like a mountain" can be applied to a multitude of diverse situations. Organizations, communities, and governments can use this process to facilitate stakeholder-based decision-making related to sustainability issues. This process engages the community in a meaningful effort to identify community and cultural values. It provides a framework for defining sustainability using a place-based approach to assessing values and addressing what matters to a community, yet it is also rooted in empirical data. Perhaps most importantly, it helps determine measures of sustainability that can inform decision-making, while also respecting the culture and history of place.

References

Chirico, J., and G. Farley (eds.). 2015. *Thinking like an island: Navigating a sustainable future in Hawaii*. Hawaii: Honolulu.

Dye, T. 1994. Population trends in Hawaii before 1778. *Hawaiian Journal of History* 28: 1–20.

Department of Business Economic Development and Tourism [DEBDT] & Hawai'i Department of Agriculture. 2012. Increased food security and food self-sufficiency strategy. Available at http://files.hawaii.gov/dbedt/op/spb/INCREASED_FOOD_SECURITY_AND_FOOD_SELF_SUFFICIENCY_STRATEGY.pdf.

Fisher, S. 2015. Hawaiian culture and its foundation in sustainability. In *Thinking like an island: Navigating a sustainable future in* Hawaii, ed. J. Chirico, G. Farely. Hawaii: Honolulu.

Hawaii Clean Energy Initiative (HCEI). 2017. Available at http://www.hawaiicleanenergyinitiative.org/.

Hawaii Green Growth (HGG). 2017. Available at https://hawaiigreengrowth.org/aloha-challenge.

Higa M. 2008. Hawaii 2050 sustainability plan: Charting a course for Hawai'i's sustainable future. Available at http://www.oahumpo.org/wpcontent/uploads/2013/02/Hawaii2050_Plan_FINAL.pdf.

Kane, H.K. 1997. *Ancient Hawaii*. Hawaii: Captain Cook.

Leopold, A. 1949. A Sand County Almanac. New York, New York.

Norton, B.G. 2005. *Sustainability: A philosophy of adaptive ecosystem management*. Chicago: University of Chicago Press.

Norton, B.G. 2015. *Sustainable values, sustainable change*. Chicago: University of Chicago Press.

Solow, R.M. 1993. Sustainability: An economist's perspective. In *Economics of the environment: Selected readings*, ed. R. Dorfman and N. Dorfman, 179–187. W. W. NOrton: New York,

Szmant, A. 2002. Nutrient enrichment on coral reefs: Is it a major cause of coral reef decline? *Estuaries* 25: 743–766.

United Nations. 1987. *Our common future—Brundtland report*. Oxford: England.

U.S. Energy Information Administration (EIA). 2017. Available at http://www.eia.gov/state/?sid=HI.

Dr. Jennifer M. Chirico is the CEO and founder of Sustainable Pacific Consulting, a Certified B Corporation that specializes in sustainability and responsible business. Dr. Chirico is an adjunct professor at Hawaii Pacific University in the Global Leadership and Sustainable Development (GLSD) master's degree program. She holds a Ph.D. in Environmental Policy, a Master's in Public Health, and a Bachelor's of Science in Management and Finance.

Chapter 6
Sustainable Change in a Fractured World

Paul D. Hirsch

Abstract Bryan Norton's body of work—with its focus on public deliberation, appropriate and fair procedures for decision-making, and epistemological humility—offers a pathway to sustainable change in the context of social diversity and scientific uncertainty. Working towards sustainable change amidst the degree of political and social divisiveness we are seeing today, however, may call for additional levels of sensitivity: to the way problems and publics are constructed, and to one's own embeddedness within the complex dynamics that comprise a problem's context. When fractures are substantial, it may be unrealistic and even unproductive to assume that deliberative efforts will yield a unified notion of the public interest. In such cases, the challenge for leaders and facilitators is to identify opportunities for the protection of shared values, and to establish democratic spaces that allow conflicts to be engaged in ways that do not require the destruction or domination of adversaries.

Keywords Polarization · Uncertainty · Leadership · Sustainability Ethics

6.1 Introduction

At both global and national scales, we are living in a political milieu shaped to a significant degree by divisiveness and polarization. Within the United States, while people's views on policy have not changed dramatically over the years, the levels of political and social sorting have. Rather than make room for difference, people holding alternative visions of the social good seem less willing to cooperate, compromise, and live alongside those who see things differently (Pew Research Center 2014). This drama of divisiveness takes place against a background of

P. D. Hirsch (✉)
Department of Environmental Studies, State University of New York College of Environmental Science and Forestry, Syracuse, NY 13210, USA
e-mail: pahirsch@esf.edu

S. Sarkar and B. A. Minteer (eds.), *A Sustainable Philosophy—The Work of Bryan Norton*, The International Library of Environmental, Agricultural and Food Ethics 26, https://doi.org/10.1007/978-3-319-92597-4_6

concern about environmental changes, at levels from the local to the global. However, like tectonic plates butting against each other and vying for the dominant position, fragmented and contending publics blanket and often obscure the singular and finite planet beneath. And although citizens in many countries remain blessed by nature's abundance, for others—be it people living on the Alaskan coastline or in drought-ridden parts of the world—the limits to that abundance loom large. The exact parameters of the limits beyond which nearly all visions of the social good will become less viable are dynamic, uncertain, and perhaps unknowable. Nevertheless, we close our eyes to them at our peril.

The work of sustainability takes place at this nexus: of social and political fractures; scientific uncertainty and ignorance; inequality of access to environmental goods and exposure to environmental ills; and dangers such as lack of access to food or lack of personal safety in conflict-ridden or hurricane-prone regions. Given the complexity of today's environmental problems, and in particular given the fractured character of a significant portion of social and political life, finding traction and achieving progress requires something more than intellectual work alone—it requires moral leadership. However, if our concern is in transcending polarization, then it won't do to look to pre-existing moral statutes, or claims of right and wrong that line up with the deepening fault-lines among existing world-views.

In the absence of agreed upon moral commitments, then, how **can** sustainable change be pursued in the context of social division, political polarization, scientific uncertainty, and various forms of inequality?

In this chapter, I engage and build on the scholarship of environmental philosopher Bryan Norton to address this question. Norton has spent his career developing concepts and approaches to help people overcome disagreements, learn and adapt in the face of complex challenges, and define a path towards environmental stewardship that transcends the boundaries of ideology to serve our deepest human values. As his Ph.D. student and ongoing collaborator, I have had the opportunity to see in action Norton's approach to sustainable change through a deliberative—and a creative—process. Recently, however, the prospects for deliberation have seemed more uncertain. The viability of a deliberative process depends on individuals and groups recognizing that they are part of a larger public by virtue of their collective interest in addressing shared problems. Given the ideological sorting that we are now witnessing in the United States, however, it seems essential to pay close attention to the ways in which problems and publics are themselves contested categories, and often get constructed in ways that gloss over important differences. Directing our attention to the fractures, then, and to the processes by which publics and problems are constructed, may highlight safer spaces for the expression of difference, as well as genuine points of communicative and collaborative potential. Moreover, if we can learn to operate amongst fractures without glossing over them, it may help to avoid more destructive conflict in the long run.

What I wish to do in this chapter is to explore how Norton's process-oriented approach to sustainable change, pursued with what can be called a *critical sensibility*, can meet the demands of the current milieu. The outline of the argument is as follows. In the first section, through focusing on the first three "procedural heuristics" defined in his 2015 book *Sustainable Change*, I characterize Norton's approach to deliberative decision-making in the context of value diversity and scientific uncertainty. In the second section, I point to some of the limits to deliberation given the polarized social contexts in which a great deal of current environmental problem solving takes place. In the third section, I point to two concepts, drawn from the critical social sciences, that can both inform deliberative problem solving and help to clarify its limits within divisive and unequal contexts. The first of these concepts is *critical reflexivity*, which entails turning the lens we generally apply to understand *other* people and phenomena towards understanding *our own* embeddedness within social landscapes characterized by various kinds of inequality, both straightforward and implicit. The second concept is political theorist Chantal Mouffe's distinction between *agonism* and *antagonism*, which speaks to the need for non-destructive ways of engaging the conflicts that arise when diverse publics fail to cohere. In the conclusion, I explore the implications of the argument for the training of environmental leaders and facilitators in the age of sustainability.

6.2 Navigating Norton's Middle Path

In helping the sustainability-minded citizen navigate a middle path between the pretenses of objectivity and the temptations of ideology, Norton relies on the philosophical tradition of American Pragmatism, particularly as exemplified by C. S. Peirce and John Dewey. As a pragmatist, Norton rejects claims to foundational "Truths" while still working to keep at bay an unproductive relativism. This is accomplished by placing emphasis on the power of reasoned deliberation, buttressed by empirical experience, through which a community can reach agreement on difficult decisions. These decisions can then be iteratively updated, and rethought if necessary, through further deliberation and experience. The upshot of his approach is a way to cut through the paralysis that results when conflicting ideologies vie for dominance and nothing tangible gets done. In his oft-cited "Convergence Hypothesis," (also oft-debated, as discussed in Minteer 2009) Norton makes the case that divergent ideologies and value orientations can nevertheless undergird shared action, and therefore people should stop arguing over ideology and try to agree on what to do rather than how to value (Norton 1991).

In the more than two decades since the publication of *Toward Unity among Environmentalists*, the book in which he initially laid out this view, Norton has grappled with the difficulties involved in doing just that. The fruits of this grappling, which include a rethinking of the dualistic notion of value that the convergence hypothesis sought to tame (Norton 2005; Minteer 2009), are most conclusively and practically laid out in his 2015 book, *Sustainable Change*. In *Sustainable Change*,

Norton offers a philosophically grounded, practically oriented guidebook for addressing environmental issues "in something like their actual complexity".

Central to Norton's approach in *Sustainable Change* are a set of *procedural heuristics*: conceptual touch-points intended to foster social learning, discovery, and adaptive decision-making in the context of value diversity and scientific uncertainty. The first of these is the *Process Rationality Heuristic*, which is Norton's foundational proposal that "the rationality of decisions depends more on finding an appropriate process than on making accurate calculations and predictions" (the Heuristics are listed in the book's appendix, on page 293 of the print copy). Derived from a distinction made by cognitive scientist and economist Herbert Simon between substantive and procedural rationality, the Process Rationality Heuristic shifts the focus away from making accurate predictions about future states (substantive rationality), towards finding an appropriate *process* for working through social and environmental changes and choices (procedural rationality). The power —and challenge—of this latter approach lies in the empowerment of those with a stake in the problem to internalize both the messiness of the uncertain science and the gravity of the difficult decisions being faced.

Norton's second, *Epistemological Heuristic,* provides additional guidance into how stakeholders and decision-makers can operate in contexts characterized by a degree of uncertainty. The Heuristic says to: "Question everything, but not all at once" and "Trust experience more than ideology". This second heuristic is personified by Norton's favorite exemplar, Aldo Leopold. In the now-canonical story, Leopold relates his interest, in the early part of the last century, in increasing the economic viability of Southwestern landscapes by encouraging more deer hunting. To manage for this interest, Leopold sought to increase the deer population by exterminating their main predator: wolves. The result included denuding of the hillsides, massive die-offs of deer, and unchecked erosion of mountaintop soils. In Leopold's words, it was "as if someone had given G-d a new pruning shears, and forbidden him all other exercise" (Leopold 1949).

One thing that is important about Leopold's story is his willingness to learn from experience, even if that experience was a result of his own profoundly harmful and deeply humbling mistake. As Norton has interpreted it (see Hirsch and Norton 2012), Leopold's essay "Thinking like a Mountain" represents a cognitive transformation after which Leopold approached decision-making with an eye to the larger ecosystems and deeper values at play. Metaphorically speaking, Leopold learned that the appropriate spatial and temporal scale from which to evaluate future decisions about deer, wolves, and hunting is the scale of the "Mountain."

In light of Leopold's experience and the lessons we can draw from it, then, "questioning everything," entails—at a minimum—questioning one's own assumptions, both regarding the values at play in a given decision context, and the relevant physical, spatial, and temporal dynamics against which those values can appropriately be understood to play out (this latter is the focus of Norton's fourth heuristic, the Scale and Boundaries Heuristic). For Norton, this questioning goes along with a healthy respect for the complexity of environmental problems and the recognition that they require a deliberated and well-thought out response. The epistemological humility called for by Norton's second heuristic is thus a starting point from which can come a healthy engagement with available knowledge and its limitations, and an approach to decision making that is truly adaptive. That is, decision-making undertaken with the specific aim of testing the veracity and robustness of available knowledge, thereby building a stronger foundation for subsequent action and adaptation, and not incidentally generating stronger support for the science that made that possible.

Leopold, of course, in facing up to a set of dynamics he did not at the outset fully grasp, was relatively alone in the mountains of the Southwest. In contrast, most of the environmental problems being confronted today inspire the interest and concern of a wide diversity of people and groups. But to what extent do they act or can they be understood as a collective, or a "public" that can act as a cohesive unit? This problem is tackled in the next section.

6.3 Fragmented Publics, Discursive Inequality, and the Limits of Deliberative Process

When I was a student in Bryan Norton's graduate classes, a sort of pattern emerged. Norton would deliver a lecture on his pragmatic approach to environmental problem solving, speaking to the possibility of forward progress available if diverse stakeholders could stop arguing over ideology and initiate a process focused on what to do. I would then raise my hand and say: "but what about power?" I was concerned with the possibility that authentic dialog would be drowned out or otherwise subverted through the more hidden operations of powerful actors. More specifically, I was worried that the power to frame the terms of debate would be exercised so as to produce only those outcomes that serve the status quo.

His response was generally to graciously acknowledge that the operations of power in certain cases highlight the limitations of his approach, which relies on open, honest, and genuine communication. He remained confident, however, that there are many contexts where authentic dialog is indeed possible and actionable, so long as the dialog is based on what to do rather than on justifying it via ideological claims that are the inevitable sites of conflict. While I was—and am—inspired and guided by his approach, I was—and am—motivated by the recognition that an ever-increasing portion of public life seems to be in the category of situation where deliberative approaches may not suffice.

It is useful here to take a critical look at Norton's third, *Normative Sustainability Heuristic*, which states that sustainable change "requires a public discourse and the emergence of a 'public' that can identify key values that must be sustained if the public interest is to be sustained." This heuristic gets at the heart of a possible limitation as to what we can hope for from deliberation. In a context that entails a high degree of political polarization and/or social fracturing, there may never be just one public; rather, there are multiple and in some cases shifting publics, fractured along the lines of political boundary, ideological position, organizational agenda, educational experience, academic discipline, or cultural world-view. In today's political context, furthermore, the social groupings one most identifies with constrain in very powerful ways how one thinks, or even can think, about problems and solutions. As social scientist Kahan (2012) points out, "people with different values draw different inferences from the same evidence." For example, the differences in people's sense of the degree to which climate change poses a threat are strongly influenced not by their knowledge of the requisite science but by a process in which people filter out information that—were they to accept it—would, in Kahan's words, "drive a wedge between themselves and their peers."

The good news embedded in this insight is that, as Kahan points out, at least some of those who downplay the cause for concern about greenhouse gases are acting rationally in the sense that they are protecting themselves against a risk that is much more immediate and salient than the long-term and diffuse threat of climate change—namely, the very real risks associated with estrangement from one's peers (which, in a fractured world, means one's ideological fellows as well). One implication of this insight is that in certain circumstances it may be preferable to acknowledge and find ways to navigate the fractured nature of public life rather than try to get everyone on the same page. This becomes even more vital to the extent that significant inequalities are present across diverse and divergent groups. In such cases, our deliberative hopes must be subjected to an especially critical eye. Inequalities can be straightforward and explicit, as when different groups hold different amounts of money or political power. They can also be implicit, as when a problem or issue is framed in a way such that marginalized actors are unable to voice their concerns (Hirsch et al. 2013).

This latter situation involves what we might call *discursive inequality*: unequal power to frame the narrative. For example, certain disciplinary perspectives may be seen as more "scientific" than others, and therefore more credible to decision-makers. Likewise, countries in the global "North" and "South" may have different levels of access to decision-making arenas, and therefore different capacities to parlay their unique interpretations of the complex dynamics that need addressing. To illustrate with a specific example, in New York State, when the debate over whether or not to open the state's natural-gas containing shale to hydraulic fracturing raged, opportunities for public comment were framed in terms of "rights" to land and mineral resources. Yet from the perspective of some members of the Onondaga Nation—members of a sovereign government whose land and water security would be impacted by this state-level choice—the legal and linguistic framing of the debate implicitly marginalized values framed in terms of collective responsibilities rather than individual rights (Manno et al. 2014). The upshot of this is that a deeper level of complexity is entailed when multiple interpretive perspectives are embedded in a social landscape characterized by disparities in the amount and types of power available to different actors. The very way an issue gets framed is itself a form of power that can be expressed to yield tangible results. Similarly, a deliberative process that unfolds according to a specific problem framing is also a form of power—one that shapes who can and can't engage, and what can and can't be said.

So what can we do to recognize and responsibly engage at the outer limits of deliberative potential? This question is taken up in the following section.

6.4 Elements of a Critical Sensibility: Reflexivity and Agonism

Herbert Simon—and later Norton—in distinguishing between substantive and procedural rationality, sought to emphasize that appropriate deliberation is a legitimate form of rationality, and may in some cases be a better basis for decision-making than a calculation that is based on abstract and a-contextual notions of value. Procedural rationality is favored to the extent that it is a superior decision model for dealing with complexities that cannot be captured by algorithmic operations. As pointed to in the last section, however, procedural rationality itself has limitations when applied in contexts that are characterized not only by difference but also by inequality, and in particular by discursive inequality—lack of parity in being able to frame the narrative and thereby to shape the way a process unfolds. Of course, the processes that characterize this additional layer of complexity are not immune to analysis and consideration. It is at least conceivable then, that the operations of discursive inequality and related issues can be made more visible, and thereby treated as part of the problem to be engaged. This requires, at a minimum, a commitment to *critical reflexivity*.

To be critically reflexive is to recognize that the construction of problems and publics isn't something that happens out there. We ourselves—as analysts, students, teachers, leaders, members of social groups—are embedded and entangled in relations of power that constrain what is possible, and for whom. While we can't magically extract ourselves (indeed, the claim of objectivity or neutrality is itself a —quite significant—form of power) we can turn a critical eye on our own perspective and think through the implications of our embedded-ness and entanglements. For example, with respect to the issue of biodiversity conservation, to be critically reflexive is to recognize the discursive inequality that may be entailed when someone living in the United States or another relatively rich country speaks to a need to protect globally significant biodiversity (Zia et al. 2011). To do so is to speak from a particular and privileged position that may serve to diminish or disregard those whose local livelihoods are affected by globally oriented conservation efforts. Practicing critical reflexivity thus means treading gingerly before claiming the higher ground that is available in the grand "we" of global society.

Foregoing the grand "we" requires a willingness to make breathing space for conflict, difference, and disagreement. The trick—not an easy one at all—is to prevent conflict from unfolding in ways that are destructive to different parties or to the social and political fabric they both rely on.

In this regard, Mouffe (2005) offers a notion of democracy that is premised on the inevitability of at least some conflict and the impossibility of reconciling it once and for all via honest and reasoned debate. From her skepticism regarding the limits of deliberation, and her commitment to avoiding destructive forms of conflict, Mouffe develops a distinction between *agonism* and *antagonism*. Like the contesting champions of ancient Greece, a relationship characterized by *agonism* involves two or more actors who might be fighting passionately for different things, but who are committed to doing so in a way that does not—as is the case in *antagonistic* relations—destroy or subsume the other. Indeed, the other is necessary for the struggle, and the struggle itself is seen as part of the process of civic and individual development.

Agonism entails championing what one values and believes even if it is not compatible, or is only partially compatible, with the dominant view. Mouffe argues that the presumption that diverse views can be reconciled—even with the best and most fair of processes—is likely to disempower one or both, or to kick a conflict down the road and add fuel to its inevitable fire. For Mouffe, glossing over of the reality of conflict and irreconcilable diversity with the veneer of consensus leads to the inevitable emergence of much more destructive forms of conflict in the long run. As I understand it, agonism also involves refraining from demonizing those on the other side of a social fracture, which risks eroding basic levels of trust and transforming agonism into antagonism.

Tolerating dissonance and being willing to engage in ongoing and non-destructive struggle is easier said than done, of course. In the next and final section, I speak to some of the implications of the argument developed thus far for the training and development of a generation of environmental leaders and facilitators who can put these ideas into practice.

6.5 Conclusion: Developing the Capacities of Environmental Leaders and Facilitators

As social and political fractures widen and planetary boundaries loom (Steffen et al. 2015), there is a need—more than ever—for people who can provide effective leadership and facilitation in the work of sustainable change. Fulfilling this need calls for careful thought as to the training and capacities that such leadership and facilitation requires. By way of conclusion, I would like to illustrate the capacities for leadership and facilitation that Bryan Norton's procedural approach to sustainable change, pursued with a critical sensibility, might encourage the development of.

Taken together, Norton's Process Rationality and Epistemological Heuristics provide important starting points for the training of environmental facilitators and leaders working amidst diversity and uncertainty. The Process Rationality Heuristic tells us that, in a diverse world, rather than bringing a group around to one's predetermined point of view, one must be able to design, implement, and support procedures that allow people with disparate experiences to communicate with each other, deliberate about shared problems, and come to at least a contingent resolution. The Epistemological Heuristic tells us that, in an uncertain world, one must learn, like Leopold, to adopt an attitude of humility, recognizing the limits of one's knowledge. This attitude, furthermore, should not be exercised so as to yield paralysis: rather, it should lead to a thoughtful expansion of the scope of consideration for evaluation and analysis, thoughtful selectivity regarding the evidence required to act, a willingness to act, and an ability to learn and adapt based on the results that ensue.

In contexts characterized by diversity, an environmental leader or facilitator must also be able to support and enable groups with different visions of the social good to cooperate while maintaining that diversity and distinctiveness. And when diversity begins to bleed into divisiveness, something more yet again may be called for—the capacity to foster agonism and to keep antagonism at bay.

Following Norton's enthusiasm for metaphor and analogy, I end this chapter with two analogies intended to shed light on what may be required of the leader or facilitator working on sustainability challenges in the context of such divisiveness. The first analogy comes from an ancient joke[1]: *a wise elder listens to two farmers, one after the other, make passionate claims for opposite (and seemingly incompatible) sides of a deeply felt argument about land. The elder listens carefully and, after each speaker has finished, proclaims: "you're right." Then, when a perplexed bystander yells out "wait a minute, they can't both be right!" the elder says: "you're right too!"*

As I interpret this joke (admittedly ruining the humor in doing so), the elder's focus is on transcending the dangers of polarization. In responding to this particular dispute, she or he recognizes an opportunity to strengthen the larger social fabric that allows for the co-existence of seemingly incompatible perspectives. To achieve this, in saying "you're right too," the elder has affirmed the legitimacy of both farmers' perspectives (which, if not affirmed, may lead the farmers to more destructive forms of conflict). The elder has also implicitly invited the two disputants, as well as bystanders like us, to adopt a stance of epistemological humility in recognition that both can in fact be right if each has only a partial view of things. From this starting point, while the conflict may by no means be resolved, its destructive potential is minimized, and a pathway for further inquiry—and struggle—is opened. One might say that the elder has skillfully moved the conflict from the antagonistic to the agonistic realm.

And what can we say about the procedures that might be called for to allow the disputants to continue on with their lives, and in doing so to avoid harming those things they both value?

---

[1]The joke seems to have Islamic origins, and was popularized in the film Fiddler on the Roof (I've also seen a version of it in a Calvin and Hobbes cartoon).

The process of supporting two people with diverging views who are in some sense both "right" is exemplified by divorce mediation (as in the joke above, divorce mediation is also a context in which the bystanders—that is, children—are in all likelihood better served if the mutual rightness of both parties can indeed be affirmed). Whereas a marriage counselor's job is to support the emergence of a unified "public" (that is, couple) the job of a divorce mediator is not focused on helping a couple heal wounds or improve their intimate relationship; rather, his or her job is to help two individuals who are growing apart to focus on the larger set of values to which they remain mutually committed, such as the well-being of kids and pets. Unlike a lawyer, whose power comes from the force of laws and judges, the power of a divorce mediator is dependent on the extent to which he or she can support a creative thinking process—one that yields solutions allowing two people's paths to diverge in ways that are compatible with the protection of values, people, and things that cannot be easily divided. This creative thinking process is in turn backstopped by the ever-present possibility for both parties of "lawyering up" if the mediator becomes aware of abuse or other activities that may negate the very possibility of thinking together. This legal—and powerful—backstop mitigates at least to a degree the influence of any ongoing and significant inequality between the two participants.

Importantly, no single person can respond to the variety of dynamics present in something like a divorce. Wise elders, marriage counselors, divorce mediators and lawyers are all part of a system designed to allow struggling people to find their appropriate mode of engagement in the spectrum from consensus to agonism, and sometimes beyond. A marriage or divorce, furthermore, is generally an order of magnitude simpler than conflicts that play out over the larger social scales of communities and countries. Furthermore, while the divorce mediator analogy conveys something important about the necessary orientation for working in a polarized—and sometimes dangerous—social milieu, it does not do justice to the scientific uncertainty that characterizes current environmental issues. What the analogy does show with regards to uncertainty, however, is that in an emotionally—or politically—charged situation, uncertainty is not just an intellectual problem, it is raw material in the struggle for dominance, and must be engaged with as such.

While we contemplate what it will take to train the elders, counselors, mediators, lawyers, and others who will navigate these conflicts, therefore, my own view—which admittedly may take the Epistemological Heuristic slightly farther than Norton intended—is that navigating the "post-fact" world may require us to recognize that there are at least some problem contexts in which uncertainty may be more appropriately called ignorance (Vitek and Jackson 2008; Stirling 2010). In such cases, questioning everything may call for an even deeper notion of humility—something akin to a secular version of religious awe. For such cases, the critically reflexive application of Norton's Epistemological Heuristic may yield a profound admission that the light of one's understanding shines dimly and partially, accompanied by a healthy respect both for the possibility that others can see what one cannot, as well as for that which remains in the dark for all.

References

Hirsch, P.D., and B.G. Norton. 2012. Thinking like a planet. In *Ethical adaptation to climate change: Human virtue of the future*, ed. Allen Thompson and Jeremy Bendik-Keyme. Cambridge, MA: MIT Press.

Hirsch, P.D., J.P. Brosius, S. O'Connor, A. Zia, M. Welch-Devine, J.L. Dammert, A. Songorwa, T.C. Trung, J.L. Rice, Z.R. Anderson, S. Hitchner, J. Schelhas, and T.O. McShane. 2013. Navigating complex trade-offs in conservation and development: An integrative framework. *Issues in Interdisciplinary Studies* 31: 99–122.

Kahan, D. 2012. Why we are poles apart on climate change. *Nature* 488: 255.

Leopold, Aldo. 1949. *A sand County Almanac and sketches here and there*. New York: Oxford University Press.

Manno, Jack P., Paul Hirsch., and Andrea M. Feldpausch-Parker. 2014. Introduction by the Onondaga Nation and activist neighbors of an indigenous perspective on issues surrounding hydrofracking in the Marcellus Shale. *Journal of Environmental Studies and Sciences* 4 (1): 47–55.

Minteer, B.A. (ed.). 2009. *Nature in common? Environmental ethics and the contested foundations of environmental policy*. Philadelphia: Temple University Press.

Mouffe, C. 2005. *On the political*. New York: Routledge.

Norton, B.G. 1991. *Toward unity among environmentalists*. New York: Oxford University Press.

Norton, B.G. 2005. *Sustainability: A philosophy of adaptive ecosystem management*. Chicago: University of Chicago Press.

Norton, B.G. 2015. *Sustainable values, sustainable change: A guide to environmental decision making*. Chicago: University of Chicago Press.

Pew Research Center. 2014. Political polarization in the American public. Available at http://www.people-press.org/2014/06/12/political-polarization-in-the-american-public/. Accessed September 8, 2017.

Steffen, W., K. Richardson, J. Rockström, S.E. Cornell, I. Fetzer, E.M. Bennett, R. Biggs, S.R. Carpenter, W. De Vries, C.A. De Wit, C. Folke, D. Gerten, J. Heinke, G.M. Mace, L.M. Persson, V. Ramanathan, B. Reyers, and S. Sörlin. 2015. Planetary boundaries: Guiding human development on a changing planet. *Science* 347 (736): 1259855.

Stirling, A. 2010. Keep it complex. *Nature* 468: 1029–1031.

Vitek, B., and W. Jackson. 2008. Introduction: Taking ignorance seriously. In *The virtues of ignorance: Complexity, sustainability, and the limits of knowledge*, ed. B. Vitek and W. Jackson, 1–17. Lexington, KY: University Press of Kentucky.

Zia, A., P. Hirsch, A. Songorwa, D.R. Mutekanga, S. O'Connor, T. McShane, P. Brosius, and B. Norton. 2011. Cross-scale value trade-offs in managing social-ecological systems: The politics of scale in Ruaha National Park, Tanzania. *Ecology and Society* 16 (4): 7. https://doi.org/10.5751/ES-04375-160407.

Paul D. Hirsch is Associate Professor in the Department of Environmental Studies at SUNY ESF: The State University of New York College of Environmental Science and Forestry. He completed a Ph.D. with Bryan Norton at Georgia Tech's School of Public Policy, an MS at the Odum School of Ecology at the University of Georgia, and a BS at Cornell University. Hirsch's scholarship and writing focus on the role of collaborative intelligence in fostering sustainable change, and on communicating effectively when both uncertain science and diverse values are at stake. He is the curriculum designer and lead instructor for ESF's online graduate certificate program in Environmental Leadership.

Chapter 7
Leadership for Sustainability

R. Bruce Hull

Abstract It is almost cliché to say this, but environmental-problems are people-problems and overcoming sustainability challenges such as climate change requires more than science and technology. Ultimately, sustainable development requires collaboration and action by individuals, teams, organizations, and institutions. Leadership is a celebrated topic in public administration and business management, but less so in environmental, engineering, and sustainability circles, perhaps because of tendencies in these fields to fixate on technical problems with technical solutions. This chapter summarizes a subset of leadership theory relevant to the wickedness of environmental and sustainability challenges. It also introduces relevant leadership tools, practices, and perspectives. Ultimately, the chapter is a homage to Bryan Norton and his prescient and powerful call for trusted, flexible, and creative processes for adapting to change as we create a desirable place to live in the future.

Keywords Wicked · Adaptive · Distributed · Collaborative · Leadership DAC

7.1 The Leadership Challenge

Humanity prospered during the stable and predictable conditions of the Holocene. Environmental conditions nurtured agriculture, urbanization, literacy, science, governance, organized religion, markets, and most other institutions that birthed civilization. Those conditions are changing. We now live in the era of great acceleration, when environmental conditions are much less predictable and much more variable (Steffen et al. 2015). These changes are likely to accelerate over

R. B. Hull (✉)
College of Natural Resources and Environment, Virginia Tech (0324), Blacksburg, VA 24060, USA
e-mail: hullrb@vt.edu

S. Sarkar and B. A. Minteer (eds.), *A Sustainable Philosophy—The Work of Bryan Norton*, The International Library of Environmental, Agricultural and Food Ethics 26, https://doi.org/10.1007/978-3-319-92597-4_7

the coming decades as humanity ends poverty, adds several billion people to the global middle class, rapidly urbanizes, and doubles energy and food consumption (Deaton 2013).

Now more than ever, we need processes that produce good decisions. Bryan Norton describes our challenge:

> [I]f one takes seriously the magnitude of acceleration of human technological change, which will, in turn, interact with background changes that will be expressed at all levels and scales of the atmosphere and the biosphere in coming decades, *then the only alternative to despair is the hope that rests on developing processes that can generate better decisions….* What is needed is a process—not one bound to narrowly defined outcomes—that keeps multiple options and opportunities open, yet is robust and flexible enough to respond to unexpected changes and to allow creative responses. Such a process cannot be rigid; it must be a process whereby its participants can learn how to learn as it faces countless new situations. (Norton 2015: 286; *emphasis added*).

The purpose of this chapter is to explain how good leadership practices support the processes Norton demands and sustainability requires. These leadership practices fall into three categories: distributed, adaptive, and collaborative. Each type of leadership practice is discussed below. But first, leadership is defined.

7.2 Leadership Defined

Leadership is an ambiguous and contested idea, perhaps even more so than "sustainability." The purpose of this section is to explain and defend a specific theory of leadership applicable to sustainability challenges, a theory that boils down to this simple statement: leadership produces direction, alignment, and commitment among stakeholders (Drath et al. 2008). Stated differently, the outcomes of leadership are direction, alignment, and commitment (DAC). This theory of leadership shifts attention away from attributes of leaders to the practices of stakeholders that promote direction, alignment, and commitment. This definition of leadership should be empowering because it means everyone can practice leadership, not just people with titles or corner offices or otherwise paid and positioned to be "leaders." That is, we can all practice leadership by helping facilitate DAC. Detailed definitions for direction, alignment, and commitment are offered below:

Direction: Strong direction exists when stakeholders agree on what they are aiming to accomplish. To paraphrase Yogi Berra: If you don't know where you are going, you'll end up someplace else. Having direction means that the collective knows what success looks like. Heifetz calls this critical task "adaptive work," which he defines as the process of learning and articulating the gap between aspirations and reality (Heifetz 1994: 22). Stakeholders must work to generate agreement on where they are and where they want to be. Generating agreement on direction is not easy. It may involve orchestrating conflict and exposing internal contradictions that mobilize people to clarify what matters most and what can be traded off. Stakeholders with weak direction are uncertain about what they should do and may feel pulled in different directions by competing goals.

Alignment: Even if stakeholders agree on goals, success is unlikely unless key actors coordinate their efforts and avoid working at cross-purposes. Stakeholders bring different resources to bear. They have different expertise, pull different levers, and exercise different types of influence. Some tasks are glamorous and thus generate too many volunteers. Other tasks are less popular because they are expensive, risky, or thankless. Stakeholders must align their time, talent, and resources so that all necessary activities occur at the right times and places. Stakeholders with weak alignment work in isolation and don't see how their tasks fit into the larger set of tasks to be done: they risk working at cross-purposes, duplicating effort, or having important work fall through the cracks.

Commitment: Even if stakeholders agree on a direction to aim and can align resources to make that happen, the effort will falter for lack of commitment. Actors must willingly sacrifice their self-interests to achieve the shared goal. Stakeholders with strong commitment feel responsible for the success and well-being of the collective effort, and know that other stakeholders feel the same. They trust one another and will stick with the effort through difficult times. Stakeholders with commitment allow other stakeholders to make demands on their time and energy. Commitment is evidenced when people buy-in and are loyal to the collective.

All three outcomes—direction, alignment, *and* commitment—must be present for leadership to have occurred. You are probably familiar with frustrating situations when one or more of these outcomes did not exist. For example, when stakeholders have direction but no alignment they agree on a mission statement but, because they are competitors for funding, will not share resources or will not do work that others can take credit for. Conversely, if stakeholders have commitment, without direction or alignment, then people head off in all directions at once, often working at cross purposes as well as risking burnout and damaging the reputation of the group. Most perversely, stakeholders can have direction and alignment but no commitment, so feet get dragged and time, money, and good will get wasted.

Another way to explain the Direction-Alignment-Commitment theory of leadership is to contrast it with what it is *not*. It is not a leader-follower theory of leadership. The leader-follower theory is probably the most well-known of all leadership theories (Bennis 2007). Its ontology has three basic concepts: a leader, followers, and a goal to achieve (as opposed to direction, alignment, and commitment). Leadership guru Ronald Heifetz (1994) offers us a thought experiment to illustrate the critical difference between these two leadership ontologies: Imagine the different responsibilities and outcomes that occur when stakeholders follow a leader's vision rather than collaborate with other stakeholders to identify and face the challenges stakeholders share. In the leader-follower model, vision of the desired future condition comes from the leader and if something goes wrong with that vision or the process leading to it, fault lies with the leader. In the DAC leadership model, goals emerge from the stakeholders and they must hold themselves accountable for both the process and the future they create.

A theory has power if it generalizes to and helps explain many situations. Direction, alignment, and commitment seem to be properties of leadership that generalize to many different cultures, contexts, and situations. Pretty much anywhere or anytime leadership occurs, the collective has achieved direction, alignment, and commitment. However, and this is an important however, the *practices* that produce DAC will be highly context, culture, and situation dependent. Hence, good leadership requires a large toolbox of practices that promote DAC in different situations. The remaining sections of this chapter review practices that work under the distributed, collaborative, and adaptive conditions typical of sustainable development challenges.

7.3 Distributed Leadership Practices

The widely-dispersed, tragedy-of-the-commons attributes of sustainability problems requires that responsibility and responses occur over vast spans of space and time. Stakeholders tend not to have direct lines of influence or authority over one another because they reside in different organizations, nations, and places. Participation will be voluntary. Most actors will neither meet nor interact. Some stakeholders will not even realize they are connected. Processes and practices are needed to coordinate disparate, widely distributed stakeholders that may never see or know one another.

Well-tested and widely applied methods exist for distributing and sustaining capacity under these conditions. Space limitations prevent full descriptions, but illustrative methods are defined here with references for follow-up.

- Collective Impact: The Collective Impact process helps diverse and distributed stakeholders produce large, systemic change. Five conditions have been found in most successful efforts: agree on a common agenda, agree on and share measurements of progress, align actions and resources, continuously communicate, and support by backbone organizations to maintain network functions (Kania and Kramer 2011).
- Scaling up: Solutions must be scaled up from multiple starting points because unique, local conditions will defy one-size-fits-all solutions. Picking the innovations to champion and scale will prove challenging. Game changing innovations will emerge unpredictably at different locations and scales of operation. Most of these innovations sputter and fail, but some, which ones we will not know in advance, will merge and grow and must be nurtured, distributed, and allowed to transform whole systems (Geels 2011; Grin et al. 2010). Intentional methods exist for scaling up innovations (see Management Systems International 2017).
- Community of Practice and Learning Networks: Networks of actors that circulate and vet best practices and build capacity of practitioners to practice their craft. These networks promote learning from practitioners, by practitioners, about their practices (Wenger and Snyder 2000).

- Roundtables: Integrating sustainability issues into global supply chains remains a critical challenge. Roundtables bring together representatives from business, civil society, and government to promote and negotiate standards for production and product certification for global commodities such as soy, corn, and livestock (Steering Committee 2012).

A key skill of distributed leadership is being able to influence others without having authority over them. Several influence strategies exist, none are foolproof, and all require nuance and practice. They include: rationalizing, asserting, negotiating, inspiring, and bridging (see Table 7.1). Most people have a preferred influence strategy and default to using it on others, which means our ability to influence others will be limited until we learn to use influence styles that are effective on others (Cohen and Bradford 2011). Most readers of this chapter will likely prefer and employ a rationalizing strategy, because scientists and scholars default to using facts and logic. But arguments of this type are ineffective on people who, for example, prefer personal relationships or inspiring stories. Likewise, if the topic of discussion concerns ethics, values, and identities, then a rationalizer offering facts and logic will seem callus and disconnected, perhaps alienating others by making them think their personal values and identities don't matter.

Table 7.1 Influence without authority

Strategy	When not to use it
Rationalizing: Use logic, facts, and reasoning to present ideas and persuade others. This influence strategy is the default for many scientists and professionals who appeal to the authority of science to convince others	It can be ineffective when it overwhelms others with detail or makes others feel the speaker values data more than principles and ethics
Asserting: Use laws, authority, and personal confidence to influence others and challenge ideas. People accustomed to being in positions of authority or enforcing regulations may default to this style	It doesn't work when people feel pressured and the actor is aggressive or heavy-handed or when stakeholders have veto-power and can sabotage the process just by withdrawing their participation
Negotiating: Use compromises, concessions, and tradeoffs to keep the process moving. Facilitators and planners may default to this style. It is ineffective when others become confused about an actor's values, because they agree with everything	Too much compromise raises concerns that the process will degenerate to the lowest common denominator people will agree to rather than a decision of consequence
Inspiring: Encourage a shared position by communicating a sense of shared mission and exciting possibility with inspirational appeals, stories, and metaphors. Mission-oriented NGO stakeholders may default to this style	It doesn't work when there is a lack of transparency on the part of the inspirer or insufficient basis for trust
Bridging: Use peers, personal relationships, and other network connections to bring influence on others	It is ineffective when stakeholders feel manipulated or forced to rely on these indirect indicators of legitimacy rather than have time to study and be persuaded by the issues

Adapted from Musselwhite and Plouffe (2012)

7.4 Adaptive Leadership Practices

A key step in problem solving is recognizing the problem. Different types of problems require different leadership practices. Sustainability problems typically fall into one of three types: chaotic, adaptive (wicked), and technical (Grint 2005; Heifetz et al. 2009; Levin et al. 2012; Norton 2015) (see Table 7.2). The bulk of this chapter is focused on leadership for adaptive or wicked types of sustainability challenges, not technical or chaotic problems.

Technical problems have far more certainty. They have been encountered before and can be solved with data, analysis, and management. Appropriate leadership activities help define the situation and identify the problem, assemble experts with tools proven to solve that problem, and allocate needed resources. Examples of technical problems include increasing the energy efficiency of residential housing to reduce carbon emissions, repairing a malfunctioning emissions control system in a fleet of vehicles to reduce pollution, or financing minority owned businesses to address poverty. Clearly, these are difficult and important problems to solve, but they require qualitatively different leadership than adaptive challenges such as global climate change, where predictions are difficult and neither the goals nor the strategies to achieve them are known. Chaotic or crisis situations present yet a different set of leadership challenges. These situations emerge when systems change rapidly, such as in disasters. Public safety may be at risk and stakeholders are willing to temporarily cede authority to a central command that will restore order.

The wickedness of sustainably challenges requires stakeholders to recognize and adjust to immense and intimidating uncertainty. Three types of uncertainty categorized in Table 7.2 are defined in more detail below: system, strategy, and goal (Norton 2015).

Table 7.2 Types of sustainability challenges and leadership vary by uncertainty

Situation type	Goal uncertainty	System uncertainty	Solution uncertainty	Leadership
Technical	Low: Goals known from past experience	Low: Causations known and predictions possible	Known. Strategies worked before	Hierarchy and Authority
Adaptive/ wicked	High: Goals emerges from collaboration and learning	High: Dynamic, nonlinear, open, and unpredictable conditions.	High: Requires continuous innovation and learning	Distributed among stakeholders
Chaotic/ crisis	High and Low: Long term goals unknowable, clear short term goals of public safety and confidence	High: Highly turbulent, perhaps unknowable	High: Look for what brings order in short term instead of right answers	Centralized command may be needed to restore order and safety

System Uncertainty: The myth of a stable or balanced nature is busted (Botkin 1990). Natural ecological systems are dynamic. Species and ecosystems evolve. Disturbance is normal. New system conditions emerge, even without human intervention. Additional complexity, openness, instability and unpredictability have been added by the tight coupling of natural ecological systems with cultural and economic systems that are also open, dynamic, unstable, and unpredictable (Steffen et al. 2015). By the early 1970s, theories about general systems, chaos, complexity, evolution, and quantum physics confirmed that human and natural systems were beyond control and that cause-effect science was insufficient. By the end of the 20th century, senior members of the American Association for the Advancement of Science acknowledged that uncertainty in human-nature coupled systems was too much for traditional science to overcome and that society's most pressing challenges had "radically outgrown" science's ability to solve them (Jasanoff et al. 1997).

Strategy Uncertainty: Because of new and emerging conditions, we won't know whether tested strategies will produce the same results they did in the past. Another, and perhaps greater, source of uncertainty emerges from the political and contested nature of strategies: we can't predict which strategies will be adopted. Strategies known to work in the past might not be used again if they are unacceptable to future stakeholders. The objections by pro-market, anti-government advocates to most any proposed tax or government regulation serve as a familiar illustration of how strategy choices are political lightning rods. Stakeholders can differ in the strategies they favor more than they differ in the goals they desire. Said differently, there can be as much conflict over means as ends. Uncertainty results because it is not possible to predict which strategies will be deemed acceptable.

Goal Uncertainty: Another source of uncertainty emerges from stakeholders not knowing what they want. Stakeholders' values and goals change as they engage and learn about a problem. Stakeholders, when they first encounter a problem, profess goals that are hypotheses that are just as refutable as hypotheses about ecological functions and engineering solutions (Norton 2005). For example, stakeholders in a community may agree to maximize wealth and freedom through real estate development. After years of pursuing these values, residents may become frustrated by traffic congestion, fossil fuel dependence, and the loss of local foods, open space, and biodiversity. That is, their goals change as they experience the landscape and lifestyle previous goals created. It is not until after we create the future that we can decide if we want to live there.

All this uncertainty creates lots of opportunity to learn by doing, so learning is critical to adaptive leadership (Senge et al. 2008). Sustainability professionals will recognize the proven four-step approach used to address these sorts of challenges: plan-do-check-fix. In the jargon of numerous environmental professions, this process is known as adaptive management. In business, the approach is called the Deming Cycle or Management by Discovery. In the planning professions, it is known as incrementalism and groping along. Project managers call it the Agile method. Advocates of innovation call it piloting and pivoting. All these approaches share Norton's pragmatist's philosophy of pursuing a "process that is corrigible through the sharing of experience and the exchange of ideas." (Norton 2015: 67). This process of questioning and revising the original problem formation is known as double loop learning (Goldstein 2009; Throgmorton 2003).

Unfortunately, admitting uncertainty threatens the legitimacy of sustainability professionals, managers, scientists, and other experts (Schon 1983). Professional identities and salaries depend upon offering advice that solves problems, not saying: "I don't know." Professionals therefore need the courage, and permission, to fail. They must transform defensiveness about failure into opportunity for inquiry and mistakes into lessons. Moreover, stakeholders need the courage to fail. They must be willing to adjust to new information, uncertainty, and unpredictably. Stakeholders, likewise, will need the persistence to do more than sporadically engage in planning processes. An open, adaptive process may require continuous monitoring and corrections.

Sense making, rather than rational planning, provides a useful way to frame the leadership challenges of dealing with uncertainty, failure, and learning. Rational planning relies on prediction and control, which become less and less effective as uncertainty mounts (Norton 2015). Sense-making is more appropriate for uncertain situations (Senge et al. 2008; Snowden and Boone 2007). Sense-making avoids analysis paralysis by emphasizing action as a means to make sense of the situation. It encourages continuously probing the system with interventions to identify patterns, generate ideas, share insights, and steer the system towards desired conditions. Sense-making encourages course corrections rather than rigidly following a charted path towards a pre-determined solution, promoting tolerance for achieving results that differ from what were aimed for. Importantly, sense-making leadership becomes a process of engaging stakeholders in deliberation and learning through repeated efforts to probe and transform the situation rather than a process of deciding, announcing, and then defending a course of action destined to fail. Efforts to defend expertise or control the situation may preempt the opportunity for key patterns to emerge and stifle trial and emergence of successful solutions.

7.5 Collaborative Leadership Practices

Stakeholders must collaborate across the vast differences of identity, culture, profession, discipline, sector, and other boundaries and barriers. Powerful silos of self-reinforcing similarity create destabilizing protectionism and polarization. Some of these silos, such as disciplines and nations, emerged because they proved effective at solving problems of the Holocene, but are increasingly less helpful and even counterproductive in the new interconnected, accelerating, and uncertain world. Moreover, no person, profession, organization, business, government, or institution possesses the power, insight, or resources to solve many of the most pressing sustainable development challenges. Collaboration is needed to align resources.

Well-known strategies for collaboration include Interest Based Negotiation (Fisher and Ury 1991), Reaching Higher Ground (Dukes et al. 2000), and Boundary Spanning Leadership (Ernst and Chrobot-Mason 2011). Rather than review these established strategies that are capably presented elsewhere, this section focuses on less well understood dimensions of collaborative leadership: embrace value plurality, make conflict useful, and build trust.

Embrace Value Plurality: Because of the newness and uncertainty of contemporary sustainability challenges, stakeholders often don't know which of their own values are relevant to the decisions they must make, let alone know the values of others with whom they need to collaborate. Stakeholders must give voice to their poorly articulated values, learn from values articulated by others, and construct a shared vision out of enormous ambiguity and plurality (Lewicki et al. 2003). Norton, in his recent book, *Sustainable Values Sustainable Change*, makes an impassioned plea for more and better understanding of values and expresses hope that his book inspires more scholarship of evaluation. He predicts that professionals skilled in evaluation will find themselves increasingly influential and valuable (Norton 2015: 286).

A first step to pluralism is buffering: people need time to recognize, articulate, and defend the values they hold. The next step is to recognize and accept that other people also hold values, that these values may differ, and that these differences produce different understandings and expectation of the world. The next step is to share and learn from the plurality of values to construct some shared understanding that can guide sustainable development. This task is more difficult than simply finding common ground where a plurality of values overlaps. It requires discovering synergies and constructing the higher ground that respects the diversity of values within a community as well as tolerating that people are fickle and values will change as a result of learning more about the situation when they engage it and others (Ernst and Chrobot-Mason 2011).

Make Conflict Useful: Cognitive conflict improves the outcomes of group processes (Runde and Flanagan 2010). It spurs innovation by introducing new expertise and questioning assumptions. If cognitive conflict gets suppressed, collaborative work suffers from group-think and/or decisions devolve to the lowest common dominator. Strategies exist to productively manage cognitive conflict. For example, to relieve the stigma of offering criticism of an idea, the group could intentionally create a time when offering convergent, critical, cognitive-conflict type thinking is expected. The purpose and expectations of comments made during that time would be different than during times when divergent, supportive, brainstorming is expected. Role playing also helps manage conflict. If people know that a person is playing the role of devil's advocate or dyadic critic, then the criticism is more likely to be interpreted as questioning the idea and not attacking the person who offered the idea.

In contrast, affective conflict is most always "bad" because it threatens identities, accentuates power imbalance, and redirects work from solving problems to blaming others. Affective conflict focuses on relationships not ideas. It violates the collaboration mantra: be hard on the problem and easy on people. Perhaps the biggest challenge to managing affective conflict is managing ourselves: when someone criticizes our ideas, we tend to assume they are attacking us personally. Our negative reaction may be triggered by motivations that we attribute to the other person. Perhaps we think they are trying to make us look bad or oppose our solution because it undermines their budget and source of power. Tools to manage affective conflict exist, but most importantly, collaborators need to be coached to recognize affective conflict, that we each have a tendency to misinterpret criticisms of our ideas as personal attacks, and that we each can strive to prevent misperceptions by carefully framing how we present our critiques of other's ideas.

Build Trust: Trust improves collaboration. Without trust, cognitive conflict degenerates into affective conflict. Trust also creates the foundation for continued problem solving by building relationships that survive differences, disagreements, time, and change. Distrust, in contrast, can lead to stalemates, withdraw, non-compliance, protests, and even violence. Distrust creates risk adverse cultures within and across organizations, curtailing the innovation, experimentation, and adaptive capacity necessary for solving to complex sustainability challenges.

Trust has four dimensions: dispositional, rational, affinitive, and procedural (Stern and Coleman 2015). Some dimensions of trust can be fostered through good leadership.

- Affinitive: Trust depends upon perceptions of shared identities and values as well as positive shared experiences. Affinity is key to getting people to the table. Affinitive trust can be built by creating opportunities for people to interact and share experiences.
- Rational: Trust depends on whether past performances delivered expected outcomes. People make a calculative evaluation: will others deliver on what is agreed? People or organizations that have delivered in the past are more trusted to deliver in the future. To promote rational trust: know what you say, say what you know, and do what you say.
- Procedural: Procedures viewed as fair, transparent, and consistent promote trust. Procedures can reduce the vulnerability of individual participants to bullies and to the actions of people in power. Procedural trust is facilitated by an open, rigorous process that tests assumptions and outcomes, encourages open debate, is persistent and ongoing, and is capable of being corrected, redirected, or reformed.
- Disposition: People enter into a situation disposed to trust or distrust based on personal histories, innate personalities, or general hearsay. There may not be much you can do to impact this dimension of trust.

7.6 Conclusion

This book is a tribute to the influence of Bryan Norton, one of America's most penetrating and insightful scholars of sustainability philosophy, policy, and practice. Norton challenges sustainability scholars and professionals to cultivate a process for decision-making in an era of immense uncertainty and dynamism. In the quote at the beginning of this chapter, he offers a sobering appraisal of the situation and I repeat one sentence here: "The only alternative to despair is the hope that rests on *developing processes that can generate better decisions*." This emphasis on process requires special efforts by stakeholders participating in the process. This chapter examined several leadership practices that support those efforts.

Leadership for sustainably is here defined as the capacity of individual stakeholders to facilitate and maintain the shared direction, alignment, and commitment that robust decisions processes must produce. This definition of leadership is more appropriate for wicked sustainability challenges than the more common definition of leadership: leaders leading followers. Importantly, everyone can practice leadership to promote DAC. We can all lead from where we are.

Direction, alignment, and commitment (DAC) are generalizable properties of leadership. Pretty much anywhere or anytime that leadership occurs, the collective has achieved direction, alignment, and commitment. However, the *practices* that produce DAC are context, culture, and situation dependent. To be good at leadership means having a large toolbox of practices that work in different situations.

This chapter reviews practices appropriate for the distributed, collaborative, and adaptive conditions typical of many sustainable development challenges. To summarize: *Distributed* leadership practices help stakeholders work across a vast, complicated, global, tele-connected systems where no one has authority over others and most actors will neither meet nor interact. *Collaborative leadership* practices help stakeholders bridge the vast differences that divide and polarize. *Adaptive leadership* practices help stakeholders make sense of, learn from, and respond to massive uncertainty and dynamism. Stakeholders must do this difficult work; it can't be done for them, enforced, or imposed if commitment is to be sustained.

"*The only alternative to despair is the hope that rests on developing processes that can generate better decisions.*" Better processes and better decisions depend, in part, on effective leadership practiced by sustainably professionals and other stakeholders.

References

Bennis, W. 2007. The challenges of leadership in the modern world: Introduction to the special issue. *American Psychologist* 62: 2–5.

Botkin, D. 1990. *Discordant harmonies*. London: Oxford University Press.

Cohen, A.R., and D.L. Bradford. 2011. *Influence without authority*. NY: Wiley.

Deaton, A. 2013. *The great escape: Health, wealth, and the origins of inequality*. Princeton: Princeton University Press.

Drath, W.H., C.D. McCauley, C.J. Palus, E. Van Velsor, P.M. O'Connor, and J.B. McGuire. 2008. Direction, alignment, commitment: Toward a more integrative ontology of leadership. *The Leadership Quarterly* 19: 635–653.

Dukes, E.F., M.A. Piscolish, and J.B. Stephens. 2000. *Reaching for higher ground in conflict resolution*. San Francisco: Jossey Bass.

Ernst, C., and D. Chrobot-Mason. 2011. *Boundary spanning leadership*. New York: McGraw Hill.

Fisher, R., and W. Ury. 1991. *Getting to yes: Negotiating agreement without hiving in*, 2nd ed. New York: Penguin.

Geels, F.W. 2011. The multi-level perspective on sustainability transitions: Responses to seven criticisms. *Environmental innovation and societal transitions* 1: 24–40.

Goldstein, B. 2009. Resilience to surprises through communicative planning. *Ecology and Society* 14 (2): 33.

Grin, J., J. Rotmans, and J. Schot. 2010. *Transitions to sustainable development: New directions in the study of long term transformative change*. NY: Rutledge.

Grint, K. 2005. Problems, problems, problems: The social construction of 'leadership'. *Human Relations* 58: 1467–1494.

Heifetz, R.A. 1994. *Leadership without easy answers*. Cambridge: Harvard University Press.

Heifetz, R.A., A. Grashow, and M. Linsky. 2009. *The practice of adaptive leadership: Tools and tactics for changing your organization and the world*. Cambridge: Harvard Business Press.

Jasanoff, S., R. Colwell, M.S. Dresselhaus, R.D. Goldman, M. Greenwood, A.S. Huang, W. Lester, S.A. Levin, M.C. Linn, and J. Lubchenco. 1997. Conversations with the community: AAAS at the millennium. *Science* 278: 2066–2067.

Kania, J., and M. Kramer. 2011. Collective impact. *Stanford Social Innovation Review* 9: 36–41.

Levin, K., B. Cashore, S. Bernstein, and G. Auld. 2012. Overcoming the tragedy of super wicked problems: Constraining our future selves to ameliorate global climate change. *Policy Sciences* 45: 123–152.

Lewicki, R., B. Gray, and M. Elliott. 2003. *Making sense of intractable environmental conflicts*. Washington, D.C.: Island Press.

Management System International. 2017. Scaling up: http://www.msiworldwide.com/approach/tools/scaling-up-framework-toolkit/. Accessed 13 Aug 2017.

Musselwhite, C., T. Plouffe. 2012. When your influence is ineffective. Harvard Business Review Blog Network. (March 28). https://hbr.org/2012/03/when-your-influence-is-ineffective/. Accessed 13 Aug 2017.

Norton, B. 2005. *Sustainability*. Chicago: University of Chicago Press.

Norton, B. 2015. *Sustainable values, sustainable change: A guide to environmental decision making*. Chicago: University Chicago Press.

Runde, C., and T. Flanagan. 2010. *Developing your conflict competence: A hands-on guide for leaders, managers, facilitators, and teams*. San Francisco: Jossey-Bass.

Schon, D.A. 1983. *The reflective practitioner: How professionals think in action*. New York: Basic Books.

Senge, P., B. Smith, N. Kruschwitz, J. Laur, and S. Schley. 2008. *The necessary revolution: How individuals and organizations are working together to create a sustainable world*. NY: Doubleday.

Snowden, D.J., and M.E. Boone. 2007. A leader's framework for decision making. *Harvard Business Review* 85 (11): 69–76.

Steering Committee of the State-of-Knowledge Assessment of Standards and Certification. 2012. *Toward sustainability: The roles and limitations of certification*. Washington, DC: Resolve Inc.

Steffen, W., W. Broadgate, L. Deutsch, O. Gaffney, and C. Ludwig. 2015. The trajectory of the Anthropocene: The great acceleration. *The Anthropocene Review* 2: 81–98.

Stern, M.J., and K.J. Coleman. 2015. The multidimensionality of trust: Applications in collaborative natural resource management. *Society & Natural Resources* 28: 117–132.

Throgmorton, J.A. 2003. Planning as persuasive storytelling in a global-scale web of relationships. *Planning Theory and Practice* 2 (2): 125–151.
Wenger, E.C., and W.M. Snyder. 2000. Community of practice: The organizational frontier. *Harvard Business Review* 78: 139–145.

Dr. R Bruce Hull is a professor and Senior Fellow at Virginia Tech's Center for Leadership in Global Sustainability based near Washington DC, which provides graduate education, leadership training, and professional development opportunities for sustainability professionals. He is President of the Board of Climate Solutions University, whose mission is to help communities adapt to climate change. He also serves on advisory councils for centers focused on Global Change and Communicating Science. He has authored and edited numerous publications, including two books, *Infinite Nature* and *Restoring Nature*.

Chapter 8
Norton Versus Callicott on Interpreting Aldo Leopold: A Jamesian View

Piers H. G. Stephens

Abstract Since Bryan Norton first advocated an American pragmatist reading of Aldo Leopold's work in 1988, he has been debating with J. Baird Callicott over interpretation of Leopold's development of the land ethic. In this chapter I give an overview of this debate, defending the general outlines of Norton's position by bringing in new interpretative work of my own. I argue firstly that Norton is correct to see a Jamesian pragmatist influence on Leopold, but maintain that this is best read as deriving from William James' own work rather than from the secondary source, Arthur Twining Hadley, and that Norton may have exaggerated the immediacy and strength of the influence. Secondly, I demonstrate how strands of evidence suggest that Leopold read James, and that elements of James' thought chimed with Leopold's own perspective sufficiently to support Norton's claim of a pragmatist influence. Thirdly, I show that aspects of Callicott's interpretation of Leopold do not exclude a pragmatist reading, and argue that Callicott has misunderstood Jamesian pragmatism in ways that undermine his case. Finally, I suggest that Jamesian pragmatic naturalism has far more radical potential than Callicott allows, and that the wider framing of the debate within the environmental philosophy community as setting pragmatist managerialism against radical ethical innovation has itself been misleading. I conclude that Norton is correct to argue that the American pragmatist tradition is a powerful positive contributor to environmental philosophy and policy, provided that the breadth of pragmatic naturalism itself is sufficiently recognized across the board.

Keywords Leopold · Pragmatism · Callicott · Truth · James · Experience

---

P. H. G. Stephens (✉)
Philosophy Department, University of Georgia, 121 Peabody Hall,
Athens, GA 30602-1627, USA
e-mail: piers@uga.edu

S. Sarkar and B. A. Minteer (eds.), *A Sustainable Philosophy—The Work of Bryan Norton*, The International Library of Environmental, Agricultural and Food Ethics 26, https://doi.org/10.1007/978-3-319-92597-4_8

8.1 An Interpretative Overview

One of the many significant landmarks in Bryan Norton's distinguished career is his multi-faceted debate with J. Baird Callicott. In this chapter I examine their differences over interpreting Aldo Leopold's environmental philosophy. The dispute's core, which pits Norton's vision of Leopold as an enlightened anthropocentrist and pluralistic land manager indebted to American philosophical pragmatism against Callicott's view of Leopold as an ethically radical nonanthropocentrist, concerns the plausibility of Norton's seeing a clear pragmatist influence on Leopold via the thought of Arthur Twining Hadley, a President of Yale University and professed admirer of William James. In what follows, I first give an overview of the debate itself. I then investigate the problems of ascribing a particular intellectual identifying label to Leopold. Having suggested that Leopold can fruitfully be read in broadly Jamesian pragmatist terms, I outline a set of relevant but often misunderstood Jamesian positions, and then explain my reasoning for a Jamesian influence upon Leopold's thought and writing, or at least for the validity of using a Jamesian frame to help capture Leopold's core ideas. I conclude that my additional argument can give Norton's interpretation, and an environmental pragmatist perspective more generally, a qualified vindication.

To start with Callicott's perspective, we should note that his carefully refined interpretation started in his earliest career, when as he puts it the "Aldo Leopold land ethic… which I had championed in my first paper in the field, 'Elements of [an] Environmental Ethic'—was being completely ignored" (Callicott, "Introductory Palinode"). His chosen task was thus to interpret and develop Leopold's land ethic, most famously expressed in the formulation that "A thing is right when it tends to preserve the integrity, stability, and beauty of the biotic community" but "wrong when it tends otherwise" (Leopold 1987: 224–5), and to provide a philosophical framework for it. For Callicott the land ethic is a holist and nonanthropocentric venture which is original and radical in its moral implications, but coherent with a philosophical background that can be reconstructed from Leopold's records and reading. He sees a reasonably clear trajectory to Leopold's intellectual transitions, with the "unifying thread through Leopold's intellectual odyssey, from resourcism (Pinchot's philosophy) toward preservationism (Muir's philosophy) to his own harmony-with-nature philosophy" taken as being "Leopold's interest in game", and he regards Leopold's move to the Midwest in 1924 as facilitating his mature concern with land health and the eventual development of the land ethic (Callicott and Freyfogle 1999: 16–17). However, Callicott sees Leopold as decisively rejecting moral anthropocentrism whilst still based in the Southwest. He interprets a key move in Leopold's paper "Some Fundamentals of Conservation in the Southwest", to which we shall return, as indicating Leopold already "thought anthropocentrism was so ridiculous that it should not be dignified by dispute" in 1923 (Callicott et al. 2009: 463). Thus in Callicott's reading, Leopold's ethical outlook swiftly becomes nonanthropocentric after his departure from Pinchot's orientation.

For further scaffolding, Callicott invokes Leopold's use of Darwinism to argue for a naturalistic understanding of ethics, involving accepted restraints on action within a community; he further sees Leopold's thought as drawing from the Humean moral sentiments tradition in which moral value is socially focused and projected by valuing subjects rather than inhering in some objective property. This latter form of focus and projection can and should be extended into attributing intrinsic value to the land community as a whole, with the moral sentiments supporting the land ethic's normative demands (Callicott 1999: 79–115). The logic is "that natural selection has endowed human beings with an affective moral response to perceived bonds of kinship and community membership and identity; that today the natural environment, the land, is represented as a community, the biotic community; and that, therefore, an environmental or land ethic is both possible—the biopsychological and cognitive conditions are in place—and necessary, since human beings collectively have acquired the power to destroy the integrity, diversity, and stability of the environing and supporting economy of nature" (Callicott 1989: 83). Callicott has made several attempts to bridge the Humean is-ought gap, most recently through a historically contextualized denial that Hume meant this distinction in the way it has been generally conceived, and that in fact Hume himself resolves the issue (Callicott 2013: 70–80).

Accordingly, Leopold's ethical scheme on Callicott's account is naturalistic, based upon sympathetic moral sentiments, holistic in orientation, and revolutionary in intent. For him, Leopold in *A Sand County Almanac* was pushing a profound transformation of our socio-ethical orientation, the "promulgation of an evolutionary-ecological worldview" with "its axiological… and normative… implications" of radical intent and importance (Callicott 2011: 516).

Norton first advanced an alternative to this reading of Leopold in his 1988 *Conservation Biology* essay "The Constancy of Leopold's Land Ethic", reprinted in the important *Environmental Pragmatism* compilation (Norton 1996). Norton's initial position was that Leopold "embraced the main philosophical elements of the land ethic early in his career" and that these included "important influences, hitherto unnoticed, derived from American pragmatism" via Arthur Twining Hadley, the Yale University president during Leopold's student days there. Leopold's policy shift from predator eradication to predator protection was for Norton motivated less by moral radicalism than by pessimism about environmental management and "a recognition that scientific knowledge is inadequate to guide gross manipulations of ecosystems" (Norton 1996: 84–5), with Leopold's key perspectival shifts occurring in the 1930s. On this account, contra Callicott, Leopold's position on anthropocentrism was ambivalent: whilst he "accepted that humans will alter the biota" and indeed must do so, he believed that managerial alterations should be geared to long term human survival and the best ecological knowledge, and further hoped that "as Americans become more aware of their interdependence with the rest of the biotic world, they will gradually develop a new conception of the world, including a moral reaction to the community of life" (Norton 1996: 98–9).

But for Norton, this hope was not paramount within Leopold's actions, nor was anthropocentrism his primary target. Rather, Norton sees Leopold as acting on the view that the long-term interests of nature and humanity harmonize once one recognizes the human species' place in the ecological community, and thus the policies mandated by enlightened anthropocentrism and nonanthropocentrism will converge. (He later developed this "convergence hypothesis" on a larger frame in Norton 1991.) As such, Leopold could deploy pragmatist pluralism and avoid needless confrontations. Although more recently Norton has expressed regret over the language of "constancy" in his 1988 essay as having "misleadingly overstated my case", his nuanced position is that "Leopold had fastened onto the Darwinian/ Pragmatic idea of testing 'ideas' by observing the outcome of actions based on those ideas—experience" by 1923 and that "the other changes Leopold underwent between 1923 and his death were built upon this central idea" (Norton 2013: 3). Thus Norton's Leopold is a practical adaptive land manager infused with pluralistic American pragmatism from an early stage, a figure in the inter-war "third way" environmentalism identified by Minteer (2006), and one whose thought aimed at ecological sustainability via experimentation and experiential social learning, always under conditions of uncertainty. Drawing on diverse ideas, Leopold in Norton's view "was an integrator" whose "guide toward the necessary integration was the pragmatic method", and who accepted "Hadley's model of longevity as the test of whether a culture has learned the 'truth' about its place, its habitat" (Norton 2005: 74, 70). This takes us to the central interpretative dispute, the significance of Leopold's Hadley citations.

In his 1988 paper, Norton relied heavily upon the one Hadley reference in Leopold's foundational essay "Some Fundamentals of Conservation in the Southwest" from 1923. In his 2005 volume *Sustainability*, however, Norton supplemented this with information that Leopold quoted Hadley elsewhere, notably in Leopold's 1918 lecture to the Women's Club of Albuquerque, where he asserted that so firmly had the "evolutionary character of Truth been established" that "President Hadley of Yale… defines the truth as 'that which prevails in the long run'". Hadley described himself as a pragmatist follower of William James, and Norton argues that Hadley's essay "The Influence of Charles Darwin on Historical and Political Science" may have significantly influenced Leopold (Norton 2005: 65–6). For Norton, Hadley's first contribution to Leopold's thought was to indicate how Darwinian natural selection could favor certain types of social institution that minimize or eliminate Malthusian scarcity. From this perspective, the capacity of a culture to successfully sustain itself in a place over the long term is the ultimate test: Leopold "borrowed from Hadley the key idea… that longevity over multiple generations provides a second, independent criterion of the success of a culture" (Norton 2005: 70).

Callicott and a group of colleagues dispute these claims and expressed their criticisms in an *Environmental Values* essay, leading to further exchanges between the parties, the last of these being Norton's *Contemporary Pragmatism* article of 2013, to which Callicott has yet to respond (Callicott et al. 2009, 2011; Norton 2011, 2013). Callicott and his team identify only four references to Hadley or his definition of truth in all of Leopold's work, these being (i) the 1918 lecture to the Women's Club of Albuquerque, (ii) the 1921 essay "The Wilderness in Forest

Recreational Policy" in the *Journal of Forestry*, minus any Hadley attribution, (iii) the 1923 essay "Some Fundamentals of Conservation in the Southwest", and (iv) the epigram for an unpublished, undated "Wilderness Bulletin", apparently written in 1924/5. Callicott et al. and Norton agree in noting that Leopold's ascription to Hadley of the view that truth is "that which prevails in the long run" is a misquotation; Hadley actually said that *right* "prevails in the long run", but given the tight relationship between the good and truth in pragmatist thought, the concepts are closely linked. I now examine each disputed instance and give my own verdicts.

8.2 Leopold and Pragmatism: Digging Deeper

In the case of (i), the 1918 lecture, there is unanimity: both parties accept the citation sincerely expresses Leopold's support for Hadley's definition of truth at the time (though he does not include the word "truth" as part of the direct quotation) and I agree. In (ii), the first article where Leopold advocated wilderness preservation and one originally undiscussed by Norton, Callicott et al. suggest that Leopold may or may not be being ironic in quoting Hadley's dictum. These Leopold statements include "truth" within the quotation marks without mentioning Hadley, and read: "It is quite possible that the serious discussion of this question [wilderness preservation] *will seem a far cry in some unsettled regions, and rank heresy* to some minds. Likewise did timber conservation seem a far cry in some regions, and rank heresy to some minds a generation ago. 'The truth is that which prevails in the long run'" (Leopold 1921: 719). Whilst acknowledging that a non-ironic reading makes sense here, Callicott et al. suggest that this could still be an ironic allusion since "truth is that which prevails in the long run and anti-conservation and anti-preservation beliefs have prevailed much longer than pro-conservation and pro-preservation beliefs" (Callicott et al. 2009: 467).

This reading draws upon their observation that Hadley upheld respecting well-established beliefs, a point also made by Norton, but here Callicott et al. apparently assume that such respect must so revere a belief's endurance in the past that its truth in the present should be constituted wholly from such longevity. As we shall see, such a reading misunderstands Jamesian truth; respect for an established body of beliefs held as true no more means that it should be unquestioningly followed than respect for the law means that all existing legislation should be treated as perfect, complete and unquestionable. There is nothing textually preceding or following the quotation to indicate ironic intent, and though I think Callicott correct to note Leopold's fondness for irony, it would be odd for a four page paper aimed at practical foresters to use scant space in a way that could easily mislead. Moreover, the definitive foresters' task that Leopold identifies before the quotation, unmentioned by Callicott et al., fits the pragmatic "third way" integrator model that Norton identifies: noting the "old conflict between preservation and use" now arising with respect to recreation, Leopold states that it is "the fundamental function of foresters to reconcile these conflicts, and to give constructive direction to these issues as they arise" (Leopold 1921: 718–9). All these points suggest a non-ironic reading, in which the meaning is that wilderness preservation *will* prevail as truth in the long run, as the only plausible interpretation.

This takes us to (iii), Leopold's most contentious use of Hadley, in his 1923 essay "Some Fundamentals of Conservation in the Southwest". The most relevant extract comes from the essay's final section, entitled "Conservation as a Moral Issue":

> Since most of mankind today profess either one of the anthropomorphic religions or the scientific school of thought which is likewise anthropomorphic, I will not dispute the point. It just occurs to me, however, in answer to the scientists, that God started his show a good many million years before he had any men for audience - a sad waste of both actors and music - and in answer to both, that it is just barely possible that God himself likes to hear birds sing and see flowers grow. But here again we encounter the insufficiency of words as symbols for realities.
>
> Granting that the earth is for man - there is still a question: what man? Did not the cliff dwellers who tilled and irrigated these our valleys think that they were the pinnacle of creation - that these valleys were made for them? Undoubtedly. And then the Pueblos? Yes. And then the Spaniards? Not only thought so, but said so. And now we Americans? Ours beyond a doubt! (How happy a definition is that one of Hadley's which states, "Truth is that which prevails in the long run"!). (Leopold, in Callicott 2013: 310).

Callicott and Norton agree that "anthropomorphic" here equates to "anthropocentric" in current terminology, but on little else. Callicott et al. maintain Leopold is saying it "is preposterous to believe that the world was devoid of beauty and goodness until the arrival of *Homo sapiens* only moments ago on the temporal scale of Earth's biography" yet this is "what anthropocentrists want us to think", and they claim that the Hadley reference must be ironic. On their argument, when Leopold asks "which man?" the purpose is to ridicule anthropocentrism by reference to ethnocentrism, since the "cliff dwellers, the Pueblos, the Spaniards and the Americans all thought that they—each respectively and exclusively—were the pinnacle of creation for whom the Southwest was made" but not all could be right. Since all these peoples manifested ethnocentrism and anthropocentrism, it evidently prevailed in the long run, and thus "by Hadley's lights, ethnocentrism is right (or true!)" (Callicott et al. 2009: 463–4).

It appears to me that here Callicott et al. have again, and in the same way, misread the pragmatist notion of respecting old truths, but I now turn to Norton's argument, which complements this complaint. Norton argues that Callicott et al. have ripped the concluding "Conservation as a Moral Issue" section from context by not considering it as following from the sections on Southwestern land degradation, and Norton sees the section's function not as "adopting a single ethic and criticizing or ridiculing all alternatives" but as "looking for a system or systems of thought that would support his criticisms of specific practices he and his colleagues engage in, and as a guide toward better policies". In keeping with this pluralism, he sees Callicott's reading as overly narrow, maintaining that Leopold's willingness to "not dispute the point" about anthropocentrism is "to accept it as a working principle, despite its unattractive aspects, not to reject it as obviously false" and furthermore that if Leopold "were rejecting anthropocentrism out of hand as ridiculous, he would have referred to evidence as to its falsity" (Norton 2013: 14, 18). So far as the key point about the Hadley quotation itself is concerned, Norton acknowledges the presence of irony in the passage but not in the Hadley aphorism's usage. In his view, the ironic point is that "all cultures seem to believe the earth is for them, and yet they disappear, and the earth survives", but the contribution to this irony from the Hadley definition is that we now know about the people that "what counts is not their ideology but… their practices which determine how long they survive"; by Hadley's definition, then, "their claimed 'nobility' is tested in their longevity, not their ideology" (Norton 2013: 15–16). By this test, Norton argues, Leopold contradicted Hadley's belief in white capitalist culture's supremacy, for by comparison with predecessors in the Southwest, the "current grazing culture cannot stand up to the test of many generations" and thus does "not have as much 'truth' as prior, indigenous cultures" (Norton 2013: 18).

My own view is that the primary focus is on temporality, and that Leopold is struggling with the evaluation problems posed by multitudinous agents' perspectives over time. I think Norton is correct to see the section as following from land degradation considerations and as loosely accepting anthropocentrism as a default option, but I also believe Leopold is grappling with the possibilities of nonanthropocentric perspectives which he has not yet learned to effectively articulate.

However, these possibilities lead us to issues about the "Conservation as a Moral Issue" section that both thinkers marginalize: Leopold's attitude to language and his use of the mystic P. D. Ouspensky's thought. Callicott largely waves away Leopold's tendencies to note the fuzziness of language to map experience, as when Leopold speaks of "intuitive perceptions" that "may be truer than our science, and less impeded by words than our philosophies" (Leopold, in Callicott 2013: 309), and though Callicott thinks Leopold managed to derive inklings of ecological hierarchy theory and a conception of Earth as a living being from it, he is disgusted at Ouspensky's influence. He even tidies up by arguing that Vladimir Vernadsky's work would have served Leopold's purposes better (Callicott 2013: 160–5, 182–9). Norton is kinder, noting that Leopold found Ouspensky's organicism "inspiring" and observing how a pluralist reading here fits with Leopold's "repeated cautions about the contingency and ambiguity of language" (Norton 2013: 17, 11). But neither fully incorporate Ashley Pryor's points about Ouspensky's influence.

Firstly, the influence illustrates "the important role that non-rational sources of knowledge, such as intuition" played in Leopold's early thought. Shading into this, Pryor follows Curt Meine in noting Leopold's reported view that "'there was a mystical supreme power that guided the universe' which was "not a personalized God" but "more akin to the laws of nature" and "kind of pantheistic" (Pryor 2011:469, 487; Meine 1988: 506). Taking this seriously, it seems Leopold sympathized with Ouspensky's aim of "the creative retrieval and synthesis of mysticism, poetry and idealistic philosophy in order to access 'the real'" (Pryor 2011: 467), which changes the perspective. For if Leopold is not speaking of a personalized deity when opposing anthropocentric science and religion to the intuition "that God himself likes to hear birds sing and see flowers grow", then the idea must be conceived as indicating knowledge through some form of immanent communion. It suggests a mystical insight, especially since Leopold follows it with a typical observation about mystical experience, that of "the insufficiency of words as symbols for realities". Consistent with such notions, Pryor observes Leopold's writing is not only "full of vivid and precise descriptions of what some of these 'other than human' beings look like and how they act" but also "what and how they *feel* and even how they *think*". She links this to Ouspensky's claim for his "higher logic" to emphasize "the role that intuition and emotions like love, reverence, compassion and wonder had in accessing the noumenal 'fourth dimension' that encompasses and unifies all nature", and notes that in relation to language, "both Ouspensky and Leopold voiced concerns about the paucity of human discursive language to adequately describe our experience" (Pryor 2011: 473, 478, fn. 22).

From this perspective, the case for Leopold's nonanthropocentrism here rests on mysticism rather than reasoning: his moral suggestion is apparently that for those with mystical sympathies there may be more than meets the anthropocentric eye and such opinions might one day prevail, but anthropocentrism is the current default position and the truth will be found as "that which prevails in the long run". Though this position is uncomfortable for all of us who reject mysticism, I think acknowledging the significance of Leopold's early mystical sympathies is needed to understand his broader perspective. On my reading here Leopold takes a pluralist position as Norton states, combining operative anthropocentrism with hope that "as Americans become more aware of their interdependence with the rest of the biotic world, they will gradually develop a new conception of the world, including a moral reaction to the community of life" (Norton 1996: 99). We should note further that Ouspensky shared a pragmatist influence with Hadley: both claimed inspiration from William James, whose views on panpsychism and on the capacity for either drugs or nature to induce mystical states are approvingly discussed in Ouspensky's *Tertium Organum* (Ouspensky 1965: 188–91, 269–75). Though vastly more sophisticated than the mystic, James shared Ouspensky's view that experienced reality is richer than any linguistic representation of it, and as we shall see, Leopold appears to have retained this Jamesian philosophical belief long after abandoning mysticism. There is thus an indirect double influence from James on Leopold at the least, and as will become clear, probably a direct influence too.

In summary, both Callicott's and Norton's readings of this Hadley reference are plausible in principle. However, I believe proper attention to the Ouspensky influence should incline us toward a pluralist reading that sees nonanthropocentrism as a possible moral future rather than a currently supported theory, and that alerts us to further Jamesian influence.

We now turn to (iv), the final instance of Leopold's using the "Truth is that which prevails in the long run" formula. This is its use as an epigram in an unpublished "Wilderness Bulletin" from 1924 or 1925. If Leopold was ridiculing the formulation in 1923 as Callicott maintains, why would he use it as an epigram a year later? Here Callicott et al. might be correct in suggesting it was a tactical reference to a different quotation in the text which alluded to Pinchot's utilitarian motto of "the greatest good of the greatest number of American citizens in the long run" (Callicott et al. 2009: 468) as encouraging wilderness preservation. But this does not logically exclude the quotation's epigrammatic use being meant sincerely for adding substantive backing to the encouragement. Indeed, as Norton notes, it is highly unlikely that Leopold "would choose a statement which he takes to be false… as the epigram for an essay" (Norton 2013: 21), least of all barely a year after ridiculing it. This seems to me to be a case of approving use of the formula.

Adding up these interpretations, it appears that only one of Leopold's uses of the formulation, that from 1923, could sensibly be read as ironic, and the plausibility of that reading recedes greatly when we consider the epigrammatic use in 1924/5. That leaves at least three, and probably all, of the four uses as supportive of the formulation. Callicott et al. may, however, can fairly point out that these direct allusions end in 1924/5 and might question the wider attribution of pragmatist ideas to Leopold, so to those issues we now turn.

8.3 Contesting the Leopoldian Heritage: On Darwin, Hume, Callicott and James

Given the exchanges so far, what criteria could establish Leopold as pragmatist? Leopold never claimed the label as Callicott et al. rightly note (Callicott et al. 2009: 480), but this does not resolve the issue, for Leopold did not habitually declare himself anyone's follower, thus making many intellectual debts opaque. Indeed, Callicott admits his own efforts are creative as well as interpretative, for he says that "Leopold had but faintly sketched the land ethic" whereas "I was determined to provide it with fully developed theoretical foundations" (Callicott, "Introductory Palinode"). It can equally be noted that Leopold never called himself a Humean: Callicott concedes that "there is no reason to suppose that Leopold was familiar with Hume's *Enquiry*" (Callicott 2013: 58), and his claim of influence rests on belief that Leopold's land ethic drew upon Humean aspects in Darwin's ethics. But Leopold's Darwin references might more probably be geared to the biology of the *Origin of Species* rather than the ethics of *The Descent of Man*, as Roberta Millstein argues (Millstein 2015), so Callicott's attempt at "fully developed theoretical foundations" from the moral sentiments school is as inferentially based as Norton's reading. Leopold's reluctance to be an open acolyte thus frustrates decisive attribution both ways. We must therefore judge by examining indications of influences, shared approaches to problem solving, consistent similarities of theme and/or underlying assumptions, and uses of language.

Accordingly, any case for pragmatist influence on Leopold will rest on consistent references, philosophical methods or habits of thought consistent with a particular pragmatist, allied to evidence of reading works from that tradition. In what follows, I attempt to build such a case by noting Jamesian themes in Leopold's work, and thus support Norton's claim of pragmatist influence. By this, I also answer a charge brought by Callicott et al. against my earlier writing that I relied "solely on the authority of Norton" in approving of Norton's claims of a James/Hadley/Leopold connection. Though they were correct that in the work concerned I offered "no corroborative evidence" to support my approval, this was due to lack of space rather than lack of evidence, which I here correct (Stephens 2007; Callicott et al. 2009: 478). However, my claim is weaker than Norton's: I will not be claiming that Leopold can be definitively labeled a philosophical pragmatist, but merely that the case for his philosophy being influenced by the pragmatist tradition is much stronger than its denial. I make the weaker claim because I believe Leopold's thought is sufficiently multifaceted to defy attempts at any simple final label. I think we can, however, establish some components as clearly present, and I start by outlining some often misunderstood aspects of William James's thought, with the Norton-Callicott debate as backdrop.

At his death in 1910, William James was America's premier philosopher and psychologist. This is sometimes forgotten, as the rise in reputations of John Dewey and Charles Sanders Peirce has obscured James's early 20th century pre-eminence, but it matters here: if Leopold was likely to read any pragmatist, it would be James, and he did so. Curt Meine notes that James was included in Leopold's reading when he began to read more literature and philosophy at home in Albuquerque from May 1917, while Callicott acknowledges that Leopold read James's *Pragmatism* and transcribed one quotation from it in his notebook (Meine 1988: 160; Callicott et al. 2009: 483). Yet no protagonist has checked the likely root of Leopold's Hadley misquotation from within *Pragmatism*. The James statement from which Hadley apparently drew his view that "the right is that which will prevail in the long run" (Hadley 1913: 98), which Leopold rendered as a truth definition, is this:

> *'The true,' to put it very briefly, is only the expedient in our way of thinking, just as 'the right' is only the expedient in our way of behaving.* Expedient in almost any fashion; and expedient in the long run and on the whole of course; for what meets expediently all the experience in sight won't necessarily meet all the farther experiences equally satisfactorily. Experience, as we know, has ways of *boiling over*, and making us correct our present formulas. (James 1968: 145, original emphases)

It is easy to see how Leopold, working from memory, could have conflated the two definitions, and since Leopold read *Pragmatism* in 1917 and gave his variants on Hadley's aphorism between 1918 and 1924, I suggest that this is exactly what he did. It would also be natural to recall the shorter formulation, especially from the President of one's alma mater. But the fact that Leopold was insistently thinking in terms of *truth* suggests James's influence was more direct than Callicott or Norton allow, so I outline James's oft misunderstood views to indicate deeper connection.

From the quotation above, we can see that James's "in the long run and on the whole" formulation is *future-orientated*: it is justified by the way new information forces revisions of our explanatory formulae, as with large scale scientific theory changes. This suggests that Callicott et al. err in supposing that instances (ii) and (iii) of Leopold's use of the formula are ironic, for in both of those cases their claims of anti-pragmatist ironic intention rested on "the long run" referring to *past* beliefs that should supposedly be retained because they had been long-lasting, whereas the Jamesian requirement is that which *will ultimately prove best* in fitting the experiential evidence. (Past beliefs have a role in Jamesian truth, but that is not it: rather, it is that when pressures of contradiction force someone to modify his stock of views, he "saves as much of it as he can" until he finds "some new idea" that "mediates between the stock and the new experience and runs them into one another most felicitously and expediently" (James 1968: 50). This clearly differs from worshipping past opinion as Callicott et al. claim Hadley does). We should also note the following further points.

i. Jamesian truth is not relativist, nor do its "expediency" and "success" concepts reduce to crass conquest or acquisition, as I have discussed thoroughly elsewhere (Stephens 2012). The theory retains realism about the physical world (James 1968: 201), but agrees with coherence theories of truth in advocating systematic coherence between propositions, thus covering facts which correspondence theories cannot meet. The dynamism of the experienced world always allows for existing truths being falsified and replaced by better ones—those that will prevail "in the long run and on the whole"—but the notion of absolute truth operates as "that ideal vanishing-point towards which we imagine that all our temporary truths will some day converge", and thus functions in intellectual enquiry as a regulative ideal and an intellectual value (James 1968: 145). Expedience is thus a necessary but not always a sufficient condition for Jamesian truth.
ii. James's notions of expediency, workability and success are not reductive as Callicott and others—probably including Hadley, given his Social Darwinism—have held. Callicott apparently assumes pragmatism's whole purpose is reductionist, with classical pragmatism starting from an epistemological subject-object gap and buying into scientific reductionism in which meaning and value can only ever be added or projected through subjectivity, with cognitive content "devoid of meaning in the Pragmatist tradition" (Callicott 2013: 179). Actually, James's radical empiricism aimed to *reconnect* multiple troubling divisions—subject and object, fact and value, humanity and nature—both to ground pragmatism in direct experience, and to rescue modern philosophy from the paucity of value and meaning with which the metaphysical underpinnings of modern science appeared to have left us. As Richard Gale observes, James "wanted to 'I-Thou'… the universe at large", and "the bifurcation of man and nature" was "his ultimate enemy" precisely because "it strips the world of any human meaning or value"; moreover, his radical empiricism's focus on lived experience in its complexity was a major inspiration in developing the phenomenology that animates much ecophilosophy today (Gale 1999: 349; Edie 1987). I have argued elsewhere that combining Jamesian radical empiricism with pragmatism (pragmatic naturalism) offers the ideal framework for environmental values that Callicott and others seek in alternative metaphysics and intrinsic value theory (Stephens 2009). As this invocation of experiential sensitivity and relational emphasis suggest, Jamesian notions of "success" incorporate the widest range of effective interactions with the world,

requiring attentiveness to the experiential flux and integrating reason with emotion as keys for the successful truth-seeker. James rejects accounts of the mind that pit reason against the passions, his account of the environmentally interactive agent is naturalistic and evolutionary but non-reductionist, and he is so concerned with aesthetics that Russell Goodman describes him as "a Romantic in his basic vision" (Goodman 1990: 59). James's scheme emphasizes relationship in the pragmatic knowledge process, and as Frankenberry notes, it is through the dynamic of feeling "that we *cognise relations*, particularly by the cognitive feelings of tendency and transition" in experience. Thus for James, "feeling *is* cognitive" (Frankenberry 1987: 92–3). Accordingly, becoming a good pragmatist truth-seeker requires effectively developing the emotional capacities in relation to the environment, and as Dooley demonstrates, the ideal of the "whole man" (or rather whole person) permeates all James's philosophy (Dooley 1974).

James's criteria for "success" in the epistemological agent are thus sophisticated. His account of truth emphasises knowledge as an interactive process, including reasons for choosing between hypotheses of equal explanatory power. James expressed these decision criteria in the *Principles*, maintaining that the "theory will be most generally believed which, besides offering us objects able to satisfactorily account for our sensible experience, also offers those which appeal most urgently to our aesthetic, emotional and active needs" (James 1918: 2: 312). The emphasis on the whole person in connecting truth to "success" is clear here, and we can thus see that James's criteria for truth ascription in progressing knowledge are coherence, correspondence to empirical reality, and aesthetic or spiritual satisfaction. Ironically, the criteria that Callicott himself recently allocated for adjudicating tenability of competing worldviews are the very same—"(1) self-consistency; (2) consistency with and comprehensive of all phenomenal experience; (3) aesthetical and spiritual satisfaction" (Callicott 2011: 510)—yet he is apparently unaware that these criteria are the keys in James's account of truth. James's emphasis on the active capacities of consciousness in generating knowledge derived from the deep influences of Darwinian evolution and British empiricism, though he saw the latter's models of the mind as insufficiently dynamic. James's project involved rethinking empiricism sufficiently to avoid Hume's sceptical conclusions, and he regarded pragmatism's link to British empiricism as strong enough to once describe Locke, Berkeley and Hume as "the first pragmatists" (Perry 1935: 2, 442).

iii. Darwin and Hume deeply influenced Jamesian pragmatism, yet Callicott apparently assumes that a school of thought linking Hume to Darwin necessarily excludes pragmatism. But Leopoldian acceptance of ideas from Hume and Darwin would not automatically demonstrate anything about Leopold's relationship to pragmatism, for Jamesian pragmatism also generated similar results from these roots. Callicott praises moral sentiments theory as it "provides a moral psychology that makes ethics a matter of… feeling as well as reason" of the sort "essential for an evolutionary account of the origin and development of ethics" (Callicott 2002: 292), but James's account also fits this bill. James repudiates the feeling/reason and fact/value divisions, and he sees the "moment one sentient being… is made a part of the universe" as creating "a chance for good and evils really to exist", for "nothing can be good or right except so far as some consciousness feels it to be good or thinks it to be right" (James 1979: 145, 147). Consciousness emerges into the world as an evolutionary product and develops through time, as the creature's attentiveness learns to select amongst stimuli to generate coherent and effective means of world interaction. James attributes consciousness generously: "My world is but one in a million alike embedded, alike real to those who may abstract them. How different must be the worlds in the consciousness of ant, cuttle-fish, or crab!" (James 1918: 1, 289). Consistent with Leopold's tendency to sympathetically imagine the perspectives of nonhumans, James devoted an entire essay, *On a Certain Blindness in Human Beings*, to lamenting our failures to recognize the inner lives of others, both human and nonhuman, and emphasizing the importance of humility, sympathy and receptivity to growth and wellbeing (James 2010). Though no holist, it is even questionable whether James's ethics are anthropocentric. Certainly his view that "any demand, however slight, which any creature, however weak, may make" has a prima facie claim to be satisfied and his correlative guideline that the "act must be the best act… which makes for the best whole, in the sense of awakening the least sum of dissatisfactions" might together provide much for a radical environmental ethic (James 1979: 149, 155; Fuller 1992; Lekan 2012). But my point here is to show that several of the features Callicott approves as "fully developed theoretical foundations" for Leopold's thought are present in the pragmatist tradition he rejects.

iv. James's radical empiricism is suspicious of excessive abstraction and verbalism. The Jamesian pragmatic naturalist does not reject abstractions but refrains from ascribing to them a higher grade of reality, for "[k]nowledge about a thing is not the thing itself" (James 1952: 478). Awareness of language's fuzziness and insistence that the real is richer than the logical are vital aspects of James's philosophy, and Leopold shares this sensibility. Callicott et al. note the presence of one James quotation from *Pragmatism* in Leopold's personal notebook—"… a spurious idea due to our inveterate human trick of turning names into things"—and promptly assert that the "phrase quoted from James… is not found in any of Leopold's other writings" (Callicott et al. 2009: 483). But though the phrase might not be found, the suspicion it expresses certainly is, and so I turn to evidence of influence.

8.3.1 Jamesian Themes in Leopold: The Evidence

The priority of concrete experience is one of the less noted features in Leopold's writing. Our tendency to label according to priorities and then fancy the label more significant than its object is persistently present however, and I shall now illustrate three ways in which Leopold criticizes such use of conceptual shortcuts in Jamesian fashion.

Firstly, the core of James's point: that the "essence of life is its continuously changing character; but our concepts are all discontinuous and fixed" and coincide with life only "by arbitrarily supposing positions of arrest therein". Accordingly, "concepts are not parts of reality… but *suppositions* rather, notes taken by ourselves" that we have drawn so that we may "*harness up* reality in our conceptual system in order to drive it the better" (James 1977: 113, 111). The more we build out our conceptual knowledge, the stronger the tendency to mistake it for the fullness of reality and the greater the dangers of dogmatism. Similarly, Leopold emphasizes differences between direct experience and representations, such as how a timber cruiser en route to Escudilla mountain "felt a curious incongruity between the remoteness of his notebook symbols and the immediacy of sweaty fingers, locust thorns, deer-fly bites and scolding squirrels" before novelty hit home as on "the far shore hung Escudilla" (Leopold 1987: 134). Such recognitions of the reductive character of linguistic mapping run alongside Leopold's insistence that merely conceptual engagement is inadequate. We recall their presence in Leopold's mystical view from 1923 that "our intuitive perceptions… may be truer than our science and less impeded by words than our philosophies" but such observations do not vanish with the loss of Ouspensky's influence. Their Jamesian flavor continue through much of his work, from acknowledgment that language has areas yet to map, such as the way that cranes represent "stages of the beautiful to value as yet uncaptured by language" to his worry that "the Ph.D. may become as callous as an undertaker to the mysteries at which he officiates" (Leopold 1987: 96, 174), but have received less attention than should be the case.

Secondly, Leopold applies this idea of how we habitually act upon the basis of created categories and sometimes allow them to then blind us to other information. He notes that whereas the "old prairie" was a whole that "lived by the diversity of its plants and animals" in their totality, ecological impoverishment began from the fact that "the wheat farmer was a builder of categories; to him only wheat and corn were useful" and hence he "failed to see the downward wash of over-wheated loam, laid bare in spring against the pelting rains". More broadly, he sees the danger of emphasizing the conceptual grasp of the "alphabetical conservationist" at the expense of direct nature experience: "Education, I fear, is learning to see one thing by going blind to another" as with how "most of us have gone blind to… the quality of marshes" (Leopold 1987: 101, 107, 158). This receptivity problem is also noted by James, who specifically advocates direct experience of nature to counter our failures of perception and empathy. "We are stuffed with abstract conceptions, and glib with verbalities and verbosities", James maintains, "and in the culture of these higher functions the peculiar sources of joy connected with our simpler functions often dry up, and we grow stone-blind and insensible to life's more elementary and general goods and joys" largely because we "of the highly educated classes (so called) have most of us got far, far away from Nature". The Jamesian solution is also the Leopoldian one: the "remedy under such conditions is to descend to a more profound and primitive level", for living "in the open air and on the ground, the lopsided beam of the balance slowly rises to the level line" (James 2010: 159–60).

The third parallel here involves freshness of experience as against abstraction and habit. Leopold, reflecting upon the shift to adulthood in our culture, wonders "whether the process ordinarily referred to as growing up is not actually a process of growing down; whether experience, so much touted among adults as the thing children lack, is not actually a progressive dilution of the essentials by the trivialities of living" for his "earliest impressions of wildlife… retain a vivid sharpness of form, color and atmosphere that half a century of professional wildlife experience has failed to obliterate or improve upon" (Leopold 1987: 120). This is again a Jamesian concern, for whilst perceiving life in its vivid directness is easier in youth, sustaining perceptual freshness is vital to Jamesian philosophical success. Ralph Barton Perry explains that there are "occasional moments when experience is most fully tasted—in the exhilaration of a fresh morning, in moments of suffering, or in times of triumphant effort, when the tang is strong, when every nuance or overtone is present" and that:

> James would arrest us at such moments and say, 'There, *that* is it. Reality is like *that*'. But our worldly minds are filled with ready-made ideas, and when we experience reality it usually has these ideas already stamped upon it. Our minds are accustomed to various short cuts, omissions, and abbreviations dictated by practical convenience, and what these omit we do not commonly apprehend. Hence the metaphysical vision, like the seeing of the painter, involves a recovery of innocence, a capture of the elusive, an unnatural access of sensitiveness. (Perry 1935: 2, 683–4)

This emphasis on receptivity links to humility. James's fallibilism harmonizes with Leopold's stresses on sensitive perception and ability to learn from mistakes. For Leopold, the "outstanding characteristic of perception is that it entails no consumption and no dilution of any resource", its promotion is "the only truly creative part of recreational engineering". It is thus "the expansion of transport without a corresponding growth of perception that threatens us with qualitative bankruptcy" and the task "of building receptivity into the still unlovely human mind" is vital. Indeed, the whole matter of capacity "to see the cultural value of wilderness boils down, in the last analysis, to a question of intellectual humility" (Leopold 1987: 173, 176–7, 200). Moreover humility, in the sense of humanity's ability to regret mistakes and learn, is the ability Leopold identifies as a key to human moral worth and learning: whereas the passenger pigeon would not have mourned our passing, we "who have lost our pigeons, mourn the loss" and it is in "this fact, rather than in Mr DuPont's nylons or Mr Vannevar Bush's bombs" that we find "objective evidence of our superiority over the beasts" (Leopold 1987: 110).

Leopold's moral psychology has been suggested as central to the land ethic (Starkey 2007), but my linkage is more modest. I share Bob Pepperman Taylor's view that humility, moderation, adventure and beauty are components of the Leopoldian good, and that Leopold draws upon tradition to promote "a particular form of American character" involving "a humility at odds with the egocentrism of modern day *Homo economicus*" in civic virtue (Taylor 2002: 173–4, 178, 183), but disagree with his viewing Leopold as anti-modern. Rather, I see him as concerned to invoke such tradition to reject narrow economic boosterism and advance a pragmatic agrarian republicanism, looking backward for political inspiration to move ecologically forward. Though Callicott downplays Leopold's 1918 stint at the Chamber of Commerce in Albuquerque which initiated his Hadley quotations, I suspect that the change of scene occasioned some valuable reflection, especially on citizenship and the longevity of human endeavours. Here is Meine's list of Leopold's May 1917 home reading, works from which Leopold started quoting: "Jefferson, John Stuart Mill, Carlisle [*sic*], Butler, Hugo, William James, Kipling, Epicurus and especially the Bible". On this eclectic list, it is striking that Jefferson, Mill, Carlyle and Butler are all renowned for reflections on the tensions between the individual and society, altruism and egoism, and the role of the virtuous citizen. Moreover, this choice of reading came directly after Leopold had helped review the Grand Canyon forest recreational facilities, a public planning operation that prompted Leopold's sad observation that there "is no getting away from the fact that the 'public interest' is sometimes cruel to individuals" and he "would rather saw many cords of wood than be on hand" when it was. We also know that in August 1917, Leopold began his hunting journals, which "would chronicle the raw data of his experience, the impressions of the natural world on a man whose particular genius was an ever expanding intimacy with that world" (Meine 1988: 160, 162), forming a feeder train to *A Sand County Almanac*. So the period before Leopold's Chamber of Commerce stint suggest a man developing new ideas and habits, including some that would last his lifetime. Marybeth Lorbiecki notes the masthead slogan Leopold gave to his quarterly civic publication *Forward Albuquerque* from 1918: "A Chamber of Commerce is a Chamber of Citizenship, Its Primary Function is to MAKE THE PUBLIC THINK" (Lorbiecki 1996: 79), and the emphasis on reflective citizenship as complementing commerce here is striking. It fits with Peter Cannavò's reading of Leopold as a pragmatic revitaliser of the Jeffersonian tradition, in which the property owner "was a trustee of the common good", but with Leopold's "conception of the common good" finally coming to be "much greener than that of the Jeffersonians" (Cannavò 2012: 866–7).

Similarly, James urged developing the "critical sense" in education to encourage critically aware citizenship and to protect democracy. Such critical sense represented the "ceaseless whisper of the more permanent ideals, the steady tug of truth and justice" which "give them but time, must warp the world in their direction"; as with Leopold, it looked to the long term, with the "feeling for a good human job anywhere, the admiration of the really admirable, the disesteem of what is cheap and trashy and impermanent, this is what we call the critical sense", being in James's eyes "the better part of what men know as wisdom" (James 1912: 311, 314).

Revolted by the rise in American commercialism and imperialism after the Spanish-American war, James vigorously opposed the excesses of an America that he believed was losing its moral character through fixation on power and profit (Stephens 2012: 33–52). James linked this to citizenship in politics just as Leopold did in ecology, and at least one recent commentator sees democratic participation as central to all James's thought (Miller 1997).

Both thinkers were thus persistently concerned about clarity and sensitivity of perception, about the tensions between private and public interest in democratic citizenship, about integrating action, feeling and reason, and about the meanings of education and knowledge. But two paragraphs perhaps illustrate the commonalities most clearly. They are from James's *Pragmatism* of 1907, which Leopold read in 1917, and from some of Leopold's final writing, the conclusion of his Foreword to *A Sand County Almanac* from March 4th 1948.

> The scope of the practical control of nature newly put into our hand by scientific ways of thinking vastly exceeds the scope of the old control grounded on common sense. Its rate of increase accelerates so that no one can trace the limit; one may even fear that the *being* of man may be crushed by his own powers, that his fixed nature as an organism may not prove adequate to stand the strain of the ever increasingly tremendous functions, almost divine creative functions, which his intellect will more and more enable him to wield. He may drown in his wealth like a child in a bath-tub, who has turned on the water and who cannot turn it off
> William James, *Pragmatism*, 1907.

> …our bigger and better society is now like a hypochondriac, so obsessed with its own economic health as to have lost the capacity to remain healthy. The whole world is so greedy for more bathtubs that it has lost the stability necessary to build them, or even to turn off the tap. Nothing could be more salutary at this stage than a little healthy contempt for a plethora of material things.
> Aldo Leopold, "Foreword", *A Sand County Almanac*, 1948.

The similarities go further than shared style and analogy. By 1907 James was alarmed at America's growing technological commercialism, just as Leopold feared the impact of the same forces on American nature forty years later, and as a reader of James it would be natural for Leopold, perhaps unconsciously, to ape the style of James's criticisms. And it looks very much as if Leopold, in introducing his crowning work, paraphrased the words of the pragmatist William James to express his fear that humanity's fixation on economic wealth and control was destroying the very foundations of the nonhuman nature that should have been its counterpoint.

In all these respects, then, we see resemblances between James's thought and Leopold's. Though I am not claiming Leopold as definitively a Jamesian pragmatist, I am claiming that his thought, assumptions and core perceptions of several issues would fit this identity. Leopold appears to *think* like a Jamesian pragmatic naturalist, even without embracing the label. Personally, I believe Leopold drew from numerous sources and that recent scholarship connecting him to virtue focused republican citizenship traditions fits best, alongside a pragmatist influence, whereas Callicott's reading over-emphasizes a purely ethical monorail. Though I sympathize with Callicott's belief in philosophy's socially transformative capacity in creating "an internal change in our intellectual emphasis, loyalties, affections, and convictions", I fear his emphasis on attitudes underestimates the Leopoldian interplay between thought, feeling, perception and action. This interplay, compatible with James's emphasis on the whole person, is expressed in such observations as that "nothing so important as an ethic is ever 'written'" but evolves from "the minds of a thinking community"; that the best definition of a conservationist "is written not with a pen but with an axe", a matter of being "humbly aware that with each stroke he is writing his signature on the face of his land"; and in his bemoaning the "ironbound taboo" that separates "the construction of instruments" as "the domain of science" from "the detection of harmony" as "the domain of poets" (Leopold 1987: 68, 153, 210, 214, 225).

Ultimately Callicott sees Leopold advocating attitudinal transformation as part of "worldview transformation", the start of a new "evolutionary-ecological worldview" that "engages the senses", "transcends the purely cognitive dimensions of human consciousness" so that they engage "the aesthetic and spiritual dimension of human experience", stimulating "the right brain as well as the left" so that you "see and hear as a propaedeutic to thinking hard", thus reintegrating many traditional divisions and engage "the whole person" while presenting "the wholeness and fullness of nature" (Callicott 2011: 521). Yet he fails to realize that many of these ingredients—the ideals of interconnecting science with aesthetics and the spiritual domain, of sensory engagement and actions undercutting the dichotomies of subject and object, of values and facts, and of all this promoting good thought against a backdrop of evolutionary awareness—*were already clearly present in James's philosophy* at the time of Leopold's writing. The focus on scientific ecology was absent from James's public writings, but the other Leopoldian elements sit easily in a Jamesian frame.

Summing up, although Leopold does not declare himself a Jamesian pragmatist or radical empiricist, he apparently thinks very much like one, probably from both primary and secondary Jamesian influences, and moreover he does so at the very stage when such ideas were part of the intellectual atmosphere within the groups with whom he worked. We cannot know which influences were in Leopold's mind when he wrote any given passage, but I see much evidence in Leopold's writings of a broadly Jamesian philosophical outlook, and contra Callicott, there is thus nothing unreasonable about reading Leopold in this light. Indeed, we could even infer an important test case from this: if Leopold's key ideas can be captured in a pragmatic naturalist framework, then the charge that pragmatists cannot articulate environmentalism's core values is false. Norton is thus vindicated in arguing for environmental pragmatism's worth in capturing environmental values; it might even be that environmental ethics' embattled divide between pragmatists and axiologists is itself misconceived. In any case, both in terms of Leopold's "in the long run" quotations and his wider intellectual habits, Norton has valuably contributed to understanding Leopold's perspective. Perhaps the simplest summary of my position on the Norton-Callicott debate over Leopold and pragmatism is that Norton is more right than wrong and Callicott more wrong than right. Accordingly, there is no need for "rescuing Leopold from the imagination of Bryan Norton". On the contrary, Bryan Norton's vision and analyses have greatly enriched and enlightened environmental philosophy throughout his career. The contributions of his intellectual imagination to this field have been immense, and neither Aldo Leopold nor anyone else needs rescuing from such a generous and inspiring influence.

References

Callicott, J.Baird. 2015. Introductory palinode. At http://jbcallicott.weebly.com/introductory-palinode.html. Accessed 28 June 2015.

Callicott, J.Baird. 1989. The conceptual foundations of the land ethic. *In Defense of the land ethic: Essays in environmental philosophy*, ed. J. Baird Callicott, 75–99. Albany, NY: SUNY Press.

Callicott, J.Baird. 1999. *Beyond the land ethic: More essays in environmental philosophy*. Albany, NY: SUNY Press.

Callicott, J.Baird. 2002. My Reply. In *Land, value, community: Callicott and environmental philosophy*, ed. Wayne Ouderkirk, and Jim Hill, 291–329. Albany, NY: SUNY Press.

Callicott, J.Baird. 2011. The worldview concept and Aldo Leopold's project of 'world view' remediation. *Journal for the Study of Religion, Nature & Culture* 5: 510–29.

Callicott, J.Baird. 2013. *Thinking like a planet: The land ethic and the earth ethic*. New York: Oxford University Press.

Callicott, J.Baird, and Eric T. Freyfogle. 1999. Introduction. In *For the health of the land*, Aldo Leopold, ed. J. Baird Callicott and Eric T. Freyfogle, 3–26. Washington DC: Island Press.

Callicott, J.Baird, William Grove-Fanning, Jennifer Rowland, Daniel Baskind, Robert Heath French, and Kerry Walker. 2009. Was Aldo Leopold a pragmatist? Rescuing Leopold from the imagination of Bryan Norton. *Environmental Values* 18: 453–86.

Callicott, J.Baird, William Grove-Fanning, Jennifer Rowland, Daniel Baskind, Robert Heath French, and Kerry Walker. 2011. Reply to Norton, re: Aldo Leopold and pragmatism. *Environmental Values* 20: 17–22.

Cannavò, Peter F. 2012. Ecological citizenship, time, and corruption: Aldo Leopold's green republicanism. *Environmental Politics* 21: 864–81.

Dooley, Patrick Kieran. 1974. *Pragmatism as humanism: The philosophy of William James*. Chicago: Nelson-Hall.

Edie, James M. 1987. *William James and phenomenology*. Bloomington: Indiana University Press.

Frankenberry, Nancy. 1987. *Religion and radical empiricism*. Albany: SUNY Press.

Fuller, Robert C. 1992. American pragmatism reconsidered: William James's ecological ethic. *Environmental Ethics* 14: 159–76.

Gale, Richard M. 1999. *The divided self of William James*. New York: Cambridge University Press.

Goodman, Russell B. 1990. *American philosophy and the romantic tradition*. New York: Cambridge University Press.

Hadley, Arthur Twining. 1913. *Some influences on modern philosophic thought*. New Haven: Yale University Press.

James, William. 1912. *Memories and studies*. New York: Longmans, Green & Co.

James, William. 1918. *The principles of psychology*, vol. 2. London: Macmillan & Co., Ltd.

James, William. 1952. *The varieties of religious experience*. London: Longmans, Green & Co.

James, William. 1968. *Pragmatism and four essays from the meaning of truth*. Cleveland: Meridian Books.
James, William. 1977. *A Pluralistic Universe*. Cambridge, Mass: Harvard University Press.
James, William. 1979. The moral philosopher and the moral life. In *The will to believe and other essays in popular philosophy*, ed. William James, 141–162. Cambridge, Mass: Harvard University Press.
James, William. 2010. On a Certain Blindness in Human Beings. In *The heart of William James*, ed. Robert Richardson, 145–163. Cambridge, Mass.: Harvard University Press.
Lekan, Todd. 2012. A Jamesian approach to environmental ethics. *Contemporary Pragmatism.* 9: 5–24.
Leopold, Aldo. 1921. The wilderness and its place in forest recreational policy. *Journal of Forestry* 19: 718–21.
Leopold, Aldo. 2013. Some Fundamentals of Conservation in the Southwest. In *Thinking like a planet: The land ethic and the earth ethic*, ed. J. Baird Callicott, 303–311. New York: Oxford University Press.
Leopold, Aldo. 1987. *A sand county almanac and sketches here and there*. New York: Oxford University Press.
Lorbiecki, Marybeth. 1996. *Aldo Leopold: A fierce green fire*. Montana: Helena and Billings.
Meine, Curt. 1988. *Aldo Leopold: His life and work*. Madison, WI: University of Wisconsin Press.
Miller, Joshua I. 1997. *Democratic temperament: The legacy of William James*. Lawrence, KS: University Press of Kansas.
Millstein, Roberta A. 2015. Re-examining the Darwinian basis for Aldo Leopold's land ethic. *Ethics, Policy and Environment* 18: 301–17.
Minteer, Ben. 2006. *The landscape of reform: Civic pragmatism and environmental thought in America*. Cambridge, Mass: MIT Press.
Norton, Bryan G. 1991. *Toward unity among environmentalists*. Oxford: Oxford University Press.
Norton, Bryan G. 1996. The constancy of Leopold's land ethic. In *Environmental pragmatism*, ed. Andrew Light, and Eric Katz. London: Routledge.
Norton, Bryan G. 2005. *Sustainability: A philosophy of adaptive ecosystem management*. Chicago: University of Chicago Press.
Norton, Bryan G. 2011. What Leopold learned from Darwin and Hadley: Comment on Callicott. *Environmental Values* 20: 7–16.
Norton, Bryan G. 2013. Leopold, Hadley and Darwin: Darwinian epistemology, truth and right. *Contemporary Pragmatism* 10: 1–28.
Ouspensky, P.D. 1965. *Tertium organum: The third canon of thought*. Translated by Nicholas Bessaraboff and Claude Bragdon. London: Routledge and Kegan Paul.
Perry, Ralph Barton. 1935. *The thought and character of William James*, vol. 2 Vols. Boston: Little, Brown & Co.
Pryor, Ashley. 2011. Thinking like a mystic: The legacy of P.D. Ouspensky's *Tertium Organum* on the development of Aldo Leopold's 'Thinking Like a Mountain'. *Journal for the Study of Religion, Nature & Culture* 5: 465–90.
Starkey, Charles. 2007. The land ethic, moral development and ecological rationality. *Southern Journal of Philosophy* XLV: 149–175.
Stephens, Piers H.G. 2007. Sustainability, democracy and pragmatism: Bryan Norton's philosophy of ecosystem management. *Organization and Environment* 20: 386–92.
Stephens, Piers H.G. 2009. Toward a Jamesian environmental philosophy. *Environmental Ethics* 31: 227–44.

Stephens, Piers H.G. 2012. The turn of the skew: Pragmatism, environmental philosophy and the ghost of William James. *Contemporary Pragmatism* 9: 25–52.

Taylor, Bob Pepperman. 2002. Aldo Leopold's Civic Education. In *Democracy and the claims of nature*, eds. Ben A Minteer and Bob Pepperman Taylor, 173–187. Lanham, MD: Rowman & Littlefield.

Piers H. G. Stephens is an environmental philosopher with a background in the history of ideas in moral and political philosophy, especially classical American pragmatism and the thought of William James. He is an Associate Professor of Philosophy at the University of Georgia and serves as the philosophy reviews editor of the international journal *Environmental Values*. He has co-edited three books, most notably *Contemporary Environmental Politics* (Routledge, 2006), and has published significantly in *Environmental Ethics*, *Environmental Politics*, *Environmental Values*, *The Pluralist*, *Ethics and the Environment*, *Organization and Environment* and other journals, as well as contributing diverse scholarly book chapters.

Chapter 9
The Problem of (with) Environmental Ethics

Daniel W. Bromley

Abstract The quest for enhanced environmental outcomes underlies the entirety of Bryan Norton's prodigious scholarship. Throughout his illustrious career, the constant thread is one of nudging ethicists to grasp the promising insights of pragmatism. I follow his lead here by arguing that the popular appeal to intrinsic value among environmental ethicists is a mistake. Much of the discussion has tended to focus on the "intrinsic" part. It is better to start by focusing on the "value" part. The idea of value requires a valuer. Every valuer will, since every valuer is unique, find different value in perceived objects, events, or phenomena. Following Peirce, it is the effects on a valuer that matter. Those who wish to attach the adjective "intrinsic" would deny the relevance of the pragmatic maxim by ascribing desirable yet unavoidable perceptual qualities to the objects of our senses. This project fails because doing so universalizes very specific and idiosyncratic sentiments. Pragmatism offers escape from this trap by insisting that all choice is informed by reasons, and sapient adults are in need of reasonable reasons. Being told that an observed object is intrinsically valuable is not a reason. Its only purpose is to render the listener an instrument of the speaker's desires. Environmental policy is most successful when skeptics can be brought around. They demand—and deserve—reasons, not moral authoritarianism.

Keywords Intrinsic value · Pragmatism · Environmental ethics

I have benefitted from helpful comments by Bryan Norton, Larry Hickman, Ben Minteer, and Sahotra Sarkar.

D. W. Bromley (✉)
Department of Agricultural and Applied Economics,
University of Wisconsin-Madison, 427 Lorch Street,
Madison, WI 53706, USA
e-mail: dbromley@wisc.edu

S. Sarkar and B. A. Minteer (eds.), *A Sustainable Philosophy—The Work of Bryan Norton*, The International Library of Environmental, Agricultural and Food Ethics 26, https://doi.org/10.1007/978-3-319-92597-4_9

9.1 The Setting

> The fundamental problem of ethics is not, therefore, What is right, but, What am I prepared deliberately to accept as the statement of what I want to do, what am I to aim at, what am I after?… It is ethics which defines this end. (Peirce 1997, p. 198)

As an economist whose career has concerned a wide range of environmental problems, I engage this pleasant task of honoring Bryan Norton as a relative outsider. In preparing for the task at hand, I have come to realize something quite interesting about environmental ethics as an epistemic community.

Someone familiar with philosophy, and with the study of ethics embedded in that broader field, would expect to discover that environmental ethics is a specialized community devoted to the comprehensive exploration of nature—its meaning under various normative approaches, its ontological possibilities, its role in the human condition, its evolving mental construction as humans moved from total immersion in nature to furtive (and often harmful) interactions with it, its instrumentality for the fulfilled life, and perhaps other deep topics. The observer might expect a reflective community offering clarity and insights in a modern age confronted by, as Leo Marx so vividly reminds us, "the machine in the garden (Marx 1964)." One would expect to encounter philosophical discourse helping the rest of us figure out what nature is, what it might be, what it means, and what those possible meanings mean. Perhaps environmental ethics will offer help to those of us who revere nature. It might also offer help to those who do not yet prize nature, but might do so under a certain set of ethical descriptions. Let me be clear that my imagined observer was not expecting a dressed up academic version of popular "environmental literature."

To my surprise, I now realize that "environmental ethics" is a term of art adopted by a community of environmental advocates who have at their disposal a suite of linguistic and conceptual instruments that, if carefully deployed, just might do a proper job of persuading the rest of us to treat nature the way environmental ethicists wish for nature to be treated. In fact, it now appears that "environmental ethics" is a movement for an "environmental ethic." For all of the apparent disdain for instrumentalism among environmental ethicists, environmental ethics—the field of study—strikes me as instrumentalism all the way down. A pragmatist wants to understand this paradox.[1]

9.2 The Hollow Allure of Nomological Foundations

I start by noting that pragmatists tend to be suspicious of tendentious behavioral dictates derived from law-like predicates. Consider Aldo Leopold's "land ethic." Aside from criticizing abusive landowners who certainly deserved severe

---

[1]I note that skepticism is not limited to pragmatists (Sober 1986).

approbation, what are we to make of Leopold's injunction? He certainly did not offer it as an invitation to deep contemplation about nature. Rather, Leopold's "ethic" was a personal commitment—in his later life, it was an essential purpose of his existence. His stirring almanac offered reasons—it was emotive (MacIntyre 1984) —and it induced what Georg von Wright and others would call a practical necessity (1963, 1971, 1983). In Leopold we see soaring rhetorical imagery well coupled with scolding, ridicule, and blunt obligation (1966). However, following MacIntyre, we also notice that Leopold's injunction is a "… linguistic survivor of classical theism" which no longer enjoys the necessary context. In short, the language of Leopold is simply a habit of mind that is no longer underwritten by external authorization. Leopold speaks alone. A doubter will ask: "Who are you to say that I must be nice to **MY** land?"

Leopold's "ethic" is nomological—law like—in a post-Nietzsche world now bereft of a plausible law giver. This renders his injunctive sentences meaningless. As MacIntyre (1984) remarks, since Nietzsche, generalized ruling injunctions—in the absence of an overarching law-giver (such as a God)—have no acknowledged and shared source of authority. The skeptic is now entitled to remark, when faced with nomological assertions, "Who says?" As readers we are of course inspired by Leopold, and we urge similar inspiration on others. But others are under no obligation to adhere to Leopold's dictates. The wrath of an all-seeing "god" is no longer to be feared, and Leopold is not automatically accorded the status of moral successor. Indeed, the dismal record of enlightened stewardship of privately owned land—the insult of greatest urgency to Leopold—is evidence of the futility of the land ethic as a guide for our actions. What is to be done when Leopold fails?

It might be argued, as Ben Minteer reminds me, that another reading would see Leopold as evoking "… older civic notions of the good community and the responsibility of the citizen as both a member of the biotic and social community." That is, perhaps Leopold is speaking "…less as a law-giver than as an articulator of the integrated ecological, evolutionary, and moral understanding of community membership and its responsibilities."[2] That is a possibility and one worth reflection. But since the behavior Leopold denounces was itself widespread throughout his geographic realm, one wonders precisely where to look for this shared "moral understanding" that Minteer finds compelling. If the moral responsibilities of Leopold's neighbors were so compelling, Leopold would have been denied his concern and outrage. And since, in those days, Leopold was largely a lonely voice, it is odd to talk of some shared commitment to nature. Indeed, the region under discussion was then known as "the pinery"—suggesting a bountiful provisioning capacity—and is today referred to as the "cutover" (reminding us of its sorrowful fate). Before Europeans showed up, Wisconsin's Native Americans were doing a plausible job of living sustainably in the North woods. That would soon change with the endless deluge of immigrants. This ought to tell us something about the alleged "shared community ethic" that simply needed to be mentioned by Leopold.

---

[2]Personal correspondence with Ben Minteer.

The only community "ethic" prominent in Wisconsin in the late 19th century was to cut and run. Friedrich Weyerhäuser was certainly not a lone actor in the pinery. Ezra Cornell, of New York university fame, was busy cutting and running.[3] Many farmers, land owners rather than timber scavengers, behaved similarly as the Depression settled harshly over the savaged landscape.

Despite Minteer's reassuring claims of "older civic notions of the good community and the responsibility of the citizen," my reading of history suggests that there was no community to whom appeals to civic responsibilities would find resonance. And perhaps it is precisely here that members of the epistemic community called environmental ethics came to the realization that something more authoritative than soaring civic appeals—and a land ethic—was required if people were to be made to behave "rightly." Moral fiber is strongest when not threated by the sharp blade of self-interest. Perhaps "there oughta be a law." Something called "intrinsic value" just might do the trick.

9.2.1 The Problematic Intrinsic Value[4]

In contemplating this problem, my reading of environmental ethics has put me in mind of Sisyphus straining beneath a large rock—on which is chiseled the words "intrinsic value." However, our Sisyphus is not alone—one can make out a number of adjectival assistants. Here is one called "subjectivist nonanthropocentric." Over there is another wearing a hat marked "objectivist nonanthropocentric." They are eying each other with suspicion and contempt. A third undernourished and spindly legged fellow is wearing a tattered shirt with "weak anthropocentric" on the back. He is unsure of himself. Then we see another figure—this one wearing a necessarily elongated sash with the words "weak anthropocentric anti-intrinsic value pragmatist." This fellow is not offering any help, and seems to be unwelcome. And then we see another assistant labeled "intrinsic value for pragmatists—Dewey says so" who seems to be glowering at the sash wearer (the "anti" fellow).

Then there is the O'Neill (1992) team with their bright day-glow numbers IV1, IV2, and IV3—they are older than the rest, seem unsure if they are wanted, and are confused about what to do. Standing off to one side, not pushing—but cheering in

---

[3]In fact, since New York state did not have any federal lands in the 1860s, Ezra Cornell's 990,000 denuded Wisconsin acres became the land endowment for New York's college of agriculture at Cornell University.

[4]There is a sense that specific names ought to be attached to each of the positions here under discussion. However, to do so requires a careful assessment of all of the writings of the various contributors to each category to make sure that there are no misattributions. The point here is not, after all, to tie specific individuals to specific labels that they may or may not readily accept. The point of this little homily is to highlight the multitude of elaborate peregrinations in the literature as a number of authors debate intrinsic value—and also seek to refine their relation to the concept, and to others. The labels are theirs, not mine. And they know who they are.

tight unison—we see the McShane (2007) contingent reminding the straining horde of the four varieties of goodness in their difficult task, and declaring just how grateful everyone will be when the summit is conquered. Standing off to one side, dismissive of all the adjectives in play, is Robin Attfield ridiculing the entire "subjectivist charade." He knows what intrinsic value is, and he knows that it does not need any help from pragmatists or any other variety of relativists. He is poised to grab the much-revered—indeed necessary—boulder from the incompetent horde and effortlessly fling it to the top. Only then will the world be completely safe from the corrosive effect of subjectivist fluff. Finally, there is an unwelcome observer. He is unwanted because his 1965 paper entitled "Intrinsic Value" has been ignored by the straining crowd. What can he possibly know? He is not even an environmental ethicist. Monroe Beardsley is smiling.

A pragmatist would be struck by this elaborate show and would suggest that any concept—and intrinsic value is simply a contrived concept by postulation (Northrop 1967)—that requires an elaborate encrustment of adjectival qualifiers is a concept without a future. Intrinsic value is also a concept without a past. It is a chimera—a conjurer's device offered up to accomplish a wanted goal when adults cannot be trusted to do what others, called environmental ethicists, want them to do. The gathered supporters of Sisyphus have their various reasons, but are they sufficient reasons? Perhaps Dewey can be counted on?

9.2.2 *Intrinsic Value for Pragmatists*

According to Ben Minteer (2001), there seems a small possibility that John Dewey authorized intrinsic value. Quoting Dewey, Minteer writes:

> If the needs and deficiencies of a specific situation indicate improvement of health as the end and good, then for that situation health is the ultimate and supreme good. *It is a final and intrinsic value* (Minteer 2001, p. 67) (emphasis in original).

Notice the tautology here: "If … improvement of health … (is indicated as) … the ultimate and supreme good, then health as the end and good … is the ultimate and supreme good." The italicized trailer—"It is a final and intrinsic value"—adds nothing of interest, and the word "intrinsic" invites several interpretations. Is Dewey using it to remind us of the obvious circularity of his argument? Notice Dewey said "it is … (an) intrinsic" rather than "it has." It seems that Minteer wants Dewey to say that improved health has intrinsic value—like stunning sunsets. Unfortunately, Dewey says improved health is a final and intrinsic value. Dewey could easily have said that improved health is a final and intrinsic goal.[5] Sarkar Sahotra asks me to consider the possibility that Dewey can be rescued here by deploying decision theory to build—through preference elicitation—a lexical

[5]Larry Hickman reminds me that for Dewey, intrinsic value simply implied something was both "unique and irreplaceable."

ordering of objectives or goals. Having done so, the "… fundamental objectives are the ones that are intrinsically valuable (not because of their contribution to any other values)."[6]

Unfortunately, this strategy has two flaws. The most obvious is the attribution of foundationalism to particular objectives that happen, through "preference elicitation," to appear at the top of some contrived list. A more serious problem is that the preference ordering of those of us living today cannot be taken as pertinent or dispositive for the preference ordering of future persons were they alive to participate in this exercise of such profound importance to their life prospects. The present always stands as a dictator over the future.

Dewey's tautology about values is not the intrinsic value environmental ethicists wish for. This is not the unavoidable, inescapable, and irresistible property of an object out in nature. This is a variety of instrumentalism in which improved health has risen to the top of a list of compelling activities whose ultimate purpose—their final cause—is a collective desire to act in a particular way. Absent the loose declaration offered by Dewey, improved health is just another admirable social value to hang on to—akin to the shared value we hold dear about free speech, and robust democratic discourse. I find Minteer's appeal to Dewey unpersuasive.

9.2.3 Nomological Deceits

Finally, we come to deeper reasons—and the place to start is to explore the possible reasons for Beardsley's amusement in the above story of Sisyphus. In his 1965 paper "Intrinsic Value" he writes:

> Many philosophers apparently still accept the proposition that there is such a thing as intrinsic value, i.e., that some part of the value of some things (objects, events, or states of affairs) is intrinsic value. John Dewey's attack seems not to have dislodged this proposition, for today it is seldom questioned. I propose to press the attack again, in terms that owe a great deal to Dewey (Beardsley 1965, p. 1).

Beardsley begins by drawing a distinction between "for its own sake" and "in itself." He shows a preference for the second version, which comes from G.E. Moore: "It is necessary to consider what things are such that, if they existed by *themselves*, in absolute isolation, we should yet judge their existence to be good" (Moore 1903, p. 187).[7] As Beardsley notes, this definition has not escaped debate. The conclusion seems to be that if anything is intrinsically valuable it is not some

[6]Personal communication.

[7]When Moore says "existed by themselves" we must wonder if Moore means a community of garter snakes in the absence of—isolated from—all possible valuers? He cannot have this in mind since valuers are necessary to undertake valuing. So this must be a world consisting of people (valuers), garter snakes, and not a single other object that might muddy up the relation between snakes, those other objects, and human valuers.

external object—nature—but an experience or psychological state somehow related to that object. This would seem to recall Peirce's Pragmatic Maxim:

> Consider what effects, that might conceivably have practical bearings, we conceive the object of our conception to have. Then, our conception of these effects is the whole of our conception of the object (Peirce 1934, p. 1).

In plain language, it is the effects of an object, and not the properties of an object that have—or do not have—value to possible valuers. Peirce rejects the idea that there is something in (or belonging to) objects that automatically triggers the unavoidable—inescapable—feeling of fondness (prizing) toward the object. The emphasis here on effects reminds us that there cannot be value in the absence of a subject to perceive those effects and then pronounce them good or useless—valuable or otherwise. To paraphrase Antonio Damasio (1999), the attribution of value to particular effects is simply a "feeling of what happens."

Arnold Berleant calls attention to the Humean distinction between sentiment and judgment:

> As Hume put it, "All sentiment is right, because sentiment has a reference to nothing beyond itself, and is always real, wherever a man is conscious of it." It is different, however, if we ascribe beauty to the landscape, for then we are referring not to our feelings but to something beyond, and our judgment must conform to that object (Berleant 2010, p. 337).

Finding a particular landscape beautiful, uplifting, inspiring—valuable—is a sentiment. Ascribing any property to that landscape—its beauty, the serenity it inspires, its magnificence, its value—is a judgment intended to bring others in line with the realized sentiment. Ascribing is to confuse the effects on the valuer with certain properties of the landscape that if but noticed by others would be immediately valued.[8] Notice that if those attributes of the landscape were intrinsic they would be immediately grasped (apprehended) as shared sentiments and would not need attribution. Beardsley concludes that:

> It can never be a necessary truth that any particular instance of aesthetic enjoyment is desirable, since later experience may reveal that it was not in fact desirable at all; therefore it can never be a necessary truth that any particular instance of aesthetic enjoyment is intrinsically desirable; therefore no general or universal proposition about the connection of aesthetic enjoyment and intrinsic value can be necessary; and therefore, there is no necessary truth about the intrinsic value of aesthetic enjoyment to be grasped by intuition (Beardsley 1965, pp. 13–14).

Notice his focus here on value as opposed to the traditional focus on objects (nature). More recently, Larry Hickman has observed that:

> … attempts by some environmental philosophers to locate intrinsic value in nature, …, are merely the obverse of … supernatural and transcendental programs … Such attempts to claim that what is valuable is primary and originary, totally apart from the constructivist

[8]See Sober (1986).

activities of human beings, suffers from the same type of *a priorism* that tends to vitiate arguments based on revelation or transcendental deductions (Hickman 2002, p. 26).

In reflecting on the nomological urge among many environmental ethicists, it bears notice that public policy in democratic societies is most coherent and fruitful precisely when appeals to law-like injunctions are not part of the necessary discourse. My position follows from the quite plausible presumption that sapient discerning adults do not much like being told what to think. Citizens have, throughout history, shown a decided affinity for arriving at their own conclusions. The problem of (with) environmental ethics is that many of its practitioners have shown a lack of trust that adults, following their own counsel, will do the right thing. Dewey's experimentalism is evidently suspect, and Leopold's lament has not worked because its implicit appeal to intrinsic value is too subtle.[9] Recent efforts to buttress it with the addition of a very explicit adjective—intrinsic—are incoherent. Perhaps a new strategy might do better.

9.3 Trust and Sufficiency

Out beyond ideas of wrongdoing and right doing, there is a field. I will meet you there (Rumi 1207–1273).[10]

It is my sense that many environmental ethicists believe that the general public is not to be trusted in its behaviors toward nature, and that it must therefore conjure a more persuasive linguistic instrument for bringing the public to heel.[11] Regardless of what I or others might say about the defects of intrinsic value as that needed behavioral club, the prior question concerns the futility of any environmental agenda predicated on the assured pronouncements of policy elites, academics, or government bureaucrats. Whether alarming warnings about climate change, the alleged necessity of many wolves among the residents of rural Finland, the imperative to protect biodiversity at all cost, the advantages of payment schemes for ecosystem services, the ritualized ceremonies of hearing from stakeholders—indeed any number of other environmental obligations—environmental policy is very often an exercise in ruling down on people (Bromley 2004, 2007). It rarely works as planned, and such authoritative pronouncements and decrees very often incite resistance. We see this at work in the aggressiveness of "climate-change deniers"

[9]I have underlined "implicit" to emphasize, as Bryan Norton reminds me, that Leopold was not making an explicit appeal to intrinsic value. He viewed the world through a lens of "good management" and was concerned of the consequences of particular land-use practices.

[10]Jalal al-Din Muhammad Rumi (1207–1273) was a Persian philosopher, theologian, poet, teacher, and founder of the Mevlevi (or Mawlawi) order of Sufism; also known as Mevlana (Our Guide). His aphorism has been taken up and modified to suit many purposes.

[11]Norton (in his many books and papers), and Minteer (2011), recognize the tension between an "ideological" environmental ethics and the democratic experimentalism central to Dewey.

and their vacuous reasons—including the curious claim of "I am not a scientist." However, in a democracy, even bad reasons must be listened to. Indeed, the essence of a democracy is precisely that of a discursive—reason giving—community (Hiedanpää and Bromley 2016).

The social purpose of epistemic communities such as environmental ethics is manifest by the apparent need, from time to time, for the public and its leaders to hear what the experts believe to be the case. Very often the proffered advice is ignored. It is too easy to scowl and ask, in disgust, "Why won't they (policy makers) do as we say?" But of course this points to the difference between an epistemic community as a source of advice, and an epistemic community as the authoritative source of ultimate truth. Pragmatism asks us to approach public policy as a domain of asking for and giving reasons. That is, policy in a democracy is best understood as the serious and honest exchange of ideas about what seems better to do now, about some particular public problem.

Those who have followed the literature on intrinsic value have seen O'Neill (1992) and Attfield argue that we need intrinsic value to rescue us from infinite regress, or to give us a "coherent non-arbitrary world (Attfield 2001, p. 152)." Just imagine, a world without intrinsic value would be incoherent—and no-one would know right from wrong. No wonder Attfield seems scared. However, to some of us, a world with only instrumental values sounds like a promising world in which human agency finally acquires the capacity to transcend foundationalist imperatives that gave us predestination, slavery, the holocaust, and other emanations of an agreeably coherent and well-ordered world before pragmatists came along to scramble secure timeless truths. Coming back to Beardsley:

> … a world in which all values are instrumental would be precisely a world in which every correct value-judgment could be supported by a reason, and so there would always be a reason for getting one thing done rather than another (Beardsley 1965, p. 12).

Imagine that—reason giving. But which reasons counts as good reasons—those on offer from a speaker? This cannot be the case since everyone who gives reasons already imagines them to be good. Something else is required. Pragmatism privileges those who ask for reasons for reasons. "Why, exactly, is that a good idea?"

In the modern world we tend to have professional reason givers—they are called scientists, experts, and perhaps even environmental ethicists. When the experts have reached a consensus on certain matters—they have arrived at Peircean "fixed belief"—Dewey called that warranted assertability. I call it warranted belief (Bromley 2006, 2008). Warranted belief emerges from within an epistemic community—a discipline—when members of such communities have arrived at settled belief about particular matters. Individual and collective action will often be informed by consulting the assertions of a particular discipline. We usually put it as seeing what the science says about a particular matter. Epistemic communities generally offer warranted belief.

While the giving of reasons is necessary and important, the opinions of those to whom warranted assertions are directed must be considered. That is, claims from an epistemic community, while perhaps warranted, must be deemed pertinent if they are to stand any chance of being acted upon. As discerning adults, those to whom warranted assertions are directed, have the obligation to consider pronouncements from scientific "experts" and decide for themselves whether they wish to act accordingly. A valuable belief is a warranted assertion that can be justified to an audience of attentive sapient agents intent on a particular action. A valuable belief is one upon which individuals are now prepared to act (Bromley 2006). In Peircean terms:

> Belief does not make us act at once, but puts us into such a condition that we shall behave in some certain way, when the occasion arises. Doubt has not the least such active effect, but stimulates us to inquiry until it is destroyed (Peirce 1957, p. 11).

9.4 Finding Reasons for Action

The elaborate straining and peregrinations among a number of environmental ethicists to bring credence to the concept of intrinsic value seems to have produced little of lasting value. I propose that environmental ethicists consider granting to the general adult population a moderate degree of sapience. It would also be useful if those who advocate on behalf of nature would communicate with that adult population in a manner that offers credible reasons for specific actions. I offer two illustrations of reason-giving in environmental policy that show what is possible when reasons are deemed reasonable.

9.4.1 Urban-Rural Environments

In the 1970s there was great anxiety about the notion of urban sprawl—beautiful, verdant and productive farmland was being swallowed up at an alarming rate. Urban interests and farming groups found common cause. There was even a sense that "prime farmland" was something definable with some measure of intrinsic value. The American Farmland Trust and other groups worked hard to slow down the disappearance of this precious natural amenity. At an instrumental level, it was asserted that farmers did not want to be pushed out by the spread of urban uses, and environmentalists wanted to protect greenbelts around cities. The argument often took on moral tones—if prime American farmland were to be lost then poor children in Africa would go hungry, and even the U.S. would run out of food. Hectoring and fear of impending starvation were advanced on behalf of farmland preservation. The American public was not fooled.

Those individuals living in Las Vegas and Phoenix were pretty sure that the ceaseless outward march of their dreadful suburbs was not swallowing up prime farmland. People in Indianapolis and Columbus and Des Moines lived with more prime farmland than they cared to think about—and it seemed to hold little intrinsic charm to them. But when the idea of "green space" as an urban amenity entered the conversation, people were ready to listen. And as a deterrent to urban sprawl, which few people seemed to revere, protecting urban green space became a compelling idea. Here were good reasons. Many cities, counties, and states began to create urban growth boundaries, and to adopt differential agricultural taxation schemes to relieve farmers in the urban shadow of high property taxes. Some states adopted agricultural zoning. No deep moral suasion was employed, and the "intrinsic value" of prime farmland never seems to have come up. But urban residents were pretty sure that they liked seeing attractive farms rather close to the city. It all seems so instrumental. William James would suggest that specific reasons had "cash value."

9.4.2 Wetlands

It is known that a very large share of total wetland areas in the U.S. have now been filled in and drained in the interest of increased agricultural production, and for the creation of ever more shopping malls at the urban fringe. The fight to protect remaining wetlands turned out to be a long and informative conversation about what, exactly, wetlands are good for. Most people figured that wetlands—traditionally called "swamps"—were just breeding grounds for mosquitoes. But then ecologists began to tell stories about filtering of rural and urban runoff to maintain water quality in downstream rivers and lakes, or wetlands as breeding habitat for birds and other charming creatures. Before long wetlands had been transformed—not in their intrinsic properties which had remained unchanged from time out of mind—but in the mental framing of many people who, upon being told about the instrumental properties of wetlands, came over to the side of environmentalists. The word "swamp"—as with "prime farmland"—seems to have disappeared from polite conversation. Once again, it was instrumental all the way down. And that instrumentalism is relational in nature—something is valuable (prized) because it does prized work. As Peirce insisted, a wetland is nothing but the sum total of the effects that a wetland has on us. Restoration ecologists helped to define those effects in a meaningful way to discerning adults. Sometimes, it seems adults can be trusted. Even courts of law can get it "right."

In a landmark case, a landowner in Marinette County, Wisconsin sought to drain and fill some wetlands (already owned by the family) for the purpose of constructing dwellings. The act of filling the wetland would impair the nutrient processing that we now know to be an important function of wetlands. The landowner was denied a permit to do this and subsequently sued the county which had denied the permit. The Wisconsin Supreme Court upheld the permitting requirement and the owner was enjoined and not compensated. The court argued that the owner had bought wetlands and after being denied the permit to drain those lands, he still "owned what he had bought." The plaintiff landowner argued that they were being made to provide the benefits of water purification in their wetlands and that they should be compensated for this service. On the contrary, the county insisted that the filtering attributes of wetlands were as old as time and the owner, if allowed to fill the wetland, would actually destroy those valuable ecosystem services. Why should the owner be paid from the public purse to prevent him destroying what nature had provided? The decision is worth quoting:

> The changing of wetlands and swamps to the damage of the general public by upsetting the natural environment and the natural relationship is not a reasonable use of that land which is protected from police power regulation … filling a swamp not otherwise commercially usable is not in and of itself an existing use, which is prevented, but rather is the preparation for some future use which is not indigenous to a swamp. Too much stress is laid on the right of an owner to change commercially valueless land when that change does damage to the rights of the public …. The Justs argued their property has been severely depreciated in value. But this depreciation of value is not based on the use of the land in its natural state but on what the land would be worth if it could be filled and used for the location of a dwelling. While loss of value is to be considered in determining whether a restriction is a constructive taking, value based upon changing the character of the land at the expense of harm to public rights is not an essential factor or controlling [Just v. Marinette County, 56 Wis. 2d 7, 201 N.W. 2d 761 (1972)].

Interestingly, we see the Court reminding the owner that nothing was taken. The Justs owned a wetland and after their inability to destroy it they still owned a wetland. Nothing was taken but their anti-social aspirations to destroy it for the sake of constructing a dwelling. Did they possess an inalienable right to destroy the wetland? It does not appear that they did. Moreover, the county's bank account, consisting solely of tax payments from its citizens, was not made available to reimburse the landowner for a destructive dream denied. Bryan Norton correctly reminds me that with the current political climate, the Wisconsin decision would most certainly have been quite different. I readily grant his point. However, if we are to take pragmatism seriously, as I intend to do, we are compelled to accept these hypothetical "setbacks" as part of the entire package. Pragmatists cannot celebrate reason giving only when the decisions predicated on those reasons are the ones we favor. We must accept all reasons and results and then—when disappointed—work hard to bring new and "better" reasons to bear. Societies are always in the process of becoming, and a suite of losing reasons today may turn out to be a suite of winning reasons tomorrow. The opposite is also possible.

It is clear that prime farmland and wetlands had little intrinsic value. The policy (and legal) landscape was altered in profound ways on instrumental grounds—reason giving. Reasons not as bludgeons about intrinsic value or so-called immutable property rights. Rather, reasons as understood by sapient adults. An honest Deweyan discourse managed to elaborate the "effects" of these two natural resources and thereby brought people over to the side of environmentalists. As Peirce would insist, it is the effects of objects that comprise their meaning (prizing) to us. And James would add that "truth happened to an idea (or two)."

9.5 Implications

> I understand pragmatism to be a method of ascertaining the meanings, not of all ideas, but only of what I call "intellectual concepts"… those upon the structure of which, arguments concerning objective facts may hinge (Peirce 1997, p. 401).

Environmental ethics is an intellectual concept—a Northropian concept by postulation (1967). Following Peirce, pragmatism is concerned to understand what speakers have in mind when they invoke the phrase "environmental ethics." What is the meaning here, anyway? Paraphrasing my opening quote from Peirce, perhaps a workable definition of environmental ethics might be:

Environmental ethics is not concerned with what is right. Rather, environmental ethics concerns the search for reasons why individuals accept particular propositions concerning what they want to do, where they seek to go, and—thus—what purpose impels them forward.

An epistemic community, of which environmental ethics is of immediate interest, has two important social roles. The first, internal to the community, concerns the quest for what Peirce would call settled belief. The second role, if the community is to connect with others in society, is external. Here, when others ask to "see what the experts have to say" (often put as "let us see what the science says"), settled beliefs (warranted assertions) are communicated beyond the boundaries of the epistemic community. The purpose of this extension is to help others figure out what to believe about complex matters. Whether those who read about or hear the warranted assertions find them dispositive (decisive for choice and action) is another matter entirely. If those pronouncements going beyond an epistemic community gain wide public assent we can think of that warranted belief as comprising valuable belief. As we know, a belief is that upon which you are prepared to act (Bromley 2006, 2008).

Environmental ethics will have its most profound impact if it can evolve into a mature and coherent epistemic community purged of its contrivances to bludgeon an unreliable public. The land ethic of Aldo Leopold, and the experiential commitments of John Muir, have successfully focused the collective mind on the created wonder of nature (Hiedanpää and Bromley 2016). The intellectual legitimacy of environmental ethics—as opposed to a community of political activists in academic disguise—depends on its willingness to engage the many publics in asking for and giving reasons.

It is precisely here that the singular contributions of Bryan Norton loom so profoundly important (Norton 2005, 2015). He has demonstrated the practical implications of engaging pragmatic philosophy at every level. In doing so, he also found intrinsic value to be a barren diversion. It is true that pragmatism can sometimes seem less "efficient"—less direct—than standard nomological decision protocols. The most compelling reason on the side of pragmatism is that democratic discourse privileges reason giving. And participants in democratic debate are pre-disposed to challenge reason givers if they find proffered reasons to be unreasonable. After all, the reason we reason is to make sure that our decisions are regarded as reasonable.

The pragmatist will always ask "why," and will expect the response to be pertinent to the question at hand. If it is not, more reasons will be asked for. The committed pragmatist regards "the reasonable" as an ideal synonym for "the rational."

References

Attfield, Robin. 2001. Postmodernism, value and objectivity. *Environmental Values* 10: 145–62.

Beardsley, Monroe. 1965. Intrinsic Value. *Philosophy and Phenomenological Research* 26 (1965): 1–17.

Berleant, Arnold. 2010. Reconsidering Scenic Beauty. *Environmental Values* 19 (2010): 335–50.

Bromley, Daniel W. 2004. Reconsidering environmental policy: prescriptive consequentialism and volitional pragmatism. *Environmental & Resource Economics* 28 (1): 73–99.

Bromley, Daniel W. 2006. *Sufficient reason: Volitional pragmatism and the meaning of economic institutions*. Princeton: Princeton University Press.

Bromley, Daniel W. 2007. Environmental regulations and the problem of sustainability: Moving beyond 'market failure'. *Ecological Economics* 63: 676–83.

Bromley, Daniel W. 2008. Volitional pragmatism. *Ecological Economics* 68: 1–13.

Damasio, Antonio. 1999. *The feeling of what happens*. New York: Harcourt Brace.

Hiedanpää, Juha, and Daniel W. Bromley. 2016. *Environmental heresies: The quest for reasonable*. London: Palgrave Macmillan.

Hickman, Larry. 2002. Pragmatic resources for biotechnology. In *Pragmatist ethics for a technological culture*, ed. Josef Keulartz, Michiel Korthals, Maartje Schermer, and Tsjalling Swierstra. Dordrecht: Kluwer.

Leopold, Aldo. 1966. *A sand county almanac*. Oxford: Oxford University Press.

McShane, Katie. 2007. Why environmental ethics shouldn't give up on intrinsic value. *Environmental Ethics* 29: 43–61.

MacIntyre, Alasdair. 1984. *After virtue*. Notre Dame: Notre Dame University Press.

Marx, Leo. 1964. *The machine in the garden*. Oxford: Oxford University Press.

Minteer, Ben A. 2001. Intrinsic value for pragmatists? *Environmental Ethics* 22: 57–75.

Minteer, Ben A. 2011. *Refounding environmental ethics*. Philadelphia: Temple University Press.

Moore, G.E. 1903. *Principia ethica*. Cambridge: Cambridge University Press.

Northrop, F.S.C. 1967. *The logic of the sciences and the humanities*. New York: Meridian Books.

Norton, Bryan. 2005. *Sustainability*. Chicago: University of Chicago Press.

Norton, Bryan. 2015. *Sustainable values, sustainable change*. Chicago: University of Chicago Press.

O'Neill, John. 1992. The varieties of intrinsic value. *The Monist* 75 (2): 119–37.

Peirce, Charles Sanders. 1934. *Collected papers*, vol. 5. Cambridge: Harvard University Press.

Peirce, Charles Sanders. 1957. *Essays in the philosophy of science,* ed. Vincent Tomas. New York: The Liberal Arts Press.

Peirce, Charles Sanders. 1997. Pragmatism, the essential peirce. Selected philosophical writings, vol. 2, ed. Nathan Houser. Bloomington: Indiana University Press.

Sober, Elliott. 1986. Philosophical problems for environmentalism. In *The preservation of species*, ed. Bryan Norton. Princeton: Princeton University Press.

von Wright, Georg H. 1963. *Norm and action: A logical enquiry*. London: Routledge and Kegan Paul.

von Wright, Georg H. 1971. *Explanation and understanding*. Ithaca: Cornell University Press.

von Wright, Georg H. 1983. *Practical reason*. Ithaca: Cornell University Press.

Daniel W. Bromley is Anderson-Bascom Professor (emeritus) of Applied Economics at the University of Wisconsin-Madison. His work focuses on the institutional foundations of economic systems, with particular reference to environmental policy, natural resource governance, and sustainable development in Africa, the Middle East, and South Asia. His 2006 book (Sufficient Reason: Volitional Pragmatism and the Meaning of Economic Institutions, Princeton University Press) offers a theory of choice and action that liberates human behavior from the fatal incoherence of rational choice theory. His latest book, with Juha Hiedanpää, is Environmental Heresies: The Quest for Reasonable (Palgrave Macmillan, 2016).

Chapter 10
Proceduralism and Expertise in Local Environmental Decision-Making

Evelyn Brister

Abstract Among Bryan Norton's most influential contributions to environmental philosophy has been his analysis and evaluation of democratic processes for environmental decision-making. He examines actual cases of environmental decision-making in their legal, political, ethical and scientific contexts, and, with contextual constraints and goals in mind, he theorizes concerning what they accomplish and how they can be improved. Informed by the political theories of both John Dewey and Jürgen Habermas, Norton's pragmatist approach holds that appropriate democratic decision procedures will generate broadly defensible decisions. Thus, his view of environmental decision-making is based in—and requires—inclusive, democratic, empirical inquiry. While accepting these criteria, I examine how, in practice, it is difficult to identify when these conditions have been adequately met. I investigate the limitations of Norton's proceduralist approach through a case study in community-based forest management in a New York State urban old-growth park. I argue that Norton's procedural priorities are too rigid given the contextual constraints of local decision-making. While they are useful for guiding an ideal, high standards sense of the decision-making process, less rigid Deweyan considerations of social learning and community engagement often provide sufficient guidelines for evaluating success.

Keywords Environmental pragmatism · Proceduralism · Deweyan inquiry
Community decision-making · Adaptive management

Bryan Norton's contributions to environmental philosophy have expanded its scope beyond ethical and metaphysical debates to address epistemological and political questions raised by the sciences that inform sustainability efforts and through the environmental policy process. He prescribes a policy process in which empirical evidence is paramount, but that is also sensitive to the values and goals of

E. Brister (✉)
Philosophy Department, Rochester Institute of Technology,
92 Lomb Memorial Dr., Rochester, NY 14623, USA
e-mail: elbgsl@rit.edu

S. Sarkar and B. A. Minteer (eds.), *A Sustainable Philosophy—The Work of Bryan Norton*, The International Library of Environmental, Agricultural and Food Ethics 26, https://doi.org/10.1007/978-3-319-92597-4_10

communities. Norton's account of environmental decision-making is a rich one, drawing on John Dewey's theory of inquiry, Jürgen Habermas's theory of deliberative democracy, and the adaptive management framework developed by C. S. Holling and natural resource managers (Dewey 1927; Habermas 1984; Holling 1978). He emphasizes flexibility and response to feedback—advocating policies that can be adapted as political situations change and as new knowledge is incorporated.

Norton develops and tests his account of environmental decision-making against cases of large-scale, high-impact environmental protection projects such as the multistate Chesapeake Bay restoration program and the preservation of critical habitat along the Platte River. He has developed these kinds of case studies because of their complexity and because the top-down bureaucratic force of large-scale conservation projects poses the most challenging test to his account (Norton 2015, 249). He believes that at more local scales, face-to-face relationships and a shared affinity for place are likely to nurture a strong motivation for democratic discussion and cooperation in decision-making (2015, 249).

This essay explores how Norton illuminates aspects of local and small-scale environmental planning decisions—decisions which, while prevalent, don't trigger the centralized, and sometimes forceful, efforts of federal agencies and state governments. I show how the Deweyan strand of his thinking corresponds especially well with the aims of local-scale conservation projects—even though his assessment of neighborly democratic communities may be overly rosy. Then, through a narrative concerning the restoration of an urban old-growth forest, I examine the relationship between scientific expertise, public opinion, and social learning, while reflecting on implications for his account of democratic inquiry in service of policy development.

10.1 Norway Maples and Local Environmental Planning

Here's a story of typical, garden-variety environmental decision-making. In 2008, an informally organized community group of neighbors and environmentalists approached the city of Rochester, NY with concerns about the management strategy for a city park containing a 25-acre remnant of old-growth oak-hickory forest. The park, named Washington Grove in 1932, had been acquired by the city in 1912 as an uncut woodlot. The forest held 200-year-old white oaks, massive black oaks, a population of relatively uncommon butternut trees, and even a rare 80-foot-tall American chestnut tree. Situated on a glacial kame moraine, this forest type was locally unique. But there had been little oak reproduction for a century and the unplanned trail system had contributed to the erosion of fragile glacially-deposited soils. Large old oaks had been falling at an alarming rate, and the neighborhood group adopted the aim of becoming stewards of this little-known city park. The city had pursued a policy of benign neglect, but some in the city government who had long admired the natural beauty of the park were receptive to the group's desire to manage the forest for ecological stability.

From an ecological standpoint, one of the most significant problems was that Norway maples had colonized the grove from nearby yards and urban streets. Norway maples, an invasive species, produce such a dense canopy of shade that they quickly exclude young oaks, sugar maples, and other native tree, shrub, and herbaceous species. In 2009, city foresters proposed eliminating Norway maples. They organized a series of community meetings to inform the public about the merits of the proposed plan, to facilitate community discussion about a shared vision for the future of Washington Grove, and to collect citizen input. Although the organized group of neighbors and environmentalists supported the plan, other park visitors expressed opposition. The park commissioners and city foresters solicited these comments, and in response they changed their presentation of the plan. Rather than describing the goal as complete removal of Norway maples over the course of several years, it was refashioned as a trial treatment to be reassessed yearly. They did not, however, scale back the first year's proposed intervention, and even though the first year's extensive cut of Norway maples was met with vocal disapproval, the second year's cut proceeded as planned. By the third year of the action plan to eliminate Norway maples, community opposition had receded. Also by that time desirable native plants like mayapple had begun to reappear spontaneously, so that managers and participants could point to this visible evidence of success.

I was an interested observer of the Washington Grove debate. As an ecologist, I supported the elimination of the invasive Norway maples, but as a philosopher, I also supported a responsive democratic process. The dilemma posed by community resistance to the conservation plan challenged my faith that a democratic process of inquiry stood a good chance of converging on a plan that satisfied current best practices for maintaining forest health. That is, the social decision-making process, as I will show in more detail later, summoned visions of a publicly shared, historic and culturally-rich natural space, but it also summoned more diverse and conflicting ideas of wildness, recreational value, aesthetic value, ownership, and the role of government. This episode of decision-making poses several questions that go to the heart of Norton's thinking: Do his procedural requirements for environmental decision-making apply in small-scale local cases? How wide is the gulf between the idealized process Norton describes and an actual, typical case, where social power comes into play and where a mix of public and private interests are involved? According to what criteria can long-term success be measured? And can Norton's account provide guidance for such an assessment only under ideal conditions, or also when the decision-making process is flawed? Attention to the political fight over cutting trees in Washington Grove may reveal reasons for the incongruence between a philosophical commitment to democratic involvement in public decisions, a view of conservation as scientifically guided, and the messy reality on the ground—while also yielding some theoretical insight.

10.2 Pragmatist Environmental Decision-Making

Bryan Norton's account of environmental decision-making is rooted in pragmatist theory via John Dewey's theory of inquiry and in pragmatic strategies via adaptive management. Both of these sources support using empirical evidence to design policy actions, which will in turn generate additional evidence to guide future improvement. Norton's account also draws on Jürgen Habermas's political theory in its commitment to procedures that promote democratic deliberation. These sources anchor Norton's account of decision-making in a scientific context. At the same time, they temper the tendency of a science-based account toward cold technocracy by being sensitive to expressed public values and passions (Norton 2005, 2015). While decision-making remains essentially empirical, both scientific knowledge and expressed public values contribute to the policy process, and both scientific managers and the public are expected to modify their values, goals, and assumptions as they learn more about particular situations. Expressed public values are "inputs" that can and do change, or at least become better justified via inquiry. The pragmatic and democratically committed roots of Norton's account ground environmental decision-making as a dynamic, transformative, deliberative process.

Norton's account develops out of his criticism of both environmental ethics and environmental policymaking: he claims that both have given too much attention to the *content* of environmental values rather than making direct use of analytic tools to trace out practical routes to problem-solving and to understand community-level processes of inquiry and deliberation. Environmental ethicists have focused most of their attention on metaphysical questions about (for instance) the source of natural value and "have made few contributions to discussions about what to actually do to improve the environment" (Norton 2005, 163). Environmental policy, for its part, has cleaved too closely to a dependence on narrow forms of economic evaluation, which is untenable because "there are some human values that will always go unexpressed in a system of cost-benefit accounting" (Norton 2005, 178). Although discussing the content of values and their role in policy decisions is not always wasted effort, Norton challenges the assumption that there is a single set of proper environmental values to be discovered. He is a value pluralist, and he is joined by other environmental pragmatists in this view (Higgs 2003; Minteer 2012; O'Neill et al. 2008; Sarkar 2012). Norton finds a Millian source for the pluralist commitment: "pluralists encourage and expect to learn from a diversity of opinion; pluralism and free expression contribute to truth-seeking" (2015, 113). However, this is not mere tolerance of difference, as we might tolerate the fantasy-ridden opinions of children. Different communities may legitimately prioritize or pursue different values because there may be more than one acceptable set of values. The role of ethicists and policymakers, then, is to assist in the process of inquiry, and the goal of inquiry is pragmatic—not to discover truth for its own sake but to transform problematic situations into more satisfying ones. Since the goal of decision-making is to agree upon actions to take, prior agreement on the fundamental values that justify the policy goal is unnecessary in those cases. These are cases where stakeholders can agree to put aside differences about the motivation for an action and still endorse the action.

Thus, the focus of Norton's account of environmental decision-making is to define a public participation process through which stakeholders engage in joint inquiry to determine a cooperative solution rather than to debate value systems. Following Dewey, Norton places ***inquiry*** at the center of environmental decision-making. Decision-making is a process of learning about the world and figuring out the best action to take in order to achieve the public interest. By way of contrast, this is not a view of policymaking as a political process of distributing power or an economic process of maximizing efficiency.[1] Rather, policymaking is a process of gathering up knowledge from experience and of using knowledge to effect change in the world in order to live well. Norton's focus on inquiry is pervasive, as here: "communities can learn and grow with respect to their values, modifying their goals en route, as well as use scientific and technical means to improve their understanding of system behavior. The incremental learning strategy also reacts to problems with experiments and actions" (2015, 11).

In the ideal case, policy actions should address the goals of many stakeholders, aiming for ***political inclusivity***. Political inclusivity is related to pluralism in two ways. First, since a commitment to value pluralism implies that the legitimacy of the value systems of various stakeholders is not prejudged, a wider public is invited to engage in the decision-making process than would be the case if dominant value systems were privileged.[2] Acknowledging value pluralism thus requires accepting a deliberative process that is inclusive and open, "compatible with, and effective within, a diverse community committed to discourse, deliberation, and cooperative action" (2005, 201). Second, a commitment to the involvement of diverse participants supports inquiry for Millian reasons: it opens the process of inquiry that supports decision-making to potentially valuable new ideas and unanticipated criticisms. Norton argues that "pluralism, as opposed to monistic, single-value systems of analysis, creates a more open and inclusive dialogue about what to do, and why" (2015, 108). Although endorsing diverse and potentially incompatible values may seem like a formula for fomenting discord, Norton is confident that in many cases the same set of policy actions can be endorsed for a variety of reasons. For example, duck hunters and birdwatchers have frequently worked together to

[1]While I will examine the outcome of focusing on the policy process as a form of inquiry rather than a contest over political authority, it's worth noting that Deweyan pragmatism has been criticized for its lack of attention to structural political inequalities (Hildreth 2009). Norton's approach, for instance, ought to be augmented by explicit attention to the distribution of credibility and epistemic authority in environmental decision-making (Brister 2012).

[2]Of course there is a presumption that participants are public-minded, tolerant, and willing to observe the conditions required for civil, democratic discourse. It is also legitimate to assume that they agree to abide by policy decisions. In theory, then, Norton's account cannot be applied in cases where self-interested stakeholders or corrupt public or private entities are involved. One point I argue later is that, in practice, it can be difficult or impossible to tell when these conditions are met, and thus difficult or impossible to set up a democratic process that protects and reproduces inclusivity.

preserve wetland habitat, even though they have different reasons for valuing the health of wild bird populations (2005, 471).

Finally, Norton's approach shifts the attention of policymakers and ethicists from value analysis to analysis of and support for a fair and open ***democratic process*** (2005, 278). In *Sustainability* (2005) the justification for the procedural emphasis is Habermasian, grounded in discourse ethics. Norton's pragmatist approach holds that appropriate deliberative procedures, so long as they are inclusive and support sincere inquiry, will generate broadly defensible decisions. Thus, environmental decision-making is characterized by a deliberative process that meets several conditions. In sum, for Norton environmental decision-making requires sincere problem-focused *inquiry,* seeks the equal and involved *inclusion* of affected stakeholders, and is assessed according to the fair and open use of *democratic procedures.*

10.3 Choosing Among Alternative Outcomes

Can we judge whether these criteria are met during an unfolding decision-making scenario? I will now call attention to some vagueness and tensions in Norton's account. I should emphasize that the case study is even now, to a degree, unfolding. It would be foolish to expect any account of a historical (much less political) process to be able to fully explain events and choices in a way that eliminates uncertainty, doubt, or the possibility of conflicting interpretations. In what follows, I draw on my personal reflections about the Washington Grove case study in order to compare a typical example of non-ideal decision-making with what should or could be expected according to Norton. This vantage point highlights the ways in which Norton's procedural approach fails to serve as a guide to resolving cases such as this one; at the same time, I identify elements of Norton's account that provide hope for identifying and promoting positive outcomes in cases of local environmental decision-making, even when the conditions for adaptive community engagement are less than ideal.

The first issue is the assumption that the public process in the Washington Grove case was, at root, a process of *inquiry*, and that the process was *politically inclusive* and fully *democratic*. This seems to imply that participants should therefore have been *equally involved* in their roles as inquirers. The process Norton outlines does not give a higher priority to participants with scientific training or to those with professional degrees or to those who live closer to the land being managed: all are afforded an equal opportunity to shape the process of inquiry.

However, there are some constraints placed on who can contribute and how. Norton, following Dewey, requires that stakeholders involved in public decision-making engage in *genuine* inquiry, meaning that they take part in the political process, respond to criticism, and agree that the goal of participation is to identify and satisfy the public interest rather than only their own private interest.[3] These requirements are intended to ensure that participants are sincere, open-minded, and willing to engage in collaborative problem-solving. In practice, however, it is difficult to identify when these conditions have been adequately met. And, as Norton acknowledges, they often are not met. Stakeholders may pose as interested in the public good while pushing only their own agenda; sometimes they use a threat to disengage as a weapon of extortion. As facilitators and negotiators, policymakers ought to seek compromise but may not be able to recognize when sincere inquiry has been abandoned. Thus, they are vulnerable to being misled by bad faith actors. And here we have an example of what is a general problem for Deweyan inquiry: that its guidelines are too weak—and too optimistic—to prescribe a direction for policy action in cases where the parties profoundly disagree or where they misbehave.[4]

At the same time that Deweyan inquiry seems too weak to discriminate which is the better of competing positions, it can also be seen as too strong, in that it can endorse *any* publicly-held value, so long as that value has been procedurally vetted and there is negligible opposition to it. This charge of relativism is a long-standing criticism of Dewey's political philosophy: critics have held that it "is vulnerable to appropriation by whatever social forces are most powerful" (Festenstein 2008, 90). Thus, there is a tension between seeing the content neutrality of Deweyan political philosophy as a strength, due to its ability to endorse any liberal public value, and as a weakness, because it is vague in pointing toward a decision procedure which is able to discriminate between conflicting visions for the public good.[5] As I watched the decision process in Washington Grove unfold, this tension became apparent. It was impossible to distinguish bad faith actors from sincere participants, or heartfelt expressions of value from pretense and selfishness.

The community expressed many objections to the proposed plan to eliminate Norway maples. The most common claim (which turned out to be accurate) was that the cut trunks and limbs would be unsightly. Others said that permanently opening up the understory would not be aesthetically pleasing to their personal tastes, no matter what was considered natural for this forest type. Some argued that the task was Sisyphean, because the only thing that would regrow would be more Norway maples—and that the costs of the management project would therefore be

[3]Sarkar (2012) raises the problem of stakeholder legitimacy explicitly, noting that as an ethical, rather than legal, issue, it has not received the amount of critical examination it deserves.

[4]This criticism of Dewey's ungrounded optimism was articulated already in the 1920s by Randolph Bourne, by Walter Lippman, and by others (cf. Misak 2013, 135–138). One *could* excuse Dewey's naiveté and idealism by claiming that the realm of theory shows us what is possible, whether or not it can be applied to real democratic processes. But Dewey's goal was always to engage with and influence real situations.

[5]The degree of content neutrality that Dewey endorses is contested; critics accuse Dewey of relativism, while supporters defend his more substantive vision of a democratic way of life. What matters here is that in cases of environmental decision-making, there are commonly conflicting values at stake—for cultural goods and for natural goods—that are endorsed by members of a democratic community.

too high, extending indefinitely into the future. One neighbor called it a form of speciesism to declare that some tree species were worth more than others or to kill today's trees for the sake of future ones; and some neighbors embraced a no-management vision they described as "letting nature take its course." Some argued that bringing management attention to the previously neglected property would increase park visitation, which would then increase erosion and other harms (although there was also the unverifiable suspicion that local social friction played a role here, since the park split neighborhoods of different demographic profiles). Finally, there was also an expression of fatalism when it was acknowledged that there had been so little oak reproduction for a century that the current character of the forest would not be maintained, and so any management decision was as good (or as bad) as any other.

10.4 Proceduralism and the Deweyan 'Problematic Situation'

In the midst of the process, participants working toward the restoration of Washington Grove could see both an ambitious, community-building activity and a process that strained trust among citizens and in the city government. Unlike Norton's prescription for success, the decision process was not laid out ahead of time and concrete measurable environmental indicators by which to measure success were not set out. At times, this procedural ambiguity drew out cynicism. Were all stakeholders being treated fairly? Were the decision-makers experienced and trustworthy? Would interventions be chosen for deliberate or arbitrary reasons?

Proceduralism is one corrective response to the opaqueness and uncertainty of this sort of political situation. A robust and well-articulated decision-making procedure can assist policy makers in identifying the public's shared values and can bring to light stakeholder knowledge that expands policy options and improves outcomes (Brister 2012, 198). The advantage of robust decision procedures is that they promise to block authoritarian and arbitrary government decisions, prevent corruption and abuse of influence, and make stakeholders' social and political inequalities irrelevant.

The procedural justice view is integrated in Norton's account of public decision-making through his endorsement of Habermasian democratic procedures conducted under appropriate communicative conditions (Norton 2005, §7.4). There is room for skepticism that clean, agreed-upon procedures can be attained in a case like the one I've described, where decisions must be made relatively quickly, without the benefit of an established and tested regulatory framework, and in a local setting.[6] On the other hand, Deweyan democratic inquiry fares better. While the case study I'll examine looks something like a failure from the perspective of Habermasian proceduralism, it appears as a success when viewed as a case of community-based democratic inquiry.

Misak's (2000) evaluation of the relationship between Habermasian political theory and pragmatist epistemology explains the difference here. Misak identifies a commonality between pragmatic inquiry and proceduralism. For both, the method of seeking right belief should be driven by reason and experience, and any such method must be open to considering different reasons and perspectives. However, as she says, "The pragmatist does not think that deliberation is always appropriate," much less deliberation that makes rule-following a priority (2000, 107). One reason is that in the course of inquiry we may change our view of what we want and of how to go about pursuing it, and during the course of such a fundamental change, we may for a time not know how to rewrite the procedural rules guiding the process, and any revision may appear ad hoc. Thus, the open-ended nature of inquiry is at odds with the predetermined guidance demanded by proceduralism. A pragmatist approach to decision-making may respect consensus building while recognizing that not every disagreement can, or should, be resolved to everyone's satisfaction.

Norton describes his approach as "procedural rationalism" (2015, 115) and as "pragmatic idealism" (2015, 118) and associates these aspects with Dewey's view of social learning. However, Dewey is not a proceduralist, except in the loose sense that he is as concerned with process as product. Norton builds an ideal account of public decision-making on the Deweyan notion of "public interest" as developed by Bozeman and Minteer (2007). He proposes a "pragmatic idealism" that evaluates the success of decisions by specifying an ideal, so that "actual processes can thus be compared with an ideal process" (2015, 118). He argues that "this approach avoids vacuity…by stating an ideal of public discourse and then characterizing a decision as 'better' insofar as it was arrived at by processes closer to the ideal" (2015, 117). He recognizes this as necessarily a form of "constrained" rather than "pure" proceduralism (2015, 116). It is constrained by the range of decisions that are open to consideration and by the imperfection of actual situations: there are instances of unequal power, knowledge, education, resources, and ability to participate.

Norton's emphasis on identifying a procedure to guide decision-making solves the evaluative problem I have posed. In providing an operative procedure for how to evaluate the legitimacy of a deliberative process, it also avoids the problem of relativism mentioned earlier, since it reinforces the content neutrality of the Deweyan approach by identifying the legitimate community values as those that emerge after a prearranged process. In addition, the Deweyan approach has a positive pragmatic benefit because for "an adaptive manager who realistically wants

[6]The procedural aspects of Norton's view likely play a more important role in cases where the federal government is involved, where deliberative procedures have been tested and routinized over time, and where there are significant differences in the magnitude of stakeholder power and access to government consideration.

to address a wicked problem…it provides a path forward, no matter how many setbacks occur" (2015, 134). In this way, it articulates a strategy to support public inquiry.

Norton's approach requires that the evaluation of decision-making be tied to evaluating how closely communicative and deliberative procedures match an ideal. I argue that this requirement is stronger than is necessary for social learning in Dewey's sense, and is perhaps impossible in situations which are strongly constrained and in flux. In such situations, it is often not possible to articulate what the ideal procedure would be, what the ultimate goal of the process is, what will definitively count as a success or failure, or what the appropriate time frame for evaluation is. And even if an ideal can be articulated, since actual processes are bound to fall short in some ways, there is no scale against which to measure how far they fall short and, therefore, when the decision-making process is no longer legitimate.

In essence, the difficulty derives from the disparity between proceduralism's idealizing method and Dewey's contextual logic. What Dewey calls "problematic situations" provide "both the impetus as well as resources for inquiry; and when settled, they function as its end" (Capps 2002, 244). Norton, too, recognizes how, on Dewey's pragmatic approach, responses to problems are embedded in local contexts that undergo a process of inquiry (not necessarily a procedure) that changes them: "Dewey…did not see 'problems' as well-formed questions awaiting decisive actions; indeed, he was quite explicit in seeing the articulation of a problem as a function of an evolving 'public'" (2015, 121). Moreover, the evaluative resources needed to adapt and improve the process of inquiry are provided internally to the context of a problematic situation, not by a universally ideal procedure. Thus, while some decision-making can proceed by designating deliberative procedures at the outset, many instances of decision-making require more on-the-fly adjustments. This does not mean they cannot be evaluated but that their measure is taken in context. They are more or less successful relative to what is revealed as possible, given local resources and expectations. Let me return to the case study to illustrate this point.

10.5 Expertise and Community Values

The possibility of managing Washington Grove to control invasive species and promote native tree regeneration had come about because of the formation of a community group. This group had spent about a year defining their aims. Separately, a group of citizens had formed to advocate for a park where dogs could be allowed off-leash. Because off-leash dogs were a de facto presence in Washington Grove, this group joined in the deliberations about reshaping the Grove's management plan. During the deliberative process, city staff definitively ruled out the possibility of changing the city code to permit off-leash dogs in a city park, and this decision caused a reconstitution of the involved public. Many of the dog-lovers lost interest in participating when their goal was taken off the table. However, it also created a counterpublic—a few participants continued to be involved in the deliberative process, pursuing obstruction as a strategy to return a discussion of dog policies to the table.

During this period when the public was being reconstituted, the city's environmental management staff altered their practices. Communication became less open, but it also became more methodical. Meetings with city staff became more difficult to obtain, and, compared to the early public meetings, later meetings focused on delivering educational content rather than on fostering interactive deliberation among citizens. If we are to judge whether the process was effectual or not, the judgment ought to take into account the political constraints that were not present initially and that constrained the emerging process. The goals of land managers could be described as having shifted from swift execution of a management plan revision to a fair consideration of the conflicting desires of stakeholders. The conditions for judging success were unavoidably local and contextual, and the measure of success had to change over time as expectations, resource constraints, and the balance of cooperative versus obstructive behavior shifted.

It's possible that Norton and I are interested in two subtly different questions. His question is: What theoretical approach will provide the type of guidance most likely to result in success? Specifically, he asks "whether (assuming good faith on the part of participants and best procedural practices represented by heuristics) communities can" participate in successful environmental decision-making (2015, 137). He says they can, and the heuristics he provides raise the chance that they'll succeed.

My question is: How can we evaluate the degree of success or failure of community attempts at environmental decision-making, given the challenges and constraints present in a situation? What I want to know is, when we're in the thick of it and the process is clearly not ideal, when the public is uncooperative or the decision-makers are naive, when the goal appears to recede and group dynamics lead to arbitrary results, does an account of public inquiry lead us to have grit and continue to pursue a "good enough" end? Is a commitment to procedurally-guided deliberation so binding that when inclusive participation is compromised, a negative judgment is compelled?

Norton has developed "a procedural account of how, if the conditions are right and the correct steps are taken, a community can learn its way toward a more sustainable future" (2015, xiv). I would add that because Norton's approach is essentially Deweyan, his heuristics and case studies also give insight for improving social learning when conditions are *not quite* right and the steps are wavering. And this is a good thing, because the time we most need insight and tenacity is when conditions are not ideal and when we can't yet enjoy the benefits of hindsight, but we *can* build consensus, constructing it using the self-correcting nature of the deliberative, evaluative process.

During the deliberative process for Washington Grove, the moment when it was most difficult to have faith in public deliberation was when some participants rejected the relevance of the scientific expertise of the foresters and of citizen participants with backgrounds in ecology and natural history. It became apparent that some participants did not trust the scientific and professional expertise of the city staff. In particular, no amount of scientific evidence would sway them to think that cutting Norway maples was a reasonable management action. These participants were not a cohesive group, but they shared a common belief: the status quo was satisfactory. At this point, we saw the local, short-term failure of Norton's "convergence hypothesis," which holds that people will tend to come together to support similar policy actions, whether their values stem from an intrinsic concern for nature or from a desire to preserve natural resources for human use. While some people supported management action in the forest—and had a wide range of reasons for doing so—others discounted any reason to alter the status quo.[7]

The process at this point was not ideal, and the time constraints were key. It was April, and cutting could only be done during a window of opportunity in May. To miss it would require waiting a year, but given the vicissitudes of government support and the dynamics of community activism, there could be many reasons other than the considered preferences of the community which might prevent the opportunity from arising again. The foresters had hoped to develop community consensus on the project through educational efforts. Some support was gained, but the consensus was not complete. Are we nonetheless justified in privileging the judgment of foresters over the values of some community members? Given that the community was split, should the decision to act or not be weighted toward maintaining the status quo, or should expertise be given greater weight?

In spite of the community's division, there are two reasons why the foresters were justified in continuing with their plan. The first reason is consistent with Norton's emphasis on adaptive management. The reason to cut invasives is to restore the natural process of oak regeneration, a need with some urgency behind it. If the invasive maples were cut (say, in a trial area), and the community did not appreciate the results, then the invasive component would regenerate quickly. On the other hand, to miss an opportunity to act would mean losing time that couldn't be retrieved.

Secondly, the foresters could draw on additional expertise which goes beyond that of laypersons. Namely, similar community conservation experiments had been performed locally and elsewhere, and a common feature of these experiments is that community dissent retreats after a relatively brief time. Thus, the city's foresters were acting not only on the basis of principles of ecological restoration but also on principles of community engagement and the accumulated experience of best practices in their profession. Perhaps this attitude can be criticized as grounded in paternalism or in scientific elitism. But at the same time it can be seen as supported

[7]I argue below that there are multiple possible explanations of the community's failure to agree upon a set of management actions. The short time frame, as well as a lack of imagination and experiential knowledge, can explain the failure without undermining a commitment to value pluralism or to Norton's convergence hypothesis. Much depends on how the context—or, as Dewey would say, the problematic situation—is defined. For a variety of empirical and theoretical responses to the convergence hypothesis, see Minteer (2009).

by a pragmatic commitment to deliberately learning from experience and as a legitimate counter to the common cognitive phenomenon known as the status quo bias. In new and challenging situations, people find it difficult to envision positive change and easy to envision failure. As it turns out, five years later, the community surrounding Washington Grove has become widely appreciative of the decision to cut Norway maples and there have been significant volunteer contributions to maintenance efforts.

10.6 Social Learning and Democratic Participation

Norton maps a reliable route for community decision-making, and he's described the journey of several communities of inquirers across rough terrain to reach successful resolutions. My reflections here show two things: first, that even for a small, local inquiry concerning an environmental problem, it's hard to scout Norton's path and hard to know when the community is on it; and second, that although the route Norton marks may be ideal, there are non-ideal paths that track right alongside it, and that likewise depend on the key elements of community inquiry, inclusivity, democratic process, adaptivity, and social learning.

There are many reasons to label the decision process in Washington Grove a failure by Norton's lights. The key to Norton's account of adaptive management is to identify measurable indicators whose improvement is the mark of success. But in Washington Grove, attempts to guide deliberation to pick indicators repeatedly devolved into discussions of values, and in particular into a split between those who valued regeneration of native species and those who wanted no change to their current experience. The process also seemed inadequate to Norton's ideal because political friction made direct, inclusive deliberation impossible and because the timeframe was inflexible. The managers' response to public dissent seemed inadequate because adjustments to the action plan were rhetorical, not substantial. Finally, the sincerity of the participants was at times in doubt—both because pursuing the public interest was not everyone's goal and because scientific evidence was deemed irrelevant, unjustified, or false. In this, my case study is hardly unique.

In spite of these shortcomings, I would not, in hindsight, judge the management action or the decision process that led to it as inadequate. Instead, I would interpret it as a moderately successful example of social learning, according to a Deweyan account, and I can imagine how Norton, too, might interpret it as only a partial failure by expanding the timeframe that defines the process.

First, the initial goal of the community group and foresters was to settle on a plan to remove Norway maples, if one was warranted. That goal was met, but not without top-down imposition of the plan for the first two years. Concrete indicators were not picked in advance, but, after several years, the foresters and citizen naturalists were able to point to the reappearance of native species and the regeneration of native trees as indicators of forest health. Thus, the success of the action plan contributed to community education in the long run. Some citizen participants who were initially unable to imagine what the forest "should" look like have now come to participate in invasive removal and native plant restoration work. Perhaps it was unrealistic to expect that a slideshow presentation on forest ecology would be a sufficient educational experience to change minds. Personal, hands-on experience over a longer period of time was necessary for community understanding of ecology to take root.

Second, in Deweyan public inquiry, the community should be open to a change in values or goals. Even the constitution of the community itself may change—and this is indeed what happened. While the process was focused in part on formulating and carrying out an invasives removal plan, city staff were also interested in encouraging the development of a citizens' group which could take on management and fundraising tasks. This was an undertaking that took several years and required the establishment of trust: the community members had to learn to trust in the city staff's goodwill and skill, and the community group had to earn the foresters' trust in their ability to build inclusive practices and navigate political discord. Over time, the constitution of the stakeholder groups shifted to include more park visitors and neighbors who had felt excluded by an earlier lawless ethos in the Grove. Thus, participant involvement grew over time and shifted to higher levels of enthusiasm and sincerity. In turn, this led to more opportunities for inquiry and community action.

The adaptive process that Norton describes "assumes that communities can learn and grow with respect to their values" (2015, 11), and it acknowledges that the most effective form of learning is through experience, when collaborative stakeholders develop a willingness to experiment and make use of scientific and technical knowledge to improve a problematic situation (2015, 57). The criteria for evaluating this case's outcome as successful is not only the achievement of a physical improvement, but also the creation of a public that continues to collaborate with the municipality on issues related to Washington Grove. These reflections show that Norton's attention to adaptive decision-making and to building collaborative publics can helpfully inform and guide participation even when constraints and obstacles lead us farther than we might like from ideal conditions.

Acknowledgements My knowledge about the case study comes from public meetings, from conversations with Brian Liberti, head of the Forestry Division for the City of Rochester, NY, and from planning documents distributed by the City of Rochester. Thanks go to William Throop for asking questions that directed my thoughts about the case, to Matthew Brown for sharing ideas about Deweyan inquiry, and to Charles Burroughs for encouragement to write about the much-loved Washington Grove. Patient and insightful audiences at Green Mountain College, the University of Waterloo, and an International Society for Environmental Ethics conference in Kiel, Germany made helpful contributions, and Ben Minteer, Sahotra Sarkar, and Paul Thompson improved my ideas and their expression immeasurably.

References

Bozeman, Barry, and Ben Minteer. 2007. Toward a pragmatic public interest theory. In *Public values and the public interest: counterbalancing economic individualism*, ed. Barry Bozeman. Washington, DC: Georgetown University Press.

Brister, Evelyn. 2012. Distributing epistemic authority: Refining Norton's pragmatist approach to environmental decision-making. *Contemporary Pragmatism* 9 (1): 185–203.

Capps, John. 2002. Achieving pluralism: Why AIDS activists differ from creationists. In *Dewey's logical theory*, ed. F. Thomas Burke, D. Micah Hester, and Robert B. Talisse. Nashville: Vanderbilt University Press.

Dewey, John. 1927/1982. The public and its problems. In *John Dewey: The later works*, vol. 2, ed. J. A. Boydston and B. A. Walsh, 1925–1927. Carbondale, IL: Southern Illinois University Press.

Festenstein, Matthew. 2008. John Dewey: Inquiry, ethics, and democracy. In *The Oxford handbook of American philosophy*, ed. Cheryl Misak. Oxford: Oxford University Press.

Habermas, Jürgen. 1984. *The theory of communicative action*, trans. T. McCarthy. Cambridge, UK: Polity.

Higgs, Eric. 2003. *Nature by design: People, Natural process, and ecological restoration*. Cambridge, MA: MIT Press.

Hildreth, R.W. 2009. Reconstructing Dewey on power. *Political Theory* 37 (6): 780–807.

Holling, C.S. 1978. *Adaptive environmental assessment and management*. New York: Wiley.

Minteer, Ben. 2009. *Nature in common?: Environmental ethics and the contested foundations of environmental policy*. Philadelphia: Temple University Press.

Minteer, Ben. 2012. *Refounding environmental ethics: Pragmatism, principle, and practice*. Philadelphia: Temple University Press.

Misak, Cheryl. 2000. *Truth, politics, morality: Pragmatism and deliberation*. New York: Routledge.

Misak, Cheryl. 2013. *The American pragmatists*. Oxford: Oxford University Press.

Norton, Bryan G. 2005. *Sustainability: A philosophy of adaptive ecosystem management*. Chicago: University of Chicago Press.

Norton, Bryan G. 2015. *Sustainable values, sustainable change: A guide to environmental decision making*. Chicago: University of Chicago Press.

O'Neill, John, Alan Holland, and Andrew Light. 2008. *Environmental values*. New York: Routledge.

Sarkar, Sahotra. 2012. *Environmental philosophy: From theory to practice*. Malden, MA: Wiley-Blackwell.

Evelyn Brister is an Associate Professor in the Philosophy Department at Rochester Institute of Technology. She has a Ph.D. in Philosophy (Northwestern University, 2002) and became involved as a community participant in the restoration of Washington Grove while obtaining an M.S. in Environmental Science at RIT (2012), where she specialized in forest ecology. Her research focuses on identifying priorities in forest management, and, more generally, on issues of knowledge, ignorance, and skepticism in scientific contexts, including climate policy and science education.

Chapter 11
Adaptive Management in Social Ecological Systems: Taming the Wicked?

Asim Zia

Abstract In many publications, Bryan Norton has proposed hierarchical systems theory to understand and manage complex environmental conservation and sustainability problems. In doing so, Norton drew from the decision science and planning theory literature to frame persistent environmental problems, ranging from Herbert Simon's un-programmed decisions to Horst Rittel and Melvin Webber's wicked problems. Norton pioneered the development of an adaptive management framework to guide a community and valued-based pragmatic and discursive approach for continually solving and re-solving the wicked environmental management problems. In this chapter, I highlight important ingredients of Norton's adaptive management framework and synthesize some of the key findings that resulted from applying some of the key elements of Norton's adaptive management framework in the field setting of addressing the "wicked" problem of tropical deforestation in Tanzania, Vietnam and Peru through a MacArthur Foundation funded project (2006–2011). The following specific problem formulation and valuation issues are explored in this chapter: space-time boundaries; identification of stakeholder values; weights on stakeholder values; and decision rule choice. We discovered that many wicked problems surrounding environmental management persist in the field settings due to the stakeholder power asymmetries, conflicting values, politics of scale across different space time horizons and institutional inertia. I propose that institutional designs and governance processes operating at different levels of the space-time hierarchy—ranging from a person's ambit to their community, city, state, country and planetary scales—must also be examined and addressed for adaptive management in social ecological systems.

Keywords Hierarchical systems theory · Adaptive management
Biodiversity conservation · International development · Social ecological systems
Multi-Criteria Decision Analysis

A. Zia (✉)
Department of Community Development & Applied Economics,
University of Vermont, 146 University Place, Morrill 208E,
Burlington, VT 05405, USA
e-mail: Asim.Zia@uvm.edu

S. Sarkar and B. A. Minteer (eds.), *A Sustainable Philosophy—The Work of Bryan Norton*, The International Library of Environmental, Agricultural and Food Ethics 26, https://doi.org/10.1007/978-3-319-92597-4_11

11.1 Introduction

In many publications, Bryan Norton has proposed hierarchical systems theory to understand and manage complex environmental conservation and sustainability problems (Norton 1994, 2005, 2015; Norton and Ulanowicz 1992). In doing so, Norton drew from the decision science and planning theory literature to frame persistent environmental problems, ranging from Herbert Simon's un-programmed decisions to Horst Rittel and Melvin Webber's wicked problems (Simon 1973; Rittel and Webber 1973). Norton pioneered the development of an adaptive management framework to guide a community and valued-based pragmatic and discursive approach for continually solving and re-solving the wicked environmental management problems. Communication and negotiations among different stakeholder groups is clearly one of the essential ingredients of Norton's adaptive management framework. Wicked issues surrounding problem formulation, value conflicts and science-policy interface challenges can be discussed and "potentially" tamed through negotiation and discussion among different stakeholders.

In this chapter, I highlight important ingredients of Norton's adaptive management framework and synthesize some of the key findings that resulted from applying some of the key elements of Norton's adaptive management framework in the field setting of addressing the "wicked" problem of tropical deforestation in Tanzania, Vietnam and Peru. During my research collaboration with Norton on an NSF funded project (2004–2008), and later on a MacArthur Foundation funded project (2006–2011), we measured stakeholder values and examined environmental valuation conflicts in variety of environmental problem domains ranging from nutrient pollution of lakes, to environmental justice issues surrounding EPA-declared hazardous waste sites and accelerating decimation of tropical forests and of the largest reserves of global biodiversity in some of the world's poorest communities. Through these research projects, we discovered that many wicked problems surrounding environmental management persist in the field settings due to the stakeholder power asymmetries, conflicting values, politics of scale across different space time horizons and institutional inertia. I present a brief overview of wicked problems and adaptive management, followed by a synthesis of findings about environmental problem formulation and valuation issues derived from three deliberative stakeholder workshops implemented in three focal tropical countries under the MacArthur funded project and draw implications for Norton's adaptive management framework. The following specific problem formulation and valuation issues are explored in this chapter: space-time boundaries; identification of stakeholder values; weights on stakeholder values; and decision rule choice. I propose that institutional designs and governance processes operating at different levels of the space-time hierarchy—ranging from a person's ambit to their community, city, state, country and planetary scales—must also be examined and addressed for adaptive management in social ecological systems.

11.2 Wicked Problems and Adaptive Management

Rittel and Webber (1973) mentioned problems of long-term planning and management as examples of wicked problems, and contrasted these with the "benign" or "tame" problems of mathematics and science. Benign problems have a unique, determinate solution. One knows when a benign problem is solved. Wicked problems, on the other hand, have no determinate solution; even the correct formulation of the problem is contested, and there is no "stopping rule". Rittel and Webber (1973) argued that many benign problems in planning and environmental protection, such as designing adequate sewer systems or timing traffic lights, have been solved by formal modeling and technological innovation but that many of the remaining problems faced by municipalities and agencies must be understood as *wicked problems*. The analysis of Rittel and Webber was prescient, for today's decision scientists readily admit that correct problem formulation is the most difficult and least understood aspect of public decision making (Keeney 1996; Keeney and Raiffa 1976; Winterfeldt and Edwards 1986; Corner et al. 2001; Coenen et al. 1998; Gregory et al. 2001; Gregory 2002; Hanne 2001; Yu 1979; Wooley and Pidd 1981; Mintzberg et al. 1976; Abaulsamh et al. 1990; Perry and Moffat 1997; Taket and White 1997). Rittel and Webber hypothesize that wicked problems have no definitive solution—and no agreed-upon formulation–because disagreements involve multiple competing interests. For wicked problems, we cannot expect "optimal" and final solutions; rather, we can only expect a negotiated and balanced outcome, a resolution that will be acceptable for a time but always open to re-negotiation as the context and power relations change in society. The question remains: how should we "resolve" wicked environmental design problems? It is a question that has intrigued Norton (2005, 2015) over many years of scholarship and research.

Various decision theorists have tackled the problem of problem formulation from three broad perspectives. First, Keeney (1988, 1992, 1996) emphasized value-focused problem formulation. Second, Simon (1955, 1982), von Neumann and Morgenstern (1944) emphasized a focus on alternatives in problem formulation. Finally, many decision scientists focused on the process of deliberation involved in problem formulation (Janis and Mann 1977; Kahneman and Tversky 1979, 1982, 1988; Kahneman et al. 1992; Gregory and Keeney 1994; Gregory et al. 1993, 2001; Gregory and Failing 2002; Henig and Buchanan 1996; Buchanan et al. 1998; Wright and Goodwin 1999). All three perspectives in decision-making require some means of setting the spatio-temporal bounds of decision problems in problem formulation. Norton (2005) acknowledged all three perspectives but in his later career (e.g., see Norton 2015) emphasized the third tack, i.e., deliberative and procedural approaches, as a focal means to adaptively manage the wicked environmental design problems. Building upon Simon's notion of substantive versus procedural rationality (Simon 1982), Norton (2015) dwelled in depth about the importance of deliberative and procedural approaches in adaptive management of environmental systems.

In the procedural context, Norton acknowledged that Rittel and Webber (1973) describe wicked problems, paradoxically, as those for which the correct problem formulation can only be established *after* the problem has been solved. Participants in the decision process—who come to the problem with very different perceptions, interests, and values—will inevitably characterize the problem in different ways. The consequence of this process is that the operational solutions devised rely on bounds established through the course of decision-making and negotiation, "after the problem has already been tamed" (Rittel and Webber 1973, p. 162). To acknowledge this consequence is to recognize the importance—and the recalcitrance—of problem formulation when the problem involves competing interests, a theme persistent in Norton's writings over time.

Repeatedly, Norton argued that *conflicting values* lie at the heart of wicked problem formulation, making goal-finding "an extraordinarily obstinate task." During the early foundational days of the field of ecological economics, in which Norton played a critical role, Norton investigated environmental value and valuation processes with multi-level perspectives derived from hierarchical systems theory, well established among ecologists and increasingly ecological economists (e.g. see Norton and Ulanowicz 1992; Norton 2003; Norton and Noonan 2007). Decision-making agents in these hierarchically nested systems face changing place attachments (e.g. see Norton and Hannon 1997, 1998), identities and sense of place (Zia et al. 2014) and changing biophysical and ecological conditions of environmental systems. Given the dynamic conditions of socio-economic and ecological systems, Norton (2005, 2015) emphasized that the space-time boundaries of an environmental problem must be understood within a hierarchical systems theoretic perspective.

While Rittel and Webber laid out 10 defining characteristics of wicked planning problems, they left the methods of their resolution as open questions for future research. Most of the follow-up research in response to Rittel and Webber (1973) is seen in computer science (DeGrace and Stahl 1990), complexity science (e.g. Zia et al. 2012) and psychology (Conklin 2001). There are more than 9600 papers and books that cite this seminal paper. The theorists of adaptive management (Norton and Steinemann 2001) and discursive planning (Sager 1997; Andrews 2002) frameworks have proposed adaptive decision making and effective communication as the most salient ways for resolving the wicked planning problems.

Norton (2005) laid out following three important characteristics of "adaptive management" to resolve wicked environmental problems:

1. Experimentalism: adaptive managers respond to uncertainty by undertaking reversible actions and studying outcomes to reduce uncertainty at the next decision point.
2. Multi-Scalar Modeling: adaptive managers model environmental problems within multi-scaled ("hierarchical") space-time systems.
3. Place-Orientation: adaptive managers address environmental problems from a "place" which means problems are embedded in a local context of natural systems but also of political forces.

Norton's adaptive management framework frames human decision makers as located within layered subsystems and supersystems, with the smallest subsystems being the fastest-changing. The larger systems change more slowly and provide the environment for adaptation by subsystems. Norton argued that this hierarchical systems theoretical approach to adaptive management allows us to associate temporal "horizons" with changing features of landscapes as is illustrated in the famous simile used by Aldo Leopold, a forest-land manager and a wildlife manager. Leopold set out to remove predators from the national forest ranges he managed in the Southwestern US. When the deer starved for lack of browse, he regretted his decision to extirpate wolves, chiding himself for not yet having learned to "think like a mountain" (Leopold 1949). Later, Hirsch and Norton (2012) extended this simile to "thinking like a planet" in the context of persistent and wicked global scale problems such as climate change and biodiversity loss.

In Norton's (2005, 2015) adaptive management approach, environmental policy decisions are modeled at two interactive levels: At the level of action, descriptive analysis is undertaken to ascertain the current state of the world, such as existing environments and policies, and the outcomes ensuing from current policies/decisions. At the meta-level of reflection, normative analysis is employed to determine the socially desirable values by which outcomes of (current and future) policy actions are measured (Norton 2005, 2015).

At the meta-level of reflection, the planners and policy designers compare the outcomes measured at the level of action with the outcomes that are deemed normatively desirable within the space-time horizon of environmental policy decisions. At this level, meta-decision choice problems are resolved through iterative experimentation and collaboration between expert and lay decision-makers. The normative analysis at the meta-level of reflection results in policy prescriptions/recommendations that aim at getting "there" from "here" given all the uncertainty, ignorance and incomplete information.

Our collaborative NSF-funded research, while diverse in its methods, scope and targets, aimed at addressing three focal questions with respect to environmental problem formulation and valuation issues that are generally confronted in adaptive environmental management:

(1) The Descriptive Question: How do human conceptions (models) of space, time, boundaries, and identity shape understanding of environmental problems?
(2) The Dynamic Question: How, and in response to what, do these conceptions change?
(3) The Normative Question: Can we offer criteria for identifying more appropriate scales at which to bound and represent a system in a given problem context?

Norton and Hannon (1997, 1998) have explicitly called for more attention to locally based values and for more study of place-based values, which are nested in larger socio-political systems. Choice of appropriate scale in formulating an adaptive management problem is the essence of the descriptive and the dynamic questions posed above. Environmental justice researchers, for example, continue to struggle with wicked spatial modeling problems. Bowen and Wells (2002), for example, highlight contradictory findings and the methodological challenges that arise due to scale selection processes in addressing environmental management problems. Anderton et al. (1994) emphasized the problem of choosing the appropriate spatial scale of analysis, but this problem continues to plague environmental justice research (e.g., Davidson and Anderton 2000; Taquino et al. 2002). Similarly, there has been increasing attention to matters of scale in ecological research and modeling. My collaborative research with Norton drew attention to a key focal point of this cross-disciplinary research—the point where geophysical modeling, social scientific study of citizen perception and values, and policy decision-making intersect—that must be better understood if disciplinary work on localism and on values related to a sense of place is to be integrated into normative decision analysis and used to improve decision making in adaptive management processes. The normative question (3) posed above was framed in this interdisciplinary context of moving the adaptive management theoretical framework from merely a natural science (and especially, ecological science) perspective to an integrated socio-economic and ecological systems perspective.

As the descriptive, dynamic, and normative aspects of environmental decisions are integrated into adaptive management science, according to Norton (2005), they become elements of public discourse, open to contradiction and to critical response. Adaptive management encompasses questions of values and goals as well as questions of means and of technique (Norton 2005). Adaptive management is a social process that attempts to achieve both cooperative action and social learning, as descriptive science is brought to bear upon human goals. Values, as well as facts, are put to challenge in the adaptive management process, the challenge of needing to give reasons and justifications for one action over another. The study of cooperative processes and what works in them may be the highest calling of descriptive and analytical environmental social science (Norton 2015). Emergent cooperation, then, is accomplished by good process, and we can study, using all the tools of the social sciences, the conditions under which cooperative action and the building of trust create a positive spiral of cooperation and expanded trust. Learning about social learning, that is, represents a wonderful, expanding role for the social sciences. This chapter examines one of these roles: how can a better understanding of how stakeholders and citizens "bound" and "scale" problems improve communication, expand areas open to empirical study, and lay the foundation for increased trust and social learning?

Bounding and scaling environmental problems is akin to taming "wicked" problems. Real world application of Norton's adaptive management characteristics, especially the characteristics pertaining to multi-scale modeling and place orientation, and integration of normative aspects of problem formulation with descriptive and dynamic aspects of problems, was one of the fundamental motivations for my collaborative work with Norton and with the large interdisciplinary team of collaborators in the MacArthur foundation funded project. This project aimed at understanding trade-offs between biodiversity conservation and economic development in three tropical countries: Peru, Tanzania and Vietnam. While the project generated numerous high impact publications (e.g. see McShane et al. 2011; Hirsch et al. 2011), three multi-day stakeholder workshops in each case study country enabled me to apply participatory and deliberative multi-criteria analysis tools to measure the stakeholder values at multiple space time scales of problem and solution bounding, consistent with Norton's adaptive management framework. Zia et al. (2011, 2015), Zia (2013) provide more details about the multi-scale modeling aspects of the valuation conflicts explored in this project in the context of real-world tropical deforestation, economic development, climate change and biodiversity loss conflicts faced by communities in these three developing countries. In what follows I provide brief context of the environmental problems in the case study areas and synthesize key findings about elicitation of stakeholder values in real-world adaptive management settings.

11.3 Application of Norton's Adaptive Management Approach in Peru, Tanzania and Vietnam

Three, multiple-day multiple stakeholder workshops were organized in Tanzania (May 2009), Peru (June 2009) and Vietnam (July 2009) to understand a variety of trade-offs involved in resolving conservation-versus-development landscape planning decisions in the sampled hotspot sites. One of the fundamental premises of case study site selection hinged on a contested hypothesis in conservation biology and international development research that biodiversity hotspots, as defined by UNEP and other UN agencies, were strongly correlated with spatial pockets of persistent poverty and poor social and human development indicators (see McShane et al. 2011 for more details). In Tanzania, for example, the eviction of pastoralists and farmers from the villages neighboring Ruaha National Park (one of the largest biodiversity hotspots in the eastern Africa) were the focus. More details on the case study site, environmental conflicts and stakeholders are presented in Zia et al. (2011). Similarly, the Vietnamese case study focused on protecting Ha Long Bay and Bai Tu Long National Park areas from a variety of economic development drivers such as mining, housing and agriculture. More details about the Vietnamese workshop and case study are available in Zia et al. (2015).

Compared to Tanzanian and Vietnamese cases, the Peruvian case was perhaps the most challenging, intransigent, and unresolved at many levels. The Peruvian case study was focused on the ongoing conflict in the Madre-De-Dios Amazonian watershed region in Peru along Brazilian and Bolivian borders. Under a continent-wide "Integration of Regional Infrastructure in South America" (IIRSA) program, a roadway project was being proposed to be built in Madre De Dios region. In addition to the roadway project, the Brazilian government had offered the Peruvian government the opportunity to design, build and operate three large scale hydropower projects in the Madre De Dios region and sell the electricity to the

Brazilians under a long-term bilateral agreement. While the Peruvian national government was interested in undertaking roadway and hydropower projects in Madre De Dios region, the local and regional stakeholders were intransigently opposed to these development projects. International stakeholders such as International Union for the Conservation of the Nature (IUCN) and World Wildlife Fund (WWF) were also not very supportive of these development projects, but these international conservation agencies did not have as much financial power as the World Bank and Inter-American Development Bank, which were financially underpinning roadway and hydropower projects in the Peruvian region.

In what follows I summarize the key findings from the three workshops vis-à-vis four deliberative decision choices (viz., space-time boundaries, identification of stakeholder values, weights on stakeholder values, and decision rule choice) that were elicited as part of the implementation of a deliberative Multi-Criteria Decision Analysis (MCDA) intervention process with multiple stakeholders in sampled case study decision making contexts. The deliberative decision process consisted of the procedural heuristic described in Table 11.1, underpinned with a simple additive weighting MCDA procedure in Tanzania and Vietnam, briefly described below (and extensively discussed in Zia et al. 2011, 2015). Peruvians did not agree upon a MCDA procedure, discussed below.

Building upon Norton and Noonan's (2007) idea of alternate development paths, we formally defined a multi-criteria expected value function V_{ik} for ith development path in a set of n development paths by kth stakeholder as in Eq. 11.1:

$$\begin{aligned} V_{ik} &= \sum_{i=1}^{n} w_{jk} x_{ijk} \\ s.t. \sum_{j=1}^{m} w_{jk} &= 1 \end{aligned} \tag{11.1}$$

Table 11.1 Procedural heuristic of deliberative MCDA

Steps	Procedures
1.	Develop a group consensus on design alternatives
2.	Develop a group consensus on criteria (mutually exclusive and typically incommensurate)
3.	Individuals assign weights on criteria
4.	Individuals assign their perceived outcome on a common scale for each alternative by each criterion
5.	Individuals participate in small group discussion to develop consensus on weights and perceived outcomes
6.	Workshop level weights and perceived outcomes are developed
7.	Workshop level weights and perceived outcomes are multiplied to evaluate design alternatives
8.	The valuation process is repeated iteratively with different set of stakeholder representatives

In this tradeoff function w_{jk} is the weight for the jth criterion in a set of m criteria by the kth stakeholder; and x_{ijk} is the value for an "outcome" or "impact" function for ith alternative on jth criteria as perceived by a kth stakeholder in a group of K stakeholders for a set of n alternatives. For an individual decision maker, the most valued development path is the one with the highest V_{ik}. A considerable challenge is how to aggregate V_{ik} across groups of multiple stakeholders for choosing a development path that reflects the pluralistic values of all affected stakeholders. Formally, this aggregation challenge is represented through the assignment of Ψ_k for aggregating V_{ik} to estimate the societal value V_i of ith development path, as shown in Eq. 11.2:

$$V_i = \sum_{k=1}^{K} \Psi_k V_{ik} \qquad s.t. \sum_{k=1}^{K} \Psi_k = 1 \tag{11.2}$$

Substituting V_{ik} from (11.1) in (11.2) yields Eq. 11.3:

$$V_i = \sum_{k=1}^{K} \sum_{i=1}^{n} \Psi_k w_{jk} x_{ijk} \qquad s.t. \sum_{j=1}^{m} w_{jk} = 1 \,\&\, \sum_{k=1}^{K} \Psi_k = 1 \tag{11.3}$$

Equation 11.3 provides one of the many possible MCDA methods to assign multi-criteria values on alternate development paths conditional upon the weights assigned to different stakeholders, the weights assigned by each stakeholder on different values in the system, as well as the impacts perceived by different stakeholders for each alternate development path vis-à-vis each value in the system. We formally stipulated that a *process* issue in aggregation refers to how a stakeholder is included or excluded from the set of K stakeholders. Furthermore, we defined that a *power* issue in aggregation refers to the problem of assigning Ψ_k weights to a kth stakeholder. In a perfectly egalitarian society, Ψ_k will be equal for all stakeholders, which is rarely the case in real societies. Power asymmetries can be explicitly represented through the asymmetric assignment of Ψ_k. Since formal MCDA cannot endogenously determine K and Ψ_k, we proposed the deployment of deliberative and softer version of MCDA application, as described in the procedural heuristic. Further, the outcome functions and x_{ijk} were scaled at three spatial scales (local, national and international) in Tanazanian case and two temporal scales in Vietnames case (Now, Future).

11.3.1 Space-Time Boundaries

In all three case study countries, stakeholders did not arrive at a consensus about the specific space-time boundaries to structure the conservation-versus-development planning problem. While there was a general agreement that the costs of the conservation are borne by local communities (such as forgone economic development opportunities and loss of local livelihoods) and the benefits of the conservation are derived by global communities (such as carbon mitigation from conserved tropical forests), there was considerable disagreement about the process to measure costs and benefits in all three countries. Despite the efforts of the workshop facilitators to draw strict spatial boundaries around the case study sites, local community stakeholders continue to protest any delineation of spatial (or temporal) boundaries. Peruvians, for example, mentioned global corporate interests as the key driving forces underpinning roadway, hydropower, and gold mining projects in Madre De Dios region as important players not bound by the case study spatial boundaries. Similarly, Tanzanians and Vietnamese stakeholders were concerned about Chinese and multi-national companies eying the natural resources in their respective case study sites. Detailed research findings about this "politics of scale" and international trade as global to local drivers of tropical deforestation are extensively discussed in Zia (2013), Zia et al. (2011). Though there was an understanding about "nested" spatial scales in all three countries (e.g. local, regional, national and international), there was considerable concern among the local community representatives about the political power and governance regimes at different levels of the nested spatial structure. In general, all mediation efforts to strictly delineate space time boundaries for the case study sites remained ineffective.

11.3.2 Values and Weights on Values

Considerable effort and time was spent in each of the three countries to elicit the values and weights on those values to compare baseline (descriptive) management scenarios with alternate (normative/prescriptive) management scenarios for the case study sites. There were significant linguistic and translation challenges in all three countries. Despite these obstacles, we were able to develop a value and weighting strategy for Tanzania and Vietnam, however these efforts were not consensually agreed upon in Peru. More specifically, Fig. 11.1 shows the values and weights on those values agreed upon in the Tanzanian workshop. There is considerable variability in the weights assigned to these values, as different stakeholders prefer different (and often conflicting) goals and values.

Similarly, Fig. 11.2 shows mean weights assigned for the values mutually agreed upon in Vietnam. A comparison of x-axes in Figs. 11.1 and 11.2 reveals that stakeholders in Tanzania and Vietnam do not necessarily choose similar values to compare alternate landscape planning scenarios. While maximization of economic

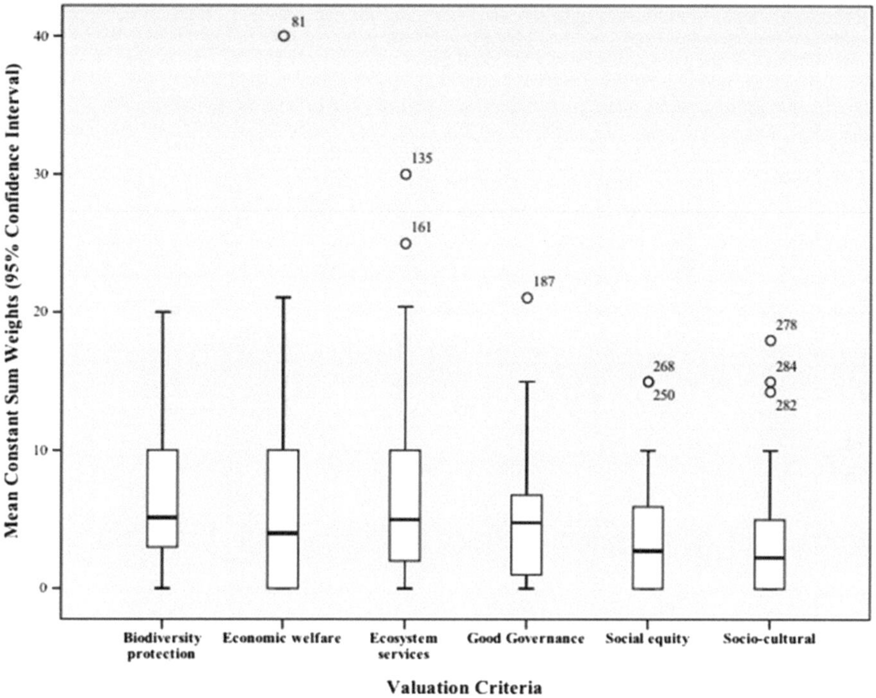

Fig. 11.1 Mean weights with 95% confidence interval on values agreed upon in Tanzania (N = 21)

welfare, protection of ecosystem services and good governance emerged as important values in both countries, Tanzanians appear to be also concerned about socio-cultural values, social equity issues and biodiversity protection. Vietnamese, on the other hand, were concerned about accessibility to the landscapes as well as the price of the land under different planning scenarios.

Compared to both Tanzanians and Vietnamese, the Peruvians instead emphasized that the value of "human rights" was considerably more important to them. More significantly, the Peruvian stakeholders did not agree upon a mutually consensual set of values to compare planning scenarios as did the stakeholders in Tanzania and Vietnam. They contested the definitions of economic welfare criteria, social welfare, ecosystem services or any other possible values that were signaled by different workshop participants. Instead, many stakeholders kept on insisting that human and animal rights were the most important considerations for them in making such planning decisions, and these values were not quantifiable on a scale. Consequently, no usable quantitative data on values or weights on values was collected during the Peru workshop.

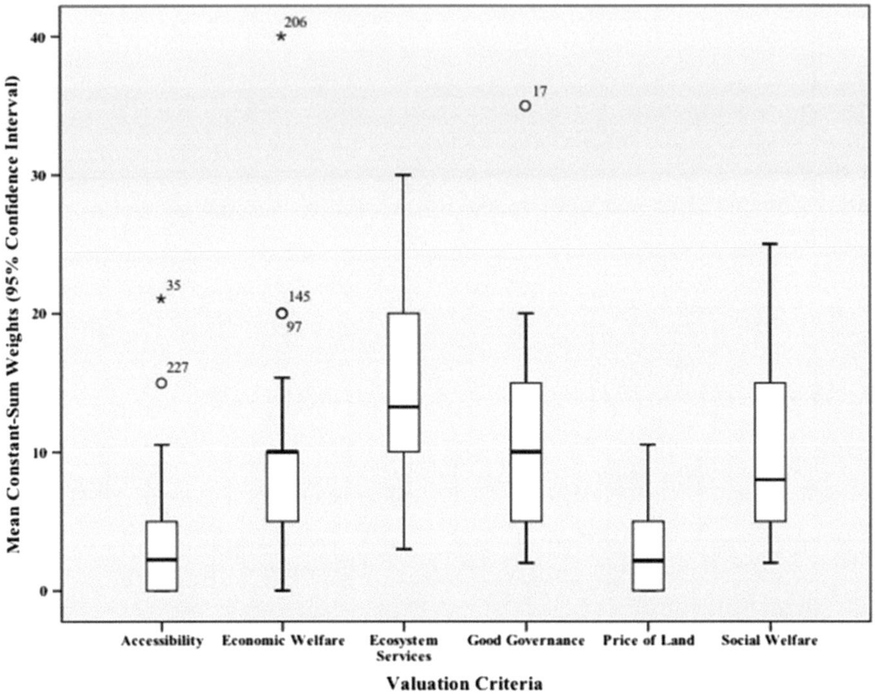

Fig. 11.2 Mean weights with 95% confidence interval on values agreed upon in Vietnam (N = 18)

11.3.3 Choice of Decision Rule

Both Tanzanians and Vietnamese agreed to use "simple additive weighting" (SAW) algorithm, as explained above for Eqs. 11.1–11.3 and Table 11.1. The SAW algorithm is one of the many algorithms that can be used for MCDA and we provided stakeholders the opportunity to choose from a sufficiently large number of algorithms (a broader review of MCDA in conservation planning processes is provided by Moffett and Sarkar 2006). The Peruvians, however, contested fundamental features of different algorithms. One fundamental feature of classifying MCDA algorithms has to do with "compensatory" versus "non-compensatory" decision rules (e.g. Hwang and Yoon 1981). Compensatory decision rules such as SAW and weighted product decision rule assume that losses on one attribute/criterion could be "compensated" by gains on another (incommensurate) attribute/criterion in a decision problem. Non-compensatory decision rules, on the other hand, do not assume the existence of such compensation across the attributes. Elimination By Aspects is an example of a non-compensatory decision rule, in which all alternatives that do not meet minimal performance standards on specific decision criteria are eliminated from consideration and only those alternatives are chosen for further consideration that meet the minimal performance thresholds. For Peruvian stakeholders, human rights was an example of an attribute, a loss on which could not be compensated by a gain in another attribute such as economic or social welfare. In essence, Peruvians rejected SAW and embraced Elimination By Aspects as a preferred decision rule, but the discussion could not move beyond that in Peruvian workshop due to the fundamental disagreements about the usage of MCDA "tools" and the basic assumptions embedded in designing mutually non-exclusive set of values, quantifying values on discrete scales and design alternatives. In contrast, both Tanzanians and Vietnamese worked through applying SAW decision rule to compare alternate conservation planning regimes in their respective case study sites.

Figure 11.3 shows mean (with 95% confidence interval) expected values elicited from Tanzanians about the five alternative planning scenarios for Ruaha National Park. The five planning scenarios included baseline national park scenario, compared with alternate scenarios of converting Ruaha National Park into a game

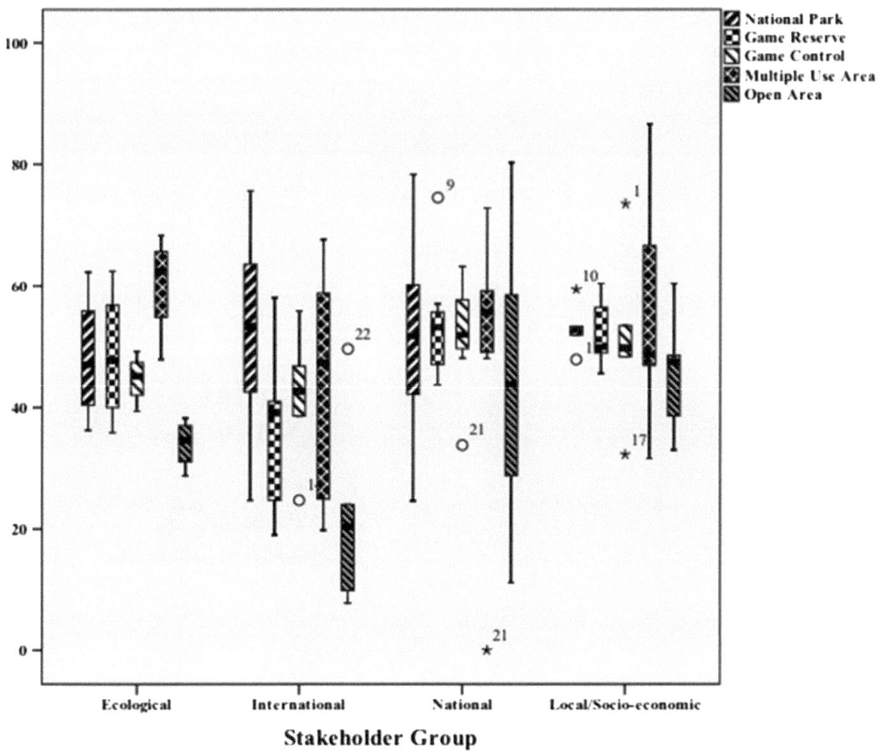

Fig. 11.3 Mean expected values with 95% confidence interval on planning scenarios elicited in Tanzania (N = 21)

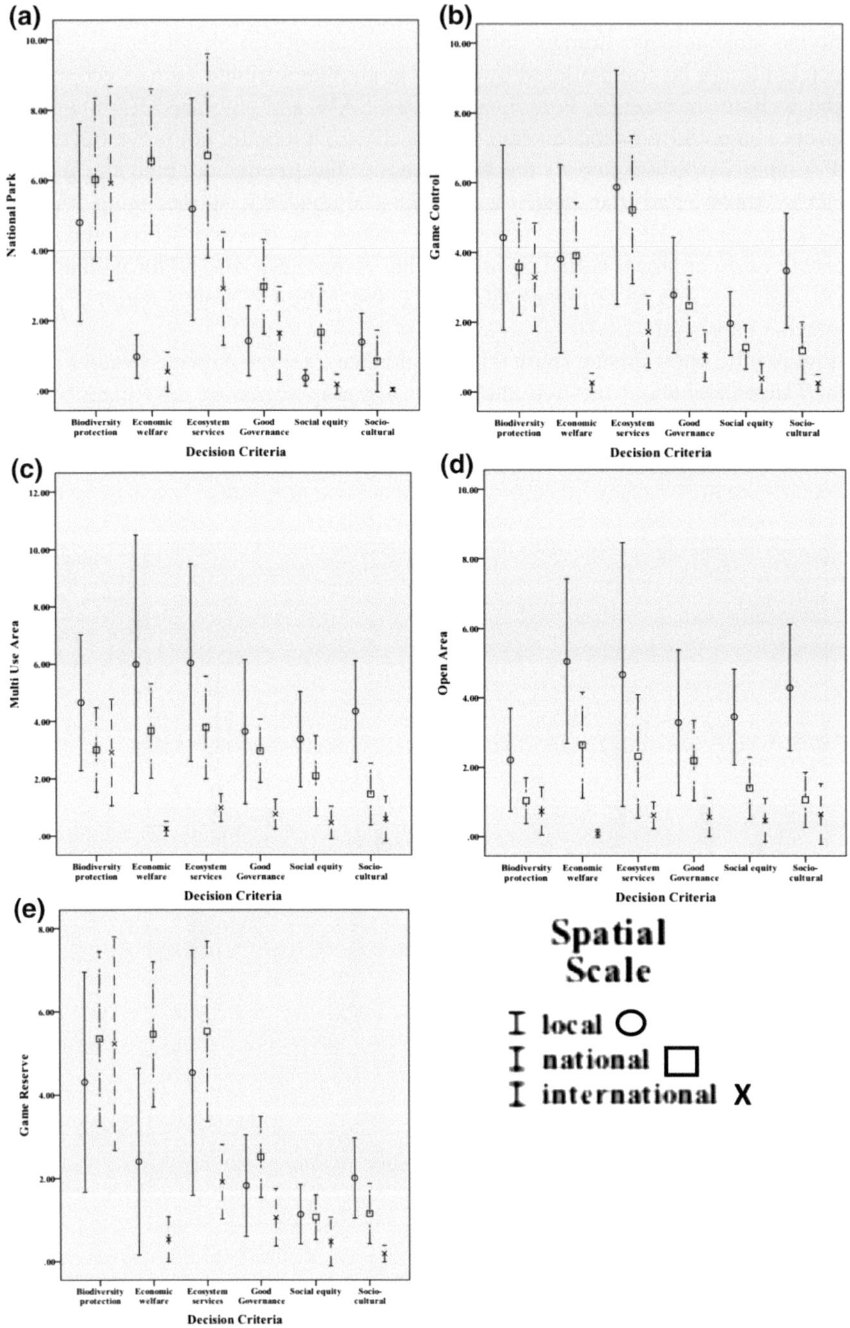
(a)
National Park
10.00
8.00
6.00
4.00
2.00
.00
Biodiversity protection
Economic welfare
Ecosystem services
Good Governance
Social equity
Socio-cultural
Decision Criteria
(b)
Game Control
10.00
8.00
6.00
4.00
2.00
.00
Biodiversity protection
Economic welfare
Ecosystem services
Good Governance
Social equity
Socio-cultural
Decision Criteria
(c)
Multi Use Area
12.00
10.00
8.00
6.00
4.00
2.00
.00
Biodiversity protection
Economic welfare
Ecosystem services
Good Governance
Social equity
Socio-cultural
Decision Criteria
(d)
Open Area
10.00
8.00
6.00
4.00
2.00
.00
Biodiversity protection
Economic welfare
Ecosystem services
Good Governance
Social equity
Socio-cultural
Decision Criteria
(e)
Game Reserve
8.00
6.00
4.00
2.00
.00
Biodiversity protection
Economic welfare
Ecosystem services
Good Governance
Social equity
Socio-cultural
Decision Criteria
Spatial Scale
local
national
international X

◂**Fig. 11.4** Pre-deliberative expected value by valuation criteria and spatial scale for national park (panel a), game control area (panel b), multiple use area (panel c), open area (panel d), and game reserve (panel e) options in Tanzania. Expected values are measured on a continuous scale from 0% (worst) to 100% (best). Error bars show 95% confidence interval around mean expected values

reserve area, game control area, multiple (mixed) land-use area, or open area available for economic development without any national or local conservation regulations. Different stakeholder groups had different scenario preferences: except for international stakeholders, national, local and ecological groups rejected the baseline national park scenario as their preferred scenario. Multiple use area emerged as the preferred scenario for these three stakeholder groups in Tanzania; however, upon further investigation with the national forestry ministry, it became clear that they were not open to changing the status of Ruaha national park to a multiple land-use area as recommended by the workshop participants. This points to the wicked problem of participation in such meetings as well as to complex power dynamics that drive decision-making in developing countries.

More interesting is the breakdown of expected values of five conservation planning alternatives for three different hierarchical scales (shown in Fig. 11.4, and extensively discussed in Zia et al. 2011). The management option of conservation provides higher value to international, followed by national and least of all local communities. In contrast, the management options of multi-use area (Fig. 11.4d) and open area (Fig. 11.4e) are perceived by the workshop participants to provide highest value on all valuation criteria at the local scale and the least value at the international scale. Essentially, we demonstrated that the proposed deliberative MCDA approach can be used to apply two out of the three characteristics of Norton's adaptive management approach, multi-scale modeling and place orientation, and elicit stakeholder values on planning alternatives at multiple scales of space (in Tanzania) and time (in Vietnam as discussed below).

Figure 11.5 shows similar mean expected values for baseline and alternate planning scenarios evaluated in the case of Vietnam. As compared with Tanzania, Vietnamese wanted to consider a "community owned" planning scenario for Bai Tu Long national park as one of the alternate planning options. This particular option, however, did not rise to the top when evaluated through SAW algorithm for the local stakeholder group as shown in Fig. 11.5.

In contrast, the community owned option was the most preferred scenario for international stakeholders present in the Vietnam workshop. For local and national stakeholders, the multi-use planning scenario appeared the most preferred option. As in the case of Tanzania, the Vietnamese policy making authorities were not too keen on changing the planning scenario to any of the proposed alternative planning regimes in Bai Tu Long despite the fact that the baseline planning scenario appeared to be the least preferred scenario for national level stakeholders who participated in the Vietnam workshop.

Figure 11.6 shows the expected value breakdown of five planning alternatives for six decision criteria at two temporal scales (now and future). We found that the management scenarios of multiple use areas and community ownership provided

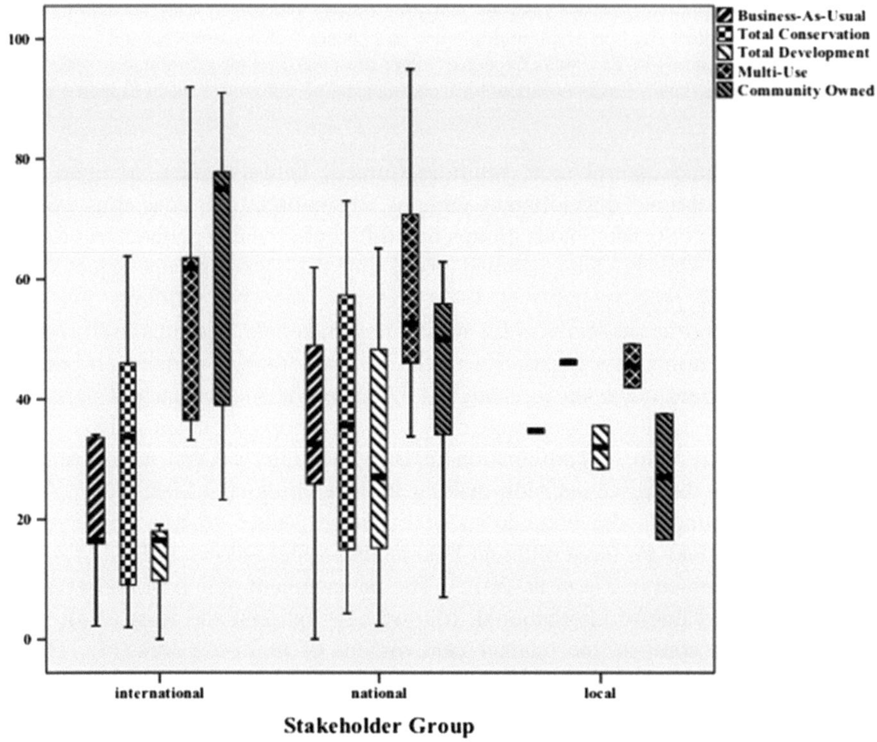

Fig. 11.5 Mean expected values with 95% confidence interval on planning scenarios elicited in Vietnam (N = 18)

relatively higher expected value for almost all valuation criteria in future time scales. On the other hand, the management scenarios of total development and business as usual dampened the expected values across the board. Finally, the management scenario of total conservation provided relatively higher expected value for the protection of biodiversity and ecosystem services but this scenario also dampened the expected values for good governance and social and economic welfare criteria. Essentially, the Vietnamese case demonstrated that deliberative MCDA approach can be used to elicit stakeholder values at different temporal scales of decision horizons.

Fig. 11.6 Pre-deliberative expected values by valuation criteria and temporal scale for business as usual (panel a), total conservation (panel b), total development (panel c), multiple use area (panel d), and community owned (panel e) management scenarios in Vietnam: expected values are measured on a continuous scale from 0% (worst) to 100% (best). Error bars show 95% confidence interval around mean expected values ▶

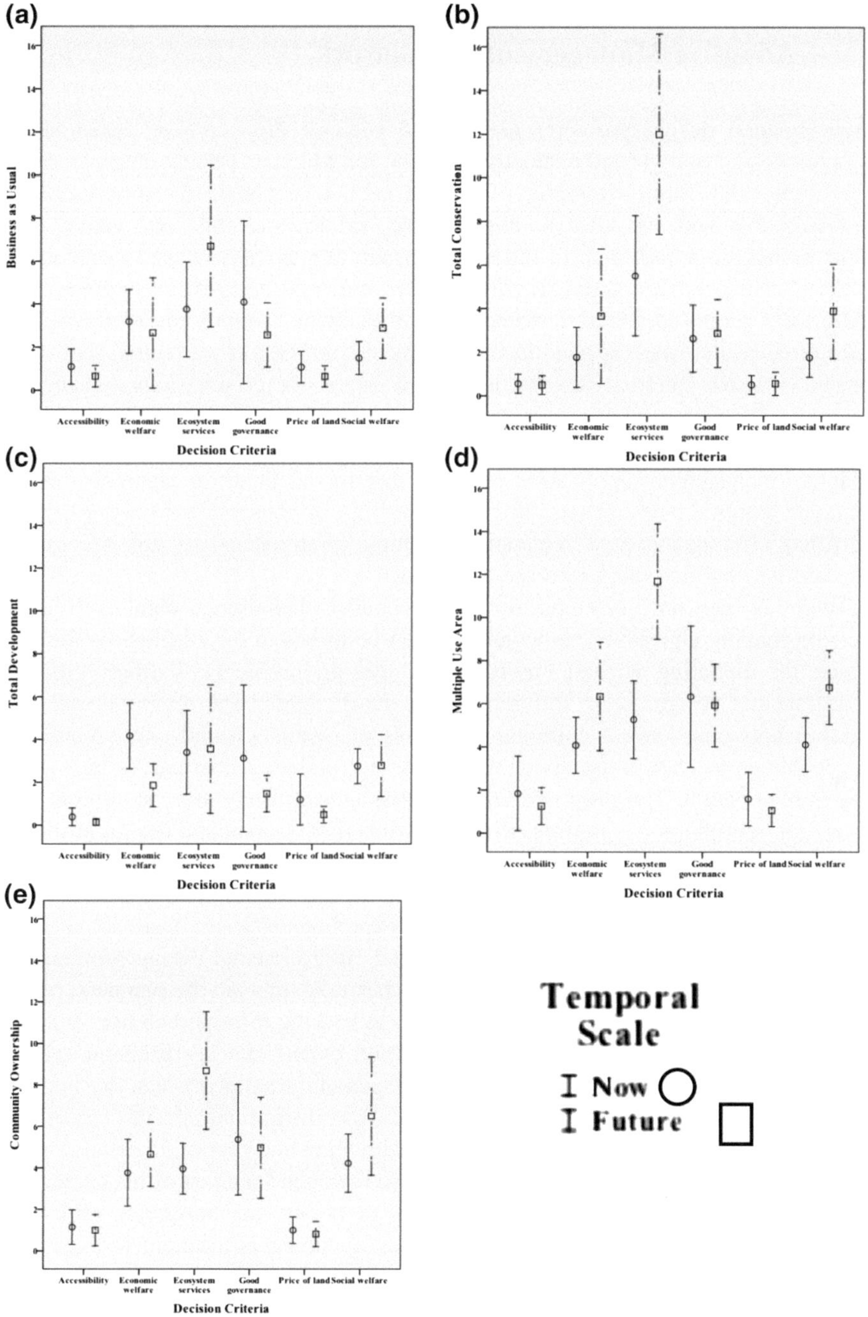
(a)
Business as Usual
(b)
Total Conservation
(c)
Total Development
(d)
Multiple Use Area
(e)
Community Ownership
Accessibility
Economic welfare
Ecosystem services
Good governance
Price of land
Social welfare
Decision Criteria
Temporal Scale
Now
Future

11.4 Concluding Reflections on Updating Norton's Adaptive Management Framework

Environmental design problems are wicked because they involve decisions on complex design options and evaluative criteria, for which no "optimal" or "correct" algorithms exist. Some examples of wicked decisions include the procedures on choosing the sets of design alternatives, values, weights on values and decision-making algorithms. In order to transparently discuss wicked decisions in public decision-making arenas involving environmental designs, two components of Norton's proposed adaptive management framework—multiscale modeling and place orientation—were applied in three tropical countries. It was found that some consensus across different stakeholder groups about wicked decisions appeared to emerge in Tanzania and Vietnam; however, Peruvian stakeholders remained divided and skeptical about choosing any specific set of decision-making algorithms. Adaptive management application revealed that mixed land-use planning options dominate extreme land use (e.g. development-only or conservation-only) planning scenarios in Tanzania and Vietnam, but these preferences do not necessarily translate into public policy in these countries.

There is no one best or optimal scenario, weighting combination, or decision-making algorithmic outcome that can be ordained for all times as the best choice for modeling wicked environmental design problems. Further, different environmental designs serve different values across spatial and temporal scales of decision horizons. Only a continuous process of evaluation and experimentation can enable us to learn more about the policy and planning alternatives in a given space-time context. The proposed adaptive management may thus be applied iteratively, at multiple points in time, while the wicked environmental design problems can at best be only temporarily resolved and re-resolved through a continuous process of dialogue, modeling, experimentation and learning. We will have to discuss and learn more, and perform more experiments, if we want to keep on improving the models for resolving the wicked environmental design problems.

The proposed deliberative decision approaches may only aid the planners, policy analysts and a broad range of decision makers in making informed choices through a process that lays bare the assumptions behind expert and lay decision-making processes. However, Norton's adaptive management framework can be updated with explicit incorporation of power dynamics and institutional design issues that characterize the landscape of biodiversity conservation and economic development, a wicked problem with which Norton remained engaged for most of his career and which has inspired many of us to continue to pursue his pragmatic approach. Institutional designs and governance processes operating at different levels of the space-time hierarchy, consistent with Norton's vision of integrating hierarchy theory in environmental valuation processes, must be examined and added as a fourth component of the adaptive management framework. This proposed refinement of Norton's adaptive management framework could potentially generate new avenues of research and application for solving wicked environmental design problems by explicitly integrating hierarchical systems theory with multi-scale societal valuation processes.

References

Abaulsamh, R.A., B. Carlin, and R.R. McDaniel Jr. 1990. Problem structuring heuristics in strategic decision making. *Organizational Behavior and Human Decision Processes* 45 (2): 159–174.

Anderton, D.L., A.B. Anderson, M. Oakes, and M.R. Fraser. 1994. Environmental equity: The demographics of dumping. *Demography* 31 (2): 229–248.

Andrews, C.J. 2002. *Humble analysis: The practice of joint fact-finding*. Westport: Praeger.

Bowen, W.M., and M.V. Wells. 2002. The politics and reality of environmental justice: A history and considerations for public administrators and policy makers. *Public Administration Review* 62 (6): 688–698.

Buchanan, J.T., E.J. Henig, and M.I. Henig. 1998. Objectivity and subjectivity in the decision-making process. *Annals of Operations Research* 80: 333–346.

Coenen, F.H.J.M., D. Huitema, and L.J. O'Toole (eds.). 1998. *Participation and the quality of environmental decision making*. Dordrecht: Kluwer.

Conklin, J. 2001. Wicked problems and social complexity. http://cognexus.org/wpf/wickedproblems.pdf. Accessed 21 Aug 2017.

Corner, J., J. Buchanan, and M. Henig. 2001. Dynamic decision problem structuring. *Journal of Multi-Criteria Decision Analysis* 10 (3): 129–141.

Davidson, P., and D.L. Anderton. 2000. Demographics of dumping II: A national environmental equity survey and the distribution of hazardous materials handlers. *Demography* 37 (4): 461–466.

DeGrace, P., and L.H. Stahl. 1990. *Wicked problems, righteous solutions*. Prentice Hall: Yourdon Press.

Gregory, R.S., and R.L. Keeney. 1994. Creating policy alternatives using stakeholder values. *Management Science* 40 (8): 1035–1048.

Gregory, R.S. 2002. Incorporating value trade-offs into community-based environmental risk decisions. *Environmental Values* 11: 461–488.

Gregory, R.S., and L. Failing. 2002. Using decision analysis to encourage sound deliberation: Water use planning in British Columbia, Canada. *Journal of Policy Analysis and Management* 21 (3): 492–499.

Gregory, R.S., S. Lichtenstein, and P. Slovic. 1993. Valuing environmental resources: A constructive approach. *Journal of Risk and Uncertainty* 7: 177–197.

Gregory, R.S., T. McDaniels, and D. Fields. 2001. Decision aiding, not dispute resolution: Creating insights through structured environmental decisions. *Journal of Policy Analysis and Management* 20 (3): 415–432.

Hanne, T. 2001. *Intelligent strategies for meta multiple criteria decision making*. Dordrecht: Kluwer.

Henig, M.I., and J.T. Buchanan. 1996. Solving MCDM problems: Process concepts. *Journal of Multi-Criteria Decision Analysis* 5 (1): 3–21.

Hirsch, P.D., and B.G. Norton. 2012. Thinking like a planet. In *Ethical adaptation to climate change: Human virtue of the future*, ed. A. Thompson and J. Bendik-Keymer, 317–344. Cambridge: MIT.

Hirsch, P.D., W.M. Adams, J.P. Brosius, A. Zia, N. Bariola, and J.L. Dammert. 2011. Acknowledging conservation trade-offs and embracing complexity. *Conservation Biology* 25 (2): 259–264.

Hwang, C.L., and Yoon. K. 1981. Methods for multiple attribute decision making. In *Multiple attribute decision making* (pp. 58–191). Springer: Berlin, Heidelberg.

Janis, I., and L. Mann. 1977. *Decision making: A psychological analysis of conflict, choice and commitment*. New York: Free.

Kahneman, D., and A. Tversky. 1979. Prospect theory: Analysis of decision under risk. *Econometrica* 47 (2): 263–291.

Kahneman, D., and A. Tversky. 1982. The psychology of preferences. *Scientific American* 246: 160–173.

Kahneman, D., and A. Tversky. 1988. *Risk and rationality: Can normative and descriptive analysis be reconciled?*. College Park, MD: Institute of Philosophy and Public Policy.

Kahneman, D., J.L. Knetsch, and V.K. Smith. 1992. Valuing public goods: The purchase of moral satisfaction. *Journal of Environmental Economics and Management* 22: 57–70.

Keeney, R.L. 1988. Building models of values. *European Journal of Operational Research* 37 (2): 149–157.

Keeney, R.L. 1992. *Value focused thinking*. Cambridge: Harvard.

Keeney, R.L. 1996. Value-focused thinking: Identifying decision opportunities and creating alternatives. *European Journal of Operational Research* 92 (3): 537–549.

Keeney, R.L., and H. Raiffa. 1976. *Decisions with multiple objectives*. New York: Wiley.

Leopold, A. 1949. *A sand county Almanac*. New York: Ballantine.

McShane, T.O., P.D. Hirsch, T.C. Trung, A.N. Songorwa, A. Kinzig, B. Monteferri, D. Mutekanga, H. Van Thang, J.L. Dammert, M. Pulgar-Vidal, and M. Welch-Devine. 2011. Hard choices: Making trade-offs between biodiversity conservation and human well-being. *Biological Conservation* 144 (3): 966–972.

Mintzberg, H., D. Raisinghani, and A. Theoret. 1976. The structure of 'unstructured' decision processes. *Administrative Science Quarterly* 21 (2): 246–275.

Moffett, A., and S. Sarkar. 2006. Incorporating multiple criteria into the design of conservation area networks: A minireview with recommendations. *Diversity and Distributions* 12 (2): 125–137.

Norton, B.G. 1994. *Toward unity among environmentalists*. Oxford: Oxford University Press.

Norton, B.G. 2003. *Searching for sustainability: Interdisciplinary essays in the philosophy of conservation biology*. Cambridge: Cambridge University Press.

Norton, B.G. 2005. *Sustainability: A philosophy of adaptive ecosystem management*. Chicago: University of Chicago Press.

Norton, B.G. 2015. *Sustainable values, sustainable change: A guide to environmental decision making*. Chicago: University of Chicago Press.

Norton, B.G., and A. Steinemann. 2001. Environmental values and adaptive management. *Environmental Values* 10: 473–506.

Norton, B.G., and B. Hannon. 1997. Environmental values: A place-based theory. *Environmental Ethics* 19: 227–245.

Norton, B.G., and B. Hannon. 1998. Democracy and sense of place values in environmental policy. *Philosophy and Geography* 3: 119–146.

Norton, B.G., and D. Noonan. 2007. Ecology and valuation: big changes needed. *Ecological Economics* 63 (4): 664–675.

Norton, B.G., and R.E. Ulanowicz. 1992. Scale and biodiversity policy: A hierarchical approach. *Ambio* 1992 (May 1): 244–249.

Perry, W., and J. Moffat. 1997. Developing models of decision making. *Journal of the Operational Research Society* 48 (5): 457–470.

Rittel, H.W.J., and M. Webber. 1973. Dilemmas in a general theory of planning. *Policy Sciences* 4: 155–169.

Sager, T. 1997. Planning and the liberal paradox: A democratic dilemma in social choice. *Journal of Planning Literature* 12: 16–29.

Simon, H.A. 1955. A behavioral model of rational choice. *Quarterly Journal of Economics* 69 (1): 99–118.

Simon, H.A. 1973. The structure of ill structured problems. *Artificial Intelligence* 4 (3–4): 181–201.

Simon, H.A. 1982. *Models of bounded rationality*. Cambridge: MIT.

Taket, A., and L. White. 1997. Wanted: Dead OR alive—ways of using problem structuring methods in community OR. *International Transactions in Operational Research* 4 (2): 99–108.

Taquino, M.C., D. Parisi, and D.A. Gill. 2002. Units of analysis and the environmental justice hypothesis: The case of industrial hog farms. *Social Science Quarterly* 83 (1): 298–316.

von Neumann, J., and O. Morgenstern. 1944. *Theory of games and economic behavior*. Princeton: Princeton University Press.

Winterfeldt, D.V., and W. Edwards. 1986. *Decision analysis and behavioral research*. Cambridge: Cambridge University Press.

Wooley, R.N., and M. Pidd. 1981. Problem structuring—a literature review. *Journal of the Operational Research Society* 32 (3): 197–206.

Wright, G., and P. Goodwin. 1999. Value elicitation for personal consequential decisions. *Journal of Multi-Criteria Decision Analysis* 8 (1): 3–10.

Yu, P.L. 1979. Second-order game problem: decision dynamics in gaming phenomena. *Journal of Optimization Theory and Applications* 27: 147–166.

Zia, A. 2013. *Post-kyoto climate governance: Confronting the politics of scale, ideology, and knowledge*. London: Routledge.

Zia, A., S. Kauffman, and S. Niiranen. 2012. The prospects and limits of algorithms in simulating creative decision making. *Emergence: Complexity and Organization (E:CO)—An International Transdisciplinary Journal of Complex Social Systems* 14 (3): 89–109.

Zia, A., B.G. Norton, S.S. Metcalf, P.D. Hirsch, and B.M. Hannon. 2014. Spatial discounting, place attachment, and environmental concern: Toward an ambit-based theory of sense of place. *Journal of Environmental Psychology* 40: 283–295.

Zia, A., P. Hirsch, H.V. Thang, T.C. Trung, S. O'Connor, T. McShane, P. Brosius, and B.G. Norton. 2015. Eliciting inter-temporal value trade-offs: A deliberative multi-criteria analysis of Vietnam's Bai Tu Long national park management scenarios. *IAFOR Journal of Sustainability, Energy and the Environment* 2 (1): 41–62.

Zia, A., P. Hirsch, A.N. Songorwa, D.R. Mutekanga, S. O'Connor, T. McShane, P. Brosius, and B.G. Norton. 2011. Cross-scale value trade-offs in managing social-ecological systems: The politics of scale in Ruaha national park, Tanzania. *Ecology and Society* 16 (4): 7. https://doi.org/10.5751/ES-04375-160407.

Dr. Asim Zia is currently serving as a Professor of Public Policy and Decision Analysis in the Department of Community Development and Applied Economics, with a secondary appointment in the Department of Computer Science, at the University of Vermont. He is Director of the Institute for Environmental Diplomacy and Security (IEDS) and Co-Director of the Social Ecological Gaming and Simulation (SEGS) lab at the University of Vermont. He has a Ph.D. in Public Policy from the Georgia Institute of Technology with Bryan Norton as his dissertation advisor; and recipient of 2004–2005 best dissertation award from Association for Public Policy Analysis and Management.

Chapter 12
Does Deliberation Promote Ecological Citizenship? The Convergence Hypothesis and the Reality of Polarization

Jozef Keulartz

Abstract This chapter will subject Bryan Norton's well-known 'convergence hypothesis' to a critical assessment with a view to improve our understanding of the possibilities and difficulties of environmental citizenship. I will argue that Norton holds overly optimistic expectations regarding the transformative force of democratic deliberation. Firstly, Norton consciously ignores the role of power relationships and strategic negotiations in political will-formation. Secondly, Norton also has a too narrow view on ethical discourse itself: he only focuses on conceptions of the good life within single communities and neglects the plurality of often incompatible conceptions between different communities. To illustrate the inevitability of bargaining on the one hand and the importance of inter-communal ethical tensions on the other, I will discuss the wolf debate in Finland that started after the country's accession to the EU in 1995.

Keywords Bryan Norton · Convergence hypothesis · Environmental citizenship
Democratic deliberation · The Finnish wolf debate

12.1 Introduction

As no other philosopher, Bryan Norton has put environmental pragmatism firmly on the map as a very promising alternative for traditional environmental ethics approaches. Already during the 1980s, long before the watershed collection of essays *Environmental Pragmatism* edited by Light and Katz (1996), Norton started to remodel environmental philosophy into a problem-oriented approach with a highly interdisciplinary character that gives priority to policy over principle, practice over theory. He thus developed into a thought-provoking philosopher who

J. Keulartz (✉)
Institute for Science in Society, Radboud University,
Heyendaalseweg 135, 6525 AJ Nijmegen, The Netherlands
e-mail: j.keulartz@science.ru.nl

S. Sarkar and B. A. Minteer (eds.), *A Sustainable Philosophy—The Work of Bryan Norton*, The International Library of Environmental, Agricultural and Food Ethics 26, https://doi.org/10.1007/978-3-319-92597-4_12

has inspired scholars and professionals from a wide range of fields dealing with all aspects of nature and the environment.

In this chapter, I will subject Bryan Norton's well known 'convergence hypothesis' to a critical assessment with a view to improve our understanding of the possibilities and difficulties of ecological or environmental citizenship.[1] The focus of this review will be on Norton's expectations about making environmental ethics more deliberate.

As Norton has argued, people with different value orientations will converge on common policies to redirect our attitudes and behavior in more sustainable directions once they start 'thinking like a mountain', or, more broadly, start 'thinking like a planet', i.e. once they realize that no human values can survive in the long run without protecting the wider ecological context in which these values evolve.

In *Toward Unity Among Environmentalists* (1991), Norton assumed that environmentalists need to work on educating the general public and improve its scientific understanding of ecological systems. If enough people will adopt the ecological viewpoint on nature, the result will be a shift in worldview, toward less consumptive and more contemplative values, for instance.

In *Sustainability. A Philosophy of Adaptive Ecosystem Management* (2005), education no longer plays such a transformative role. This role has been taken over by the idea of 'democratic deliberation', which in contrast to education implies a two-way communication between science and the general public. This shift from education to deliberation reflects a change in Norton's view on science that took place under the influence of the idea of 'post-normal science' launched by Funtowicz and Ravetz (1993).

The most prominent characteristic of post-normal science is the extension of the peer community to include stakeholders with varied interests and viewpoints. To answer the question how to develop and preserve procedural norms that encourage fairness, openness, and mutual respect among the members of the extended peer community, Norton relies heavily upon Jürgen Habermas' work on 'discourse ethics' that is at the core of his influential theory of deliberative democracy.

In Norton's most recent book, *Sustainable Values, Sustainable Change. A Guide to Environmental Decision Making* (2015), deliberative procedures still play an important transformative role, although he now believes that they must be supplemented with changes in the incentive structures under which people make decisions—what Thaler and Sunstein (2009) have called 'nudging'.

I will argue that Norton holds overly optimistic expectations regarding the transformative force of democratic deliberation. Norton's lack of realism becomes apparent when we compare his conception of deliberative democracy with Habermas' original conception. Norton significantly deviates in two respects from Habermas' 'process model of deliberative democracy'. On the one hand, Norton

[1]Following Dobson (2003, 68), I consider ecological and environmental citizenship as complementary because "they can both plausibly be read as heading in the same direction: the sustainable society."

consciously ignores the role of power relationships and strategic negotiations in political will-formation. On the other hand, Norton also has a too narrow view on the role of ethical discourse in political will-formation: he only focuses on conceptions of the good life *within* single communities and neglects the plurality of often incompatible conceptions *between* different communities.

To illustrate the inevitability of bargaining on the one hand and the importance of inter-communal ethical tensions on the other, I will extensively discuss the wolf debate in Finland that started after the country's accession to the EU in 1995.

12.2 The Convergence Hypothesis

As noted, Bryan Norton has been a leading figure in the pragmatist wave that gained traction in the mid-1990s. He is inspired by American pragmatism's founder Charles Sanders Peirce, and especially by his best-known student John Dewey and his method of reconstructive thinking.

Pragmatism's primary concern is to facilitate the solving of problems and the settlement of conflicts emerging out of our joint activities and practices in order to improve cooperation and enable peaceful cohabitation. Because of its focus on conflict resolution for the sake of further cooperation, pragmatism has always been interested more in the *process* of moral inquiry than in its ready-made *products*, such as moral rules and regulations (see Caspary 2000, 153). In order for the process of moral inquiry to be successful, pragmatists argue, we should abandon some long-standing philosophical principles that form obstacles to productive problem solving and satisfactory conflict resolution.

Pragmatists first and foremost reject every form of foundationalism, and have embraced fallibilism instead. According to Dewey, rightly hailed as America's major prophet of anti-foundationalism, the 'quest for certainty,' which philosophy has celebrated since Descartes, ought to be given up permanently because it is illusory. Without exception, all our convictions are provisional and open to repeal or review. The preoccupation with general and abstract truths is often counterproductive because it distracts attention from concrete problems and conflicts, which actually call for tact, flexibility, and context sensitivity.

It is this anti-foundationalism that undergirds Dewey's and his fellow pragmatists' rejection of dualistic thinking, a way of thinking which is deeply rooted in Western philosophy, judging by the importance of the law of the excluded middle that was already formulated by Aristotle. This principle asserts that every judgment is either true or false (not true); there is no third possibility: *tertium non datur*! This 'law of the excluded third' is in line with a whole host of dichotomies that characterize Western thought, such as theory and practice, fact and value, body and mind, and reason and emotion. These dichotomies encourage black-and-white thinking that does not allow for shades of gray or for a middle course but brings conflicts to a head and leads debates to degenerate into unproductive boundary disputes.

Pragmatists' alternative to foundationalism and dualism is contextualism, a concept that features prominently in Dewey's work. Human activities do not develop in a vacuum, but are situated within a cultural and natural milieu, that is subject to constant change; they are embedded in a dynamic context of time and space. Neglect of context, Dewey ventures to assert, "is the greatest single disaster which philosophic thinking can incur" (Dewey 1960, 92).

12.2.1 Anthropocentrism Versus Nonanthropocentrism

Throughout his entire career, Norton has taken great pains to bring Dewey's reconstructive thinking to bear on debates about environmental ethics and environmental policy, debates that are more often than not riddled with intense controversy. As is well known, the earliest controversy that has bitterly divided the environmental movement is the classic 'conservation versus preservation debate' that was sparked at the beginning of the last century by the question whether the city of San Francisco should be allowed to increase its water supply by building a dam in Hetch Hetchy Valley, located within the boundaries of Yosemite National Park. This question provoked a seven-year struggle between conservationists, who held that natural resources should be used in a wise, sustainable manner to benefit society, and preservationists, who believed that nature should be protected from human interference at all costs.

The conservationist faction was led by Gifford Pinchot, the first head of the U.S. Forest Service. He borrowed the maxim of his wise-use version of conservation from the founder of modern utilitarianism, Jeremy Bentham, who considered 'the greatest good for the greatest number' as the ultimate measure of right and wrong. Pinchot adopted this principle of utility with the salient addition: 'in the long run'. Modern utilitarianism became the basis for welfare economics and its most important tool: cost-benefit analysis. In arguing in favor of the dam, Pinchot and his allies followed this economic line of thinking as he told Congress in 1913 that the costs of "substituting a lake for the present swampy floor of the valley…is altogether unimportant compared with the benefits to be derived from its use as a reservoir" (quoted in Nash 1982, 161).

Pinchot's opponent was John Muir, the preservationist faction's spirited leader. Muir, who was co-founder and first president of the Sierra Club, denounced the idea that nature exists for the sake of humans as a sign of human arrogance and vigorously proclaimed that nature should be protected for its own sake. He praised the Hetch Hetchy Valley as "one of Nature's rarest and most precious mountain temples", and consequently condemned the damming of this valley as a sacrilege. He denounced the supporters of the dam as "devotees of ravaging commercialism", who "seem to have a perfect contempt for Nature, and, instead of lifting their eyes to the God of the mountains, lift them to the Almighty Dollar" (Muir 1997, 817). Despite Muir's appeal to President Roosevelt to save Hetch Hetchy, Pinchot's view eventually prevailed. Congress approved the damming of the valley in 1913. Muir died a year later, some say of a broken heart.

The Hetch Hetchy controversy reflects a deep divide between two conflicting worldviews: whereas utilitarian-minded conservationists maintain an anthropocentric position that assigns only instrumental value to non-human nature, preservationists are committed to a nonanthropocentric perspective in which all living things have intrinsic value. The struggle between these seemingly incompatible worldviews runs like a red thread through the recent history of conservation and later, environmentalism. We can witness an ongoing succession of new editions of the classic 'conservation versus preservation debate'.[2]

12.2.2 Thinking like a Mountain

The reason that the struggle between those who value nature for its own sake and those who value nature for the sake of human welfare is rather persistent and flares up time and again seems simple: this dichotomous constellation can only lead to deadlock because it leaves no room for any middle ground: you are either an anthropocentrist or a nonanthropocentrist—*tertium non datur*!

In order to break such an impasse and open up space for negotiation and deliberation among different and sometimes diverging perspectives, pragmatists suggest that we should replace thinking in terms of binary oppositions by the idea of broad continua, thus turning 'either-or' choices into 'more-or-less' ones. In Norton's own words: "Pragmatists believe that dichotomies and dualisms usually stand in for disguised continua and are best avoided—dissolved—when possible" (Norton 2005, 507). For such a pragmatist view on the 'conservation versus preservation debate', Norton turned to Aldo Leopold, whose 'land ethic' he considers to be inspired by classic American pragmatism.[3]

Leopold started his career as a forester and resource manager in the conservationist tradition of Pinchot. He took part in a predator eradication program during his early years as a U.S. Forest Service assistant in Arizona. The eradication of wolves, mountain lions and other large predators was meant to provide a maximum supply of deer. Over time, Leopold was forced to reconsider this management policy. The large scale killing of wolves and other predators eventually became linked to an explosion of deer populations (such as on Arizona's Kaibab Plateau in

---

[2]In *Toward Unity Among Environmentalists*, Norton devotes much attention to the 'deep ecology versus shallow ecology debate' that was initiated by Norwegian philosopher Naess (1973). Another debate that echoes the 'conservation versus preservation debate' is the 'parks versus people debate', that emerged at the end of the previous century (Miller et al. 2011). The latest edition of the classical debate is the 'traditional conservation versus new conservation debate' (Kareiva et al. 2012).

[3]Norton's pragmatist reading of Aldo Leopold is not uncontroversial. For an overview of the debate, see Piers Stephens in this volume.

the 1920s), which in turn led to overgrazing and overbrowsing of shrubs and trees, and eventually to soil erosion and the degradation of the local rimrock mountain landscape.

Leopold poetically described his conversion from a wolf eradicator to a wolf protector in a brief, 3–4 page essay—'Thinking Like a Mountain'—from his posthumously published book *A Sand County Almanac*. Leopold recounts how he and another forester, after eating lunch on a high rimrock, shot a she-wolf from a distance. He approached the mortally wounded wolf in time to watch "a fierce green fire dying in her eyes". This experience marked a turning point in his perception of the wolf. He had always been convinced that no wolves would mean hunters' paradise, but after seeing the green fire die, he "sensed that neither the wolf nor the mountain agreed with such a view." Leopold now saw wolves as integral part of natural systems, as members of 'biotic communities' that, in the long run, thrive far better with wolves than without wolves. He had learned to value wolves ecologically and to think 'like a mountain'—in the recognition that "only the mountain has lived long enough to listen objectively to the howl of a wolf" (Leopold 1949, 129–133).[4]

Time and again throughout his work Norton points out that Leopold, in his essay 'Thinking Like a Mountain', has metaphorically introduced what Norton, in the spirit of Dewey's reconstructive philosophy, has called a 'contextualist' approach to environmental management. Leopold recognized, Norton argues, that activities designed to maximize the production of resources may damage the larger ecological context in which these activities are embedded. According to Norton, Leopold's work has significantly anticipated the application of Hierarchy Theory in ecology, a discipline which focuses on the complexity and internal organization of ecological systems. Hierarchy Theory models nature as a set of nested systems, with large-scale and slow-changing systems acting as environment, or 'context', of smaller-scale and faster-changing subsystems. The larger systems determine the range of choice available in smaller systems, that are constrained by the larger systems of which they are the parts.

Following in the footsteps of Leopold and his 'protohierarchy theory' (Norton 2015, 98), Norton is convinced that the dichotomy between the anthropocentric values of the conservationists and the nonanthropocentric values of the preservationists is a false one. It should be replaced by the idea of a shared spectrum of values ranging from the short-term values embodied in the smallest systems to the long-term values embodied in the largest systems. What once appeared as a rancorous dispute between two factions with opposed worldviews now appears as two environmental strategies that stress different values in different situations: conservationists emphasize the short-term economic values, such as cost-efficiency, that guide the resource-use decisions concerning areas devoted to human use, while

[4]Recently, in 2009, some new Leopold family letters surfaced, including a 9-page letter to his mother, that contains evidence that the wolf-killing incident took place in 1909, some 20 years before Leopold's conversion (Rolston 2015, 46).

preservationists focus on the long-term values, such as health and integrity, of the larger ecological systems that determine the room for maneuver of the resource-producing activities in those areas. The division between conservationists and preservationists only occurred because they disagreed about whose values are to prevail with respect to the management of a given area (see Norton 1991, 191–2).

As an alternative to the divisive characterization of environmentalism as split between anthropocentric and nonanthropocentric approaches, Norton has offered his convergence hypothesis, stating that "provided anthropocentrists consider the full breadth of human values as they unfold into the indefinite future, and provided nonanthropocentrists endorse a consistent and coherent version of the view that nature has intrinsic value, all sides may be able to endorse a common policy direction" (Norton 1997, 87).

12.3 From Education to Deliberation as Transformative Force

In order to move past black-and-white thinking and toward convergence by seeking to bridge the perceived anthropocentrism/nonanthropocentrism rift, Norton relies heavily on science. In *Toward Unity Among Environmentalists*, he argues that his convergence hypothesis has a dual status: it is not only meant as a very general empirical hypothesis, but also as an 'article of faith' of environmentalists—the faith that growing scientific understanding, as yielded by ecology, will lead to similar policy goals among environmentalists with very different value commitments. If they reach policy consensus it is "mainly because they agree about science, not because they agree about values" (Norton 1991, 203). In order to be successful, environmentalists need to work on educating the general public and improve its scientific understanding of ecological systems. If enough people will adopt the ecological viewpoint on nature, the result will be a shift in worldview, toward less consumptive and more contemplative values, for instance (ibid., 194).[5]

12.3.1 Post-normal Science

However, education does no longer play such a transformative role in Norton's 2005 book *Sustainability. A Philosophy of Adaptive Ecosystem Management*. This role has been taken over by the idea of 'democratic deliberation', which in contrast to education implies a two-way communication between science and the general

[5]Earlier on, especially in *Why Preserve Natural Variety* (1987), Norton attributed transformative value to direct contact with and (aesthetic) experience of wild species and unspoiled ecosystems. See also Minter and Rojas, this volume.

public. This substitution of education by deliberation reflects a change in Norton's view on science that took place under the influence of Funtowicz and Ravetz (1993).

These philosophers have claimed that the image of science as an objective and impartial provider of empirical facts and rational explanations that can safely steer the course of policy and politics has fallen to pieces. Especially when very complex problems are involved like climate change, this traditional image of science does no longer correspond to reality. As a consequence of the growing complexity within many scientific disciplines, uncertainties are also increasing, not only with respect to technical and methodological issues, but also with regard to epistemological and ethical questions. At the same time the decision stakes are becoming higher and higher, reflecting conflicting purposes between stakeholders. Under these conditions the puzzle-solving strategies of normal science (in the Kuhnian sense) are no longer appropriate. According to Funtowicz and Ravetz, a new science is called for: 'post-normal science.'

The most prominent characteristic of post-normal science is the extension of the peer community. The recognition of the plurality of legitimate perspectives and ways of knowing and the inclusion of an ever-growing set of stakeholders, Funtowicz and Ravetz claim, has important implications for both society and science. "With mutual respect among various perspectives and forms of knowing, there is a possibility for the development of a genuine and effective democratic element in the life of science" (Funtowicz and Ravetz 1993, 740–1). With the emergence of post-normal science, the practice of science "is becoming more akin to the workings of a democratic society, characterized by extensive participation and toleration of diversity" (ibid., 754). In short: "Post-normal science can provide a path to the democratization of science" (ibid., 739).

Norton has embraced this new approach to science that he believes to be "compatible with and even evocative of the practical epistemology of pragmatists" (Norton 2005, 114). Like Funtowicz and Ravetz, Norton considers as key aspect of post-normal science the extension of the peer community "to include not just scientists from established disciplines, but also parties, stakeholders who have varied interests and viewpoints and who express their viewpoints in an open and public process" (ibid., 115).

12.3.2 Discourse Ethics

To answer the question how to develop and preserve procedural norms that encourage fairness, openness, and mutual respect among the members of the extended peer community, Norton relies heavily upon Jürgen Habermas and his work on 'discourse ethics' that is at the core of his version of deliberative democracy (Norton 2007, 14–15). Norton believes that this ethics tradition, that is strongly influenced by American pragmatism—Norton speaks of 'third generation pragmatism'—offers a new and more productive approach to the field of environmental ethics.

The cornerstone of Habermas' procedural approach is formed by his 'discourse principle'. According to this principle, only those norms can claim validity to which all potentially affected persons could agree as participants in rational discourses. To count as rational a discourse must meet the following conditions to ensure that the outcomes are determined by no force other than the force of the better argument: "(i) that nobody who could make a relevant contribution may be excluded; (ii) that all participants are granted an equal opportunity to make contributions; (iii) that the participants must mean what they say; and (iv) that communication must be freed from external and internal coercion" (Habermas 1998, 44).

As Norton points out, Habermas distinguishes three types of rational discourse: pragmatic, moral, and ethical discourses. *Pragmatic* discourses aim at seeking and justifying suitable means for realizing goals and preferences that are already given; *moral* discourses focus on questions of justice and examine what is equally 'good for all'; *ethical* discourses aim to reach self-understanding and gain clarity about what course of action is 'good for us' as persons and communities. Ethical judgments cannot claim universal validity, like moral judgments, but only particular validity, because these judgements are all about ideals of the 'good life' that evolve in the context of the life histories of concrete individuals and communities.

Norton considers it an advantage of Habermas' discourse ethics that it allows ample room, not only for the universalistic procedural claims of morality, but also for the more particularistic and substantive choices that shape the ideals of good life of communities and its members. Norton calls such substantive values "constitutive values"—they "*constitute* the community by giving it unity across generations; they are all tied up with the community's sense of itself and its members' sense of themselves as individual members of that community; they give meaning to the life of individuals and the community" (Norton 2005, 388).

He also calls these values 'sense-of-place values'; they are an expression of a community's 'authentic' relationship with its natural habitat. If these values are lost, the community's sense of self is diminished together with the integrity of the place that community members call their home. This explains why it are precisely these substantive values that are at stake "when groups protest major changes in the landscape of an area, as when opponents of chip mills protest clear-cutting of hardwoods and their replacement with plantation pines, or when it is claimed that certain projects destroy the 'character' of the place" (ibid., 289).

Such local resistance clearly demonstrates that the fundamental communal values, that tie generations together in multigenerational communities, involve a commitment to protect the integrity of the cultural and natural context in which a community's identity and sense of self have evolved. It is because of this deeply felt commitment, Norton argues, that ethical reasoning is ideally suited to guide members of a community toward consensus through open and inclusive processes of deliberation and cooperative decision making, in which they can try to persuade one another to change crucial beliefs with the help of what Habermas (1998, 37) has called the "unforced force of the better argument" only.[6]

12.3.3 From Land Ethic to Earth Ethic

In his most recent book, *Sustainable Values, Sustainable Change. A Guide to Environmental Decision Making* (2015), published a decade after *Sustainability. A Philosophy of Adaptive Ecosystem Management*, Norton turns his attention to so-called 'third-generation' environmental problems. He refers to the work of sociologist Robert Cameron Mitchell (1989), who has distinguished two types of environmental problems. Until the 1960s, conservation organizations were predominantly concerned with land and wildlife issues. These first-generation issues usually involved threats to particular areas or species. A new generation of environmental problems started to emerge in the late 1950s. These second-generation problems are not necessarily site- or species-specific but tend to spill over local boundaries and affect entire regions. In particular, Rachel Carson's book, *Silent Spring* (1962), called for awareness of the new type of problems by focusing on the impact of DDT.

Today, we are faced with environmental problems such as climate change, biodiversity loss, and ozone depletion, that are global in scale and long-term in scope. As Norton points out, the impact of these third-generation problems transcends Leopold's land ethic that seems to be more suited to second-generation problems. In light of third-generation issues, 'thinking like a mountain' and its ecosystem is no longer sufficient, but need to be complemented by 'thinking like a planet' and its atmosphere.

As Norton himself has acknowledged, by complementing the land ethic with an earth ethics there seems to arise a conflict between, on the one hand, his community-based approach to adaptive management with its focus on public deliberation and local involvement, and, on the other hand, the growing need to address environmental problems that cross local and regional boundaries and encompass the whole earth and its atmosphere. Norton is convinced that this conflict can be reconciled and that third-generation problems can be addressed by developing policies mainly from the bottom up. "Despite serious constraints on

[6]Norton has so much confidence in communal deliberation that he envisions a future nondualistic world, in which "the convergence hypothesis will wither away for lack of polarized interests to be brought together" (Norton 2005, 510).

local activities from power structures at larger scales, local communities can catalyze action if they are organized and pursue effective processes" (Norton 2015, 279).[7]

12.4 Does Deliberation Encourage the Emergence of Green Values?

The question now arises as to whether Norton holds overly optimistic expectations regarding the transformative force of democratic deliberation: can local deliberation actually lead to global convergence on constitutive or sense-of-place values that point in the direction of a more sustainable development? In *Sustainable Values, Sustainable Change*, Norton has discussed this question in response to Manuel Arias-Maldonado's book *Real Green. Sustainability after the End of Nature* (2012), in which this political scientist raises serious doubts that deliberative methods encourage the emergence of green values and promote ecological citizenship. According to Arias-Maldonado, "There is no guarantee that ecological values will be embraced as a result of free and equal deliberations (...) It is not reasonable to expect that citizens will dramatically change their preferences through deliberation" (Arias-Maldonado 2012, 144).

12.4.1 The Role of Incentives in Deliberation

In answering this criticism, Norton has argued that Arias-Maldonado can so easily downplay the role of deliberation in the transformation of people's preferences in the direction of sustainability because he seems to maintain a too narrow view of deliberation that includes only information and argumentation. Abstract reasoning alone will not do the job but must be embedded within a broader adaptive management process that combines procedural rationality with empirical knowledge on which measures will work and which are likely to fail.

With respect to this kind of knowledge, Norton refers to the work of social psychologist Thomas Heberlein. In his book, *Navigating Environmental Attitudes* (2012), Heberlein argues that information and education alone is seldom successful in changing behaviors that have detrimental environmental effects. What is needed to achieve significant behavioral change is a structural change in the social environment that, in particular, affects the incentives under which people make

[7]In his 2013 book *Thinking like a Planet: The Land Ethic and the Earth Ethic*, Baird Callicott has also made an attempt to extent Leopold's ethic into an earth ethic. He argues, contra Norton's pragmatist approach, that such an ethic should be based a holistic nonanthropocentric theory (see Callicott in this volume). For a reply, see (Norton 2015), 265–273.

decisions. As Norton now explains it, true deliberation can lead citizens to develop and introduce incentive systems such as tax breaks that will actually shift human behavior toward more environmental protection as suggested by Heberlein.

Norton claims that it is this recourse to Heberlein's empirical work that differentiates his approach to democratic deliberation from that of Arias-Maldonaldo. However, this claim is somewhat problematic because Arias-Maldonado also insists on the need to introduce proper incentives to steer people in more sustainable directions. He cites Richard Thaler and Cass Sunstein's seminal book, *Nudge. Improving Decisions about Health, Wealth and Happiness* (2009). Under the seemingly contradictory label of 'libertarian paternalism' they have launched the idea that citizens can be *nudged* by incentives such as green taxes or cap-and-trade systems to take better decisions regarding environmental protection in ways that do not violate their freedom of choice. Thaler and Sunstein believe that incentive-based approaches are more efficient and effective, and less expensive and intrusive than command-and-control approaches.

"But," Arias-Maldonado asks, "who nudges the nudger? In other words, who is to decide which kind of decisions should be promoted by arranging choice architecture in a particular manner?" In answering this question he argues that "nudges should reflect the state of the debate around sustainability in each given moment", and that they "must respond to the outcome of democratic procedures, which in turn reflect social change" (Arias-Maldonado 2012, 171).

In a similar vein, Norton has argued that participants in communal deliberations must be willing to make decisions regarding changes in incentives that affect their own and other's behavior. This is because changes in attitudes alone are unlikely to lead to behavioral change. But on the other hand, without changes in attitudes in the direction of sustainability citizens will not be able to change the choice architecture of their community in environmentally beneficial ways. And these changes in attitudes are conceived by Norton as outcome of deliberative processes.

In other words, although deliberative reasoning is not sufficient to lead to behavioral change, it is nevertheless a necessary precondition to reach such a change—which brings us back to the question whether Norton asks more of deliberation in an age of climate change than it can deliver: can local deliberation lead to global convergence on sustainable management?

A quick look at the vicissitudes of Natura 2000, one of the most ambitious supranational initiatives for nature conservation worldwide, clearly reveals that the empirical evidence for Norton's assumption that communal deliberation can lead to sustainable management at larger, regional or global, scales, is certainly disputable.[8]

---

[8]Minteer and Manning (2000) have found support for the empirical validity of the convergence argument in their study of the Vermont public's general environmental commitments and their specific attitudes toward national forest policy. The question, of course, is whether and to what extent the findings of a single empirical study can be extrapolated to other cases.

12.4.2 Natura 2000 and the Shift from Top-Down to Bottom-Up

Natura 2000 constitutes the cornerstone of European nature conservation policy; stretching over 18% of the EU's land area and almost 6% of its marine territory, it represents the largest network of protected areas in the world. Its aim is to protect Europe's most valuable and threatened species and habitats, listed under the 1979 Birds Directive and the 1992 Habitats Directive. These directives are legally binding texts that require the individual Member States to convert them into national legislation. If a Member State fails to pass the required national legislation, or if the national legislation does not adequately comply with the requirements of the Directives, the European Commission can initiate legal action through the European Court of Justice.

Although significant successes have been achieved, reflecting the largely uncontested character of the overall goals of Natura 2000, increasing implementation problems emerged in the various European Union (EU) Member States. Almost half of the 15 'old' EU Member States (i.e. states granted EU membership before May 2004)—including Denmark, Finland, France, Germany, Greece, Ireland and the Netherlands—have been taken before the European Court of Justice by the European Commission for failure to submit lists of designated sites in accordance with the Habitats Directive (Paavola 2004).

In implementing Natura 2000, the EU strongly relied on classic top-down measures and traditional 'command-and-control' regulations. On the assumption that it is up to scientific experts and not to ordinary citizens and politicians to determine the direction of nature policy, the starting point of the entire process was ecological knowledge about the various ecosystems and the environmental conditions in which they are viable. However, as soon as it became clear that the interests of many local stakeholders would be substantially affected, the implementation process almost came to a standstill.

Everywhere in Europe, the implementation process with regard to nature policy provoked similar conflicts with similar outcomes. In response to the protests of farmers, fishermen, foresters and other local residents, governments gradually abandoned their centralist, top-down approach and increasingly switched over to more participatory forms of decision-making.

However, the results of the shift from a top-down to a bottom-up approach were far from satisfactory. Everywhere in Europe this shift went hand in hand with a 'dilution' of the original nature goals, both qualitatively—from deeper to lighter shades of green—and quantitatively—in terms of hectares: almost all final national lists of designated Natura 2000 network sites, prepared for submission to the European Commission, were reduced to about half of their original size, and dropped below the minimum of 10% asked for by European legislation (Keulartz and Leistra 2008).

The trials and tribulations of the implementation of Natura 2000, the most important legal framework in the field of nature conservation in Europe, give ample reason to severely doubt the realistic character of Norton's expectations with respect to democratic deliberation. I will argue Norton invokes such unrealistic expectations because his conception of deliberative democracy is too simplistic compared to Habermas' own original conception.

12.4.3 Habermas' Process Model of Deliberative Democracy

Habermas lets his 'deliberative process model of rational political opinion- and will-formation' start with the *pragmatic* justification of general policy proposals and programs. Such plans have the form of hypothetical recommendations that largely rely on empirical knowledge about the efficiency of the technical and strategic means for realizing some chosen goals. However, this knowledge is naturally fallible and generally far from value-neutral and thus not uncontested. Already with the political evaluation of expertise and counter-expertise, normative viewpoints come into play that begin to compete openly in the second stage of the political opinion- and will-formation. At this point Habermas' deliberative model branches off into three trajectories.

First and foremost, policy plans need to be judged in the light of the justness of the resulting action norms. Such assessment is conducted in *moral* discourses that center on the question what without exception is equally 'good for all'. With reference to Kant's categorical imperative, Habermas argues that "a norm is just only if all can will that it be obeyed by each in comparable situations" (Habermas 1996, 161). This universalization principle demands that we leave the ethnocentric perspective of particular collectivities behind us and take the comprehensive perspective of an unlimited communication community. An agreement with respect to questions of justice can only be considered a *rational motivated consensus* if moral discourses are organized in such a way that the outcomes are determined by no force other than the 'force of the better argument'.

In complex societies, however, it is often impossible to reach such a consensus by lack of generally accepted starting points. In that case, two alternatives are open: when we are faced with non-universalizable interests we can switch to *strategic negotiations* that are central to the liberal model of democracy, in which the democratic process takes place exclusively in the form of compromises between competing interests; when we are faced with non-universalizable value orientations we can resort to *ethical discourses* that form the centerpiece of the republican model of democracy, in which the focus of democratic will-formation is on the process of a community's ethical self-understanding and political self-organization (Fig. 12.1).[9]

[9]The final element of Habermas' model are legal discourses, in which it is determined whether the proposed policy plans can be integrated into the existing legal system. Like the pragmatic discourses that open the deliberative process model, legal discourses are best understood as matters requiring expert opinion.

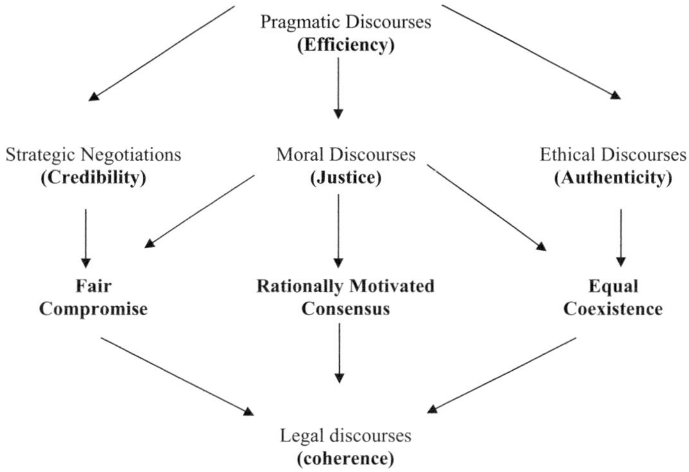

Fig. 12.1 Process model of political will-formation (adapted from Habermas 1996, p. 168)

At this point, Norton significantly deviates in two respects from Habermas' deliberative process model. Firstly, Norton explicitly ignores the role of strategic negotiations. Habermas calls such neglect an "ethical curtailment" or an "ethical foreshortening" of political discourse that is typical of the republican model of democracy (Habermas 1994, 104; 1998, 244).[10] Secondly, Norton also has a too narrow view on ethical discourse itself: he only focuses on conceptions of the good life *within* single communities and neglects the plurality of often incompatible conceptions *between* different communities. To illustrate the inevitability of bargaining on the one hand and the importance of inter-communal ethical tensions on the other, I will zoom in on the problematic implementation of Natura 2000 regarding the gray wolf population in Finland.

12.5 The Inevitability of Bargaining

In Habermas' model, rational political will-formation appears as "a network of discourses *and* negotiations that can be linked to one another via multiple pathways" (Habermas 1996, 167; italics added). For Habermas, negotiations are an essential part of his process model. Arguing must give way to bargaining in situations in which social power relations can only be disciplined, not neutralized, through procedurally correct agreements among power holders. Whereas rational argumentation is subject to the criterion of *validity*, the outcome of bargaining

[10]Both expressions are a translation of the German phrase used by Habermas: "Ethische Engführung politischer Diskurse".

processes is largely determined by the degree of the *credibility* of the threats and promises of the bargaining parties in light of their extra-political resources such as money and manpower. A compromise that bargainers can achieve, can be called a *fair compromise* if their negotiations are regulated by procedures that ensure that all relevant interests are given equal consideration and all parties have an equal opportunity to influence one another during the bargaining process. Through these fairness conditions of compromise formation, the moral viewpoint of justice can be brought to bear at least indirectly on bargaining processes. "The procedural conditions under which actual compromises enjoy the presumption of fairness must be justified in moral discourses" (ibid.).

Unlike Habermas' deliberative process model, strategic negotiations have no place at all in Norton's philosophy of adaptive ecosystem management. He has consciously ignored the role of political and economic power relationships in policy formation and management, even though he is "well aware of this confounding variable that effects official decisions so pervasively" (Norton 2005, p. xiv). However, as Habermas has pointed out, in complex societies it is not—even under ideal conditions—always possible to settle controversies by argumentative means only. Whenever the proposed regulations affect the various interests each in a different way without any generally accepted common interest, there simply is no alternative for bargaining.

Bargaining plays a significant role in the resource conflicts that have arisen in many EU-countries during the implementation of Natura 2000. Nature protection measures for endangered species have triggered negative consequences for legitimate societal interests such as fishing, farming, and forestry. This is especially true with respect to protection measures for large carnivores that have made a substantial recovery in recent times—there are currently around 17,000 bears, 10,000 wolves and 10,000 lynx in Europe (Boitani and Linnell 2015).

The wolf debate in Finland can serve as an illustration of the nature and magnitude of conflicts over the implementation of Natura 2000 Directives, particularly considering that Finland as EU's 5th largest and most sparsely populated country hosts a population that has fluctuated between 120 and 250 wolves over the past two decades, which are very small numbers when compared with a country such as Spain, for example, where the wolf population is estimated at 2000–3000 individuals.

12.5.1 The Wolf in Finland

By the 1920s, the gray wolf had been hunted to near extinction in Finland. Its recovery started during the 1970s, initially based on individuals migrating from the Soviet Union. As Finland's wolf population steadily increased, it caused a lot of problems, especially in rural areas. The return of the wolf raised concerns among parents over the safety of their children in their home yards and on their way to and back from school. The wolf's reappearance also posed a severe threat to livestock farming, with reindeer as the main victims of wolves, followed by sheep and cattle. Hunters experienced problems with wolves because they compete for the same prey, mainly moose and white-tailed deer, but also because of the loss of their hunting dogs that are attacked and killed by wolves (Hiedanpää et al. 2016b).

After Finland's accession to the EU in 1995, the wolf became included in Annex IV of the Habitats Directive as a species in need of strict protection, with the exception of the reindeer husbandry area in the northernmost region of Finland, where wolves are listed in Annex V, allowing for a more flexible regime. From the start, the EU has shown growing impatience with Finland's failure to meet the requirements of strict protection for wolves. In 2001 the European Commission initiated an infringement procedure in order to improve the level of protection of the wolf. The Finnish government showed its commitment to wolf protection by preparing a wolf management plan that was made public in 2005. Proposed changes included reforming the system of damage compensation and damage prevention and formulating a clearer interpretation of the conservation status of wolves. Despite this effort, the Commission called Finland to the European Court of Justice in 2005 for being too generous in its policy of granting wolf hunting licenses. The Court rendered its judgment in 2007 and ordered the government of Finland to rectify this policy under penalty of a fine. After this judgment, the Finnish government took some measures, including the instruction to the regional game districts to follow a tighter line in granting wolf hunting licenses. The Commission was satisfied with these measures and closed the case in 2008.

Of course, the purpose of the government's adjustments was to increase the size of the wolf population, but the opposite happened: after having peaked at approximately 250 individuals in the winter of 2006/2007, the wolf population started to decline in the winter of 2007/2008. The only plausible explanation for this decline was that wolves had been illegally killed. "As the government and the wildlife administration improved their actions in the face of EU pressure, the rural communities reacted and started to take the right into their own hands. The rural communities addressed the wolf problem by managing the population size by their own means" (Hiedanpää et al. 2015, 39).

In 2011 the legislation was revised once again by taking the authority to grant wolf hunting licenses from the regional game districts and giving it to a new central organization, the Finnish Wildlife Agency. Moreover, the criminal sanctions for illegal hunting were significantly increased, with a maximum sanction of four years imprisonment and a maximum fine of 9100 euro. However, these measures didn't stop the decline—in 2013 the wolf population further decreased to 120–135 individuals, its lowest level recorded in the 2000s.

The criminal sanctions were hardly effective in preventing illegal hunting because with fewer than 10% of all hunting crimes being solved the risk of getting caught was very low (Borgström 2012). But, more importantly, the effectiveness of the sanctions was severely undermined because the whole wolf conservation policy was not perceived as legitimate by the affected local communities who complained that their concerns were not taken seriously enough. Nationwide surveys indicated that the percentage of Finnish citizens who were afraid of wolves increased from 32 to 47% between 2009 and 2013 (Hiedanpää et al. 2016a). The lack of legitimacy led to disobedience which manifested itself in strong community support for illegal hunting (Pohja-Mykrä and Kurki 2014). The conflict over Finland's wolf policy escalated in 2013 when a group of twelve angry locals in the rural municipality of Perho took the law in their own hands and killed three wolves out of a pack of five.[11]

To improve the legitimacy of the wolf conservation regime, the government was forced to make some concessions. In response to the complaints of thirteen parent associations, the government passed the so-called 'yard wolf decree' in 2013. This decree allowed for up to five extra licenses for hunters to be exempt from the strict protection for wolves that regularly visit human settlements and may potentially be of danger to human safety (Hiedanpää et al. 2016a).

Another major concession was made by the government when it incorporated wolf hunting as a tool for population management in the new Finnish Wolf Management Plan that came into force on January 2015. For the first time since 2007, when the European Commission condemned Finland for breaching the EU rules for strict wolf protection, the government embarked on a policy of granting wolf hunting licenses on the basis of certain predetermined quotas. This policy started with a test phase of two years with quota of max. 29 and 46 wolves allowed to be killed for population management purposes in 2015 and 2016 respectively.

The drafting process of the new wolf management plan had an open and interactive character, and included online discussions, surveys, workshops, seminars and local events all across Finland (Hiedanpää et al. 2016b). This process carried elements of bargaining, because social power relationships have unmistakably influenced the final outcome, with the state power to provide a threat of sanctions and a promise of compensations on the one hand and the counter power of non-compliance and disobedience of rural comminutes on the other. It is questionable and requires further examination whether these power relationships were sufficiently disciplined through procedures that provide all parties with an equal opportunity to influence one another to call the new plan a fair compromise that balances the conflicting interests.

The wolf debate in Finland clearly demonstrates that one would be ill advised, following Norton's example, to ignore the role of power relationships and the need for bargaining in nature protection and conservation conflicts. This debate can also highlight another shortcoming of Norton's philosophy of adaptive ecosystem management: his too narrow view on ethical discourse.

---

[11]This wolf hunt that took place during a whole weekend was well planned and organized, using snowmobiles and mobile phones. One of the poachers held a high position in the local game preservation society. https://www.theguardian.com/world/2016/jan/21/finland-approves-wolf-hunt-trial-cull-illegal-poaching.

12.6 The Importance of Inter-communal Ethical Tensions

In contrast to moral discourses that center on the question what without exception is equally 'good for all', ethical discourses turn on the issue of the good life and on the question of what is 'good for us' as members of a specific nation, members of a local community, inhabitants of a region, and so on. In ethical discourses the attempt is made to reveal, through critical reflection, the deeper consonances in a common form of life that can bridge differences of opinion. Unlike a moral consensus such an ethical consensus is not (and cannot be) entirely rationally motivated. Participants in ethical discourses can never work themselves free of the place- and time-bound context of the cultural form of life in which they de facto find themselves (Habermas 1996, 163).

In addition, an ethical consensus can only exist *within* a single community. In modern pluralistic societies, however, discussions take place primarily *between* different communities, which often hold competing views of the good life. In that situation an ethical consensus is impossible, and because ethical convictions differ from negotiable interests a fair compromise is not suitable either; in ethical matters one cannot bargain nor make concessions without compromising oneself and one's integrity (Bohman 1996).

So the question becomes how to handle deep-seated value conflicts if the possibilities for consensus and compromise are excluded. Habermas suggests that we then should aspire to an "equal coexistence" of different ethical convictions. Because of the requirement of equality, ethical debates in pluralistic societies ought to be subjected to procedures that guarantee that all parties have an opportunity to express their ethical views and to effectively influence the outcome of these debates. Like a fair compromise, an equal coexistence cannot stand on its one moral feet: here as well, the procedural conditions under which an actual coexistence can count as an equal one must be justified in moral discourse.

The ideal of equal coexistence requires a certain attitude of the conflicting parties. They have to appreciate the fact that they are competing for primacy within the same universe of discourse with others who cannot beforehand be branded as unreasonable. Such reflexive awareness rejects the naivety of dogmatic beliefs, recognizes its own fallibility and leaves room for 'reasonable disagreement' as the basis of equal and peaceful coexistence of different ethical worldviews. Only then can there be a balanced debate in which one party, without renouncing its own claim to validity, is able to respect the other parties as allies in the common quest for genuine truths (see e.g. Habermas 1994, 94; 1998, 66).

Viewed through the lens of Habermas' account of ethical discourse, it indeed appears overly optimistic and unrealistic to expect, like Norton, that people worldwide will learn to think 'like a planet' and change their behavior in more sustainable directions merely by deliberations on the interpretation and application of constitutive or sense-of-place values that give meaning to a local community's life and its authentic relationship with the natural environment. Again, the Finnish wolf debate can make it clear that Norton's faith in the transformative power of communal deliberation is blind as long as he overlooks the frequent occurrence of rational or not so rational disagreements over the value of competing life styles in ethical disputes over nature protection and conservation policies.

12.6.1 Wolf Haters Versus Wolf Lovers

The stated goal of the new Finnish wolf management plan to grant derogations from strict protecting and allow for managerial wolf hunting was to raise acceptance of wolves and to prevent illegal poaching. It is, however, questionable whether the concerns of the rural population were sufficiently addressed by this concession. Recent research into wolf policies in Wisconsin and Michigan has shown that allowing wolf culling is substantially more likely to increase poaching than reduce it: “Liberalizing killing appears to be a conservation strategy that may achieve the opposite outcome than that intended” (Chapron and Treves 2016, 5).

On the other hand, for some parts of the population, especially from urban areas, this concession went much too far. Immediately after the new wolf management plan was put into force on 22 January 2015, hundreds of people took to the streets in Helsinki, Tampere and Joensuu. The demonstrators voiced their anger about what they called a coordinated plan to exterminate the Finnish wolf. They feared that under the new plan, half of the wolf population might be killed, while the population should be at least twice as big for it to be genetically healthy.[12] The demonstrations were under strong police protection after hunters and militant anti-wolf activists announced counter-demonstrations on the internet. The pro-wolf activists accused the government to have acted under strong political pressure from the far right “True Finns” party which has made the ‘protection of people from wolves’ one of their main topics in the Finnish wolf areas.

So it seems that the new wolf management plan may have widened rather than narrowed the deep rift between proponents and opponents of wolf protection and conservation. The conflict between the ‘wolf-lovers’ and ‘wolf-haters’ reflects the increasing tensions between urban and rural communities in Finland and elsewhere in Europe. Rural depopulation and economic stagnation have shifted the balance of power to what is perceived as the ‘urban elite’ who are generally committed to an environmental ethos and consider hunting as a brutal anachronism. Rural communities, for their part, experience the wolf as symbolic for the projects of the urban elite which threaten pastoral rural livelihoods and lifestyles already forced onto the defensive. Because hunting is a key tenet of the way of life in rural areas of Nordic

[12]They argued that, except for the regular hunting of wolfs, one has to take into account the illegal poaching of about 30 wolfs per year, the unlimited amount of exempt licenses in the reindeer husbandry areas, and the police termination warrants to kill wolves that come closer than 100 m to buildings.

countries, a ban on hunting is seen as an attack on rural identity (Von Essen et al. 2015, 203).[13]

People from urban and rural areas increasingly foster different and often diverging conceptions of the good life. City dwellers have generally adopted a cosmopolitan attitude; they see themselves as 'citizens of the world' who embrace diversity and welcome immigration. Rural residents, by contrast, usually maintain a more parochial attitude; they stress the importance of strong local attachments and have a territorial sense of belonging. Urban and rural people can, in shorthand, be differentiated from each other with the help of two of the three central 'ethics' that can coexist within any culture around the world, according to cultural anthropologist Richard Shweder (1991): urbanites prefer a more liberal 'ethic of autonomy', in which rights, justice, fairness and freedom are seen as important moral goods, whereas rural people favor a more conservative 'ethic of community', with duty, respect, loyalty, and interdependence as key virtues (cf. Haidt 2012).[14]

It goes without saying that these different conceptions of the good life go hand in hand with different conceptions of nature. Rural people living in close spatial proximity to a natural resource tend to favor its economic exploitation, whereas urban dwellers living at a greater distance tend to favor its conservation. Again, in shorthand, it could be argued that urbanites generally prefer a 'wild' concept of nature, while rural residents opt for a 'functional' or an 'Arcadian' concept of nature (Keulartz 2004).

In an ethnographic study of conflicts over wolf protection in Norway, Olve Krange and Ketil Skogen have noted that wolves simply did not fit into the image of nature of the rural population that counts many avid hunters: "The appearance of wolves was associated with urban life and city people, and an urban concept of nature. In the hunters' minds, that is a romantic view, based on a dream-like glorification of untouched nature. Through this construction, the wolf becomes an icon of urbanity… The wolf is not a part of real nature; it is an urban implant" (Krange and Skogen 2011, 477).

It is obvious that, while thinking like a mountain or like a planet, urban and rural residents increasingly envision different mountains and different planets. Instead of a convergence we witness a growing polarization of constitutive or sense-of-place values. The possibility of reasonable disagreements as basis of a peaceful coexistence of different ethical convictions is becoming more and more overshadowed by the reality of vitriolic disagreements.

---

[13]Anti-wolf sentiments are by no means restricted to Nordic countries. In East Germany, for instance, several anti-wolf groups have been set up recently. A good example is 'Bündnis gegen den Wolf' (Alliance against the wolf) that was set up in Mecklenburg-Vorpommern, where Angela Merkel's Christian Democrats were beaten into third place by the anti-immigrant and anti-Islam Alternative for Germany (AfD) party in the State election in September 2016. On November 22, 2016, the first anti-wolf demonstration was held in Bautzen, a town in eastern Saxony. It seems to be no accident that a month before, Bautzen was the scene of a far-right anti-migrant rally.

[14]Shweder's third ethic is the 'ethic of divinity'.

12.7 Concluding Remarks

If one wants to promote ecological or environmental citizenship, it is important to realize that one should not only address the disproportionate burden that communities suffer from environmental hazards, such as air and water pollution, and severe rainfall or drought, but that one should also try to prevent or mitigate the harm that communities suffer from the adverse social and economic consequences of the transition to a more sustainable society. It is for instance, evident that the transition to renewable energy affects the interests of different communities very differently. Hence, the need to strive towards achieving a balance of conflicting interests through compromise formation under fair bargaining conditions.

However, to reach beyond the traditional environmental constituencies and engage other sectors of society in developing environmental citizenship requires more than a fair balancing of interests. It also requires a political culture in which often widely diverging lifestyles and worldviews can compete with one another on an equal footing. As Habermas (1998, 220–226) has argued, an equal and peaceful coexistence can only be achieved on the condition that the conflicting parties abstain from dogmatically asserting themselves against one another. A dogmatic attitude towards their own culture will prevent its ongoing renewal and adaptation to constantly changing conditions. Visions of the good life owe their viability and vitality to their ability to self-transformation, by sketching out alternatives to the status quo and by integrating external impulses—they don't prosper through seclusion but through exchange.

In this chapter, I have argued that a community-based model of deliberation that ignores or downplays power differentials and ethical frictions *between* different communities raises too optimistic expectations regarding the prospects of convergence on common policies in more sustainable directions. Deliberation is after all a more complex phenomenon than Norton seems to suggest and asks for a more comprehensive repertoire than ethical reasoning only. But fortunately, Norton's toolbox contains a set of heuristic devices, such as scenario development (what Dewey has called 'dramatic rehearsal'), boundary work, social learning, and experimentation, that could be adapted to facilitate and foster communication and cooperation across borderlines that separate communities with different and sometimes diverging interests and lifestyles.

References

Arias-Maldonado, M. 2012. *Real Green. Sustainability after the End of nature*. London and New York: Routledge.

Bohman, J. 1996. *Public deliberation. Pluralism, complexity, and democracy*. Cambridge, MA: The MIT Press.

Boitani, L., and J.D.C. Linnell. 2015. Bringing large mammals back: Large carnivores in Europe. In *Rewilding European Landscapes*, ed. H.M. Pereira and L.M. Navarro, 67–84. Berlin: Springer International Publishing.

Borgström, S. 2012. Legitimacy Issues in Finnish Wolf Conservation. *Journal of Environmental Law* 24 (3): 451–476.
Callicott, J.B. 2013. *Thinking like a planet: The land ethic and the earth ethic*. New York: Oxford University Press.
Caspary, W.R. 2000. *Dewey on democracy*. Ithaca & London: Cornell University Press.
Chapron, G., and A. Treves. 2016. Blood does not buy goodwill: allowing culling increases poaching of a large carnivore. *Proceedings of the Royal Society B* 283: 20152939. https://doi.org/10.1098/rspb.2015.2939.
Dewey, J. 1960. Context and thought. In *On experience, nature and freedom*, edited, with an introduction, ed. Richard J. Bernstein. New York: The Bobbs Merrill Company, Inc.
Dobson, A. 2003. *Citizenship and the environment*. Oxford, NY: Oxford University Press.
Funtowicz, S., and J.R. Ravetz. 1993. Science for the post-normal age. *Futures* 25: 735–755.
Habermas, J. 1994. *Justification and application: Remarks on discourse ethics*, trans. C. Cronin. Cambridge, MA: MIT Press.
Habermas, J. 1996. *Between facts and norms: Contributions to a discourse theory of law and democracy*, trans. W. Rehg. Cambridge, MA: MIT Press.
Habermas, J. 1998. *The inclusion of the other: Studies in political theory*. Cambridge, MA: MIT Press.
Haidt, J. 2012. *The righteous mind. Why good people are dived by politics and religion*. Allen Lane.
Hiedanpää, J., M. Matti Salo, and J. Kotilainen. 2015. Teleodynamics and institutional change: The hardship of protecting the Amur tiger, big-leaf mahogany, and gray wolf. *Journal for Nature Conservation* 26: 36–44.
Hiedanpää, J., J. Pellikka, and S. Ojalammi. 2016a. Meet the parents: Normative emotions in Finnish wolf politics. *Finnish Journal for Human-animal Studies* 2: 4–27.
Hiedanpää, J., H. Kalliolevo, M. Salo, J. Jani Pellikka, and M. Luoma. 2016b. Payments for improved ecostructure (PIE): Funding for the coexistence of humans and wolves in Finland. *Environmental Management* 58: 518–533.
Kareiva, P., M. Marvier, and R. Lalasz. 2012. Conservation in the Anthropocene. Beyond Solitude and Fragility. http://thebreakthrough.org/index.php/journal/past-issues/issue-2/conservation-in-the-anthropocene.
Keulartz, J. 2004. Concepts of nature as communicative devices: The case of Dutch nature policy. *Environmental Values* 13 (1): 81–99.
Keulartz, J., and G. Leistra (eds.). 2008. *Legitimacy in European nature conservation policy—Case studies in multilevel governance*. Heidelberg: Springer.
Krange, O., and K. Skogen. 2011. When the lads go hunting: The "Hammertown mechanism" and the conflict over wolves in Norway. *Ethnography* 12 (4): 466–489.
Leopold, A. 1949. *A sand county Almanac*. London: Oxford University Press.
Light, A., and E. Katz (eds.). 1996. *Environmental pragmatism*. London: Routledge.
Miller, T., B. Minteer, and L.-C. Malan. 2011. The new conservation debate: The view from practical ethics. *Biological Conservation* 144: 948–957.
Minteer, B., and R. Manning. 2000. Convergence in environmental values: an empirical and conceptual defense. *Ethics, Place and Environment* 3 (1): 47–60.
Mitchell, R.C. 1989. From Conservation to Environmental Movement: The Development of the Modern Environmental Lobbies. In *Government and Environmental Politics: Essays on Historical Developments since World War Two*, ed. M.J. Lacey. Washington, DC: Wilson Center Press.
Muir, J. 1997. *The Yosemite* (1912); reprinted in *John Muir: Nature Writings*. New York: Library of America.
Naess, A. 1973. The shallow and the deep, long range ecology movement. *Inquiry* 16 (1–4): 95–100.
Nash, R. 1982. *Wilderness and the American Mind*. New haven, CT: Yale University Press.
Norton, B. 1987. *Why Preserve natural Variety?*. Princeton, NJ: Princeton University Press.
Norton, B. 1991. *Toward unity among environmentalists*. Oxford, NY: Oxford University Press.

Norton, B. 1997. Convergence and contextualism: Some clarifications and a reply to Steverson. *Environmental Ethics* 19: 87–100.
Norton, B. 2005. *Sustainability. A philosophy of adaptive ecosystem management*. Chicago: The University of Chicago Press.
Norton, B. 2007. The re-birth of environmentalism as pragmatic, adaptive management, Invited Lecture. State University of NY at Buffalo. 8 March 2007.
Norton, B. 2015. *Sustainable values, sustainable change*. Chicago: The University of Chicago Press.
Paavola, J. 2004. Protected areas governance and justice: Theory and the European Union's Habitats Directive. *Environmental Sciences* 1: 59–77.
Pohja-Mykrä, M., and S. Kurki. 2014. Strong community support for illegal killing challenges wolf management. *European Journal of Wildlife Research* 60: 759–770.
Rolston, Holmes I.I.I. 2015. Rediscovering and rethinking leopold's green fire. *Environmental Ethics* 37: 45–55.
Shweder, R.A. 1991. *Thinking through cultures: Expeditions in cultural psychology*. Canbridge, MA: Harvard University Press.
Thaler, R., and C. Sunstein. 2009. *Nudge. Improving decisions about health, wealth and happiness*. London: Penguin.
Von Essen, E., H.P. Hansen, H. Nordström Källström, N. Peterson, and T.R. Peterson. 2015. The radicalisation of rural resistance: How hunting counterpublics in the Nordic countries contribute to illegal hunting. *Journal of Rural Studies* 39: 199–209.

Jozef Keulartz is emeritus Professor of Environmental Philosophy at the Radboud University Nijmegen, and senior researcher Applied Philosophy at Wageningen University and Research Centre. He has published extensively in different areas of science and technology studies, social and political philosophy, bioethics, environmental ethics and nature policy. His books include *Die verkehrte Welt des Jürgen Habermas* [*The Topsy-Turvy World of Jürgen Habermas*, 1995] and *Struggle for Nature—A Critique of Radical Ecology* (1998). He is co-editor of *Pragmatist Ethics for a Technological Culture* (Kluwer, 2002), *Legitimacy in European Nature Conservation Policy* (Springer, 2008), *New Visions of Nature* (Springer, 2009), *Environmental Aesthetics. Crossing Divides and Breaking Ground* (Fordham University Press 2014), *Old World and New World Perspectives in Environmental Philosophy* (Springer, 2014), and *Animal Ethics in the Age of Humans* (Springer, 2016).

Chapter 13
Deliberative Decisions and Formal Multicriteria Analysis: Addressing Norton's Skepticism

Sahotra Sarkar

Abstract Norton has argued for the salience of deliberative strategies for making environmental decisions which is supposed to be preferable to formal decision analysis. This paper argues that formal multicriteria decision analysis, when deployed with care, has the flexibility to absorb the advantages of deliberative decision making. It can also be used for decision support during a deliberative process. This feature of decision analysis is illustrated using a case study from Merauke in Papua province of Indonesian New Guinea.

Keywords Conservation area networks · Decision theory · Objectives hierarchy Merauke · Mulicriteria analysis · Systematic conservation planning

13.1 Introduction

A fundamental tenet of environmental pragmatism, one that Norton and I share, is that environmental philosophy should direct its attention to practical problems, particularly towards socially relevant public policy decisions. From this perspective, metaethical problems, such as the source of environmental values or even their ultimate justification, typically fade into irrelevance in most contexts. Norton has focused on sustainability for the last few decades. However, in his early work (e.g., 1986, 1987) he focused on biodiversity conservation which, through the methodology of systematic conservation planning (Margules and Pressey 2000; Margules and Sarkar 2007; Sarkar and Illoldi-Rangel 2010), has been one of my major interests. Central to systematic conservation planning is a decision procedure of which the most basic part consists of the identification of a set of notional

S. Sarkar (✉)
Departments of Philosophy and Integrative Biology, University of Texas at Austin, 2210 Speedway, Austin, TX 78712, USA
e-mail: sarkar@austin.utexas.edu

S. Sarkar
Presidency University, Kolkata, India

S. Sarkar and B. A. Minteer (eds.), *A Sustainable Philosophy—The Work of Bryan Norton*, The International Library of Environmental, Agricultural and Food Ethics 26, https://doi.org/10.1007/978-3-319-92597-4_13

conservation areas to be managed for the persistence of biodiversity. This part has always (and without any residual controversy) been algorithmically solved because of the large size of the relevant data sets (Margules et al. 1988; Sarkar 2012a); it is the easy part and will not concern us any further here.

However, the delineation of notional conservation area networks is embedded within a broader decision context of systematic conservation planning. Over the years, it has become clear that the decisions that have to be made routinely require tradeoffs amongst multiple incompatible criteria beyond the representation of biodiversity in conservation area networks (Sarkar 2005, 2012b). My collaborators and I have been promoting the use of formal multi-criteria analysis (MCA) for this problem for over almost two decades (Sarkar et al. 2000, 2017); while we have acknowledged potential contributions from a variety of techniques for MCA (Moffett and Sarkar 2007), by and large, we have gravitated towards multiattribute value theory (MAVT) (Sarkar et al. 2017). In promoting such formal approaches we run afoul of Norton's (e.g., 2005) skepticism about the value of formal decision analysis, particularly those methods that are reducible to algorithms and typically incorporated into software packages. Norton (2005, and elsewhere) implicitly emphasizes a contrast between the use of such protocols and the democratic participatory process of deliberative decision making that he endorses. Beyond Norton's political reservations about the use of these methods (which I share), there seems also to be an epistemological worry about problematic commitments that these methods require. This paper is intended to address both though, like Norton, it makes no attempt to keep them artificially separate.

I am sympathetic to Norton's skepticism and am perhaps even more sensitive (as an advocate of formal methods) than him to the danger of mechanically relying on algorithms to make decisions. This problem was of serious concern even in the days of the dominance of classical cost-benefit analysis (see Norton 1987) but technological developments since then have made it even more pernicious. The ubiquitous use of GIS technology today underscores this point. It facilitates the fast mechanical production of beguiling spatial maps (for instance, during land use planning) that are visually impeccable but entirely opaque about the processes of their production, the assumptions and approximations that were made, and all the uncertainties that were elided from the final product. Yet, these are the images that are the baseline results used in almost all spatial environmental planning today. We need not even embrace procedural rationality to acknowledge the glaring pitfalls of blind reliance on GIS-generated results. Without a reflective critical attitude, these results are of dubious value because they give no indication of where they are inaccurate, imprecise, or otherwise uncertain. I am only emphasizing the use of GIS technology because of its likely familiarity; other computer-aided planning and decision-analytic technologies are often equally problematic. Even in contexts where GIS representations have been used as part of deliberative decision making sessions (and there have been several of these), I am not aware of a single published case in which the inherent biases and limitations of GIS as a technology has been explicitly addressed (and I welcome counter-examples).

Moreover, many formal decision-theoretic methods presume that the goal is to obtain an optimal solution given the feasible alternatives and the modeled constraints. (The problem to be solved is supposed to be one of "constrained optimization.") For instance, a variety of software tools can be used to identify a set of conservation areas that optimally represent biodiversity features in minimal areas (Sarkar et al. 2006). However, the search for optimality may take a computer a very long time because the formal problems that have to be solved are known to be computationally complex; that is, they are NP-hard which means that they belong to a set consisting of the hardest-known computational problems (Sarkar et al. 2006). Moreover, the algorithms cannot take account of many relevant features such as the political feasibility of proposed solutions which may repeatedly change during discussions between stakeholders. It is often the case that several good though formally sub-optimal solutions are politically feasible but the optimal one is not. In practice—and recall the context of environmental pragmatism—the optimum is irrelevant. The focus should move instead to good though possibly sub-optimal but politically feasible solutions. Deliberation must override computation. (I leave for another occasion a discussion of the question whether the search for optimality should even be considered rational in such contexts, only noting here that the strategy being endorsed here is in agreement with Norton's emphasis on satisficing over the mechanical pursuit of optimality.)

There are many other similar situations that strongly suggest that formal methods should be deployed only to the extent that they are deemed to be justified in the informed judgement of stakeholders (without assuming that recourse to these methods is itself a relevant criterion). To make the same point in a different way: the use of formal methods for decision analysis is justified so long as they are viewed and used as decision support resources and not as decision-making tools. The problem is that they are easy to use, and that they can be easily used without critical attention and reflection, typically by merely clicking on buttons in a user-friendly (preferably "idiot-proof") interface. In other words, we should share Norton's skepticism. To end with a slogan: people should run programs, not the other way around.

Nevertheless, I will argue here that formal methods can and should be profitably used for complex multicriteria environmental decisions and that the process, done properly, not only can but must give a central role to the deliberative process. Complexity in environmental decisions arises from many sources; the three that are most amenable to formal analysis are uncertainty, multiple criteria, and multiple agents. Given the space constraints of this volume there will be no discussion of uncertainty in this contribution though formal analysis, especially those based on Bayesian methods, which are increasingly credibly being used to accommodate uncertainty in environmental decisions. In concordance with the general emphasis on deliberative processes here, it will be assumed that multiple agents will act as a single entity making decisions using results from internal deliberations. Not only does this reflect one of the standard methodologies used in environmental decisions, viz., stakeholder consultation, it also allows us to avoid well-known pitfalls and paradoxes of formal group decision analysis (Sarkar 2012b). This chapter will be restricted to multicriteria decisions that ignore uncertainty and are made by single agents.

The perspective advocated here will be illustrated using a recently published example of systematic conservation planning from Merauke, Indonesia. The next two sections will provide, first, a very brief discussion of that methodology and, second, a more detailed description of how MCA fits into systematic conservation planning. The example being used will then be introduced and analyzed. Subsequent sections will take up the strengths and weaknesses of formal approaches to MCA. It is a pleasure to dedicate this paper to Norton's long and illustrious career and to acknowledge his influence on my work.

13.2 Systematic Conservation Planning

The problem of appropriately selecting sets of areas to be managed for the persistence of biodiversity—what have since come to be called conservation area networks—emerged in the 1970s when it became clear that there are multiple legitimate demands on land and seascapes and that, therefore, not all areas of potential importance for biodiversity can be set aside for conservation (Margules and Sarkar 2007). Areas must be prioritized for different types of habitat management including habitation and production besides biodiversity conservation. Systematic conservation planning, which was initially developed in Australia (Margules et al. 1988; Margules and Pressey 2000), is a structured stage-wise approach for efficiently mapping conservation area networks for the representation of biodiversity and its management for persistence, with feedback, revision, and reiteration, where needed, at any stage of the process—see Margules and Sarkar (2007) for a detailed treatment. (Pressey and Bottrill (2009) provide an influential alternative formulation while Sarkar (2012a) provides a history).

The stages of systematic conservation planning are shown in Fig. 13.1 (of which there are many variants); the caption of the figure provides more detail on each stage of the process. What deserves emphasis here is that the process is not a linear progression of stages and that the envisioned feedback, revision, and reiteration, especially as conservation area networks are created and managed, place the process under the rubric of adaptive management (Sarkar 2005).

The planning process begins with the delineation of a planning region, the selection of the biodiversity units (the "constituents"), and the identification of legitimate stakeholders, in a set of stages that intimately interact with each other. The problem of selecting biodiversity units is that of defining biodiversity; most philosophers, following Norton, accept that biodiversity constituents must reflect both descriptive features of the natural world and embody normative social commitments about what should be preserved (in this context see, also, the discussion of normativity in conservation biology by Callicott et al. (1999) and Takacs (1996) for sociological background). I have recently reviewed the problem of defining biodiversity (Sarkar 2017) acknowledging the critical role played by Norton's early work (summarized in Norton 2003).

All three of these stages of systematic conservation planning require deliberation between stakeholders as does the next stage of setting goals for biodiversity and other features including quantitative targets for biodiversity representation; the last stage typically also requires more technical input from conservation scientists and other professionals who can provide advice on what representation targets are appropriate. Very often resource constraints mean that the biodiversity constituents cannot be adequately mapped within the time horizon of required decisions and a technical problem of identifying adequate measurable "surrogates" for them must be broached using a variety of techniques developed in the 2000s. In most cases, data must be refined and treated, for instance, should the biodiversity surrogates include species, their spatial distributions must typically be modeled (and, ideally, corroborated in the field).

Using the output of all these stages, existing conservation areas must be assayed for their performance at representing biodiversity, and new ones selected to satisfy known constraints of economy and also the biodiversity representation targets and other relevant criteria. Economy in representation is implicitly incorporated into the solution of the two problems that are used to delineate notional conservation area networks: (i) the "minimum area" problem: how to satisfy the representation targets for biodiversity surrogates in a minimal set of conservation areas; and (ii) the "maximum representation" problem: given a maximum total area that can be set aside for conservation, how to choose a set of conservation areas within this budget constraint so as to maximize the representation of biodiversity surrogates up to the required targets (Sarkar et al. 2006). A defining aspect of systematic conservation planning is that these problems are approached algorithmically using a wide variety of software decision support tools (Sarkar et al. 2006; Margules and Sarkar 2007; Moilanen et al. 2009). Sets of prioritized areas must also be refined using what is known about the biological and socio-political vulnerabilities of different potential conservation areas. This means that, at this stage, some assessment must be made whether putative conservation areas are at tangible risk due to local ecological issues, global changes, or socio-political threats such as inappropriate transformations for development. If so, these areas must be removed from a notional conservation area network and the area selection process reiterated.

Multiple criteria may enter into the initial prioritization of areas or be subsequently incorporated by generating large portfolios of notional conservation area networks that each satisfies biodiversity representation targets that can then be ranked using the other criteria. These have been called "iterative stage" and "terminal stage" protocols (respectively) (Moffett and Sarkar 2007). Both have been used in practice. The next section will discuss MCA in more detail.

Two points deserve emphasis here. First, in all practical applications, multiple criteria will be relevant. Since conservation is a land or seascape use choice, it is a matter of social policy and that makes sociopolitical criteria relevant along with considerations of biodiversity representation and persistence (Borgerhoff-Mulder and Coppolillo 2004). However, even if we restrict ourselves only to

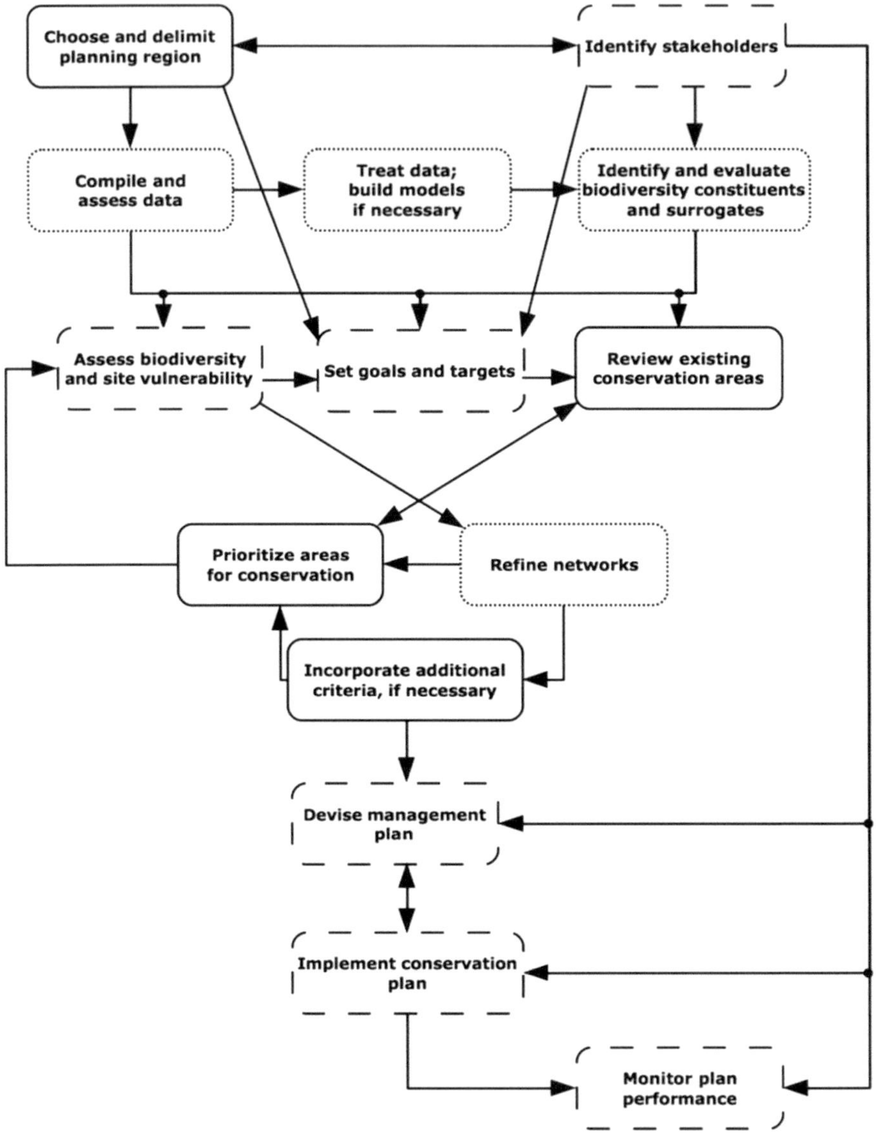

biodiversity-related issues (as an academic exercise), we are still faced with multiple criteria. For any notional conservation area network, criteria of adequacy must include spatial criteria that influence the persistence of biodiversity such as the size and shape of individual units, their dispersion across land and seascapes, their connectivity, their alignment with habitat types, and the number of replicates within the network of assemblages of each biodiversity component. Second, these criteria are not all mutually compatible and, as Norton and a few others began emphasizing

◂**Fig. 13.1** Systematic conservation planning (after Sarkar 2014). Arrows indicate which stages directly influence others. A bidirectional arrow indicates interaction between stages, that is feedback potentially in both directions. (Only major influences are shown as there is potential for interaction between almost any two stages.) Boxes with solid borders indicate that a stage is relatively well-understood (in terms of scientific capabilities) at the time or writing (2017); dotted borders indicate those that are fairly well-understood; and dashed borders indicate the least understood stages which merit further exploration. *Choose and delimit planning region*: precise geographical boundaries of the planning region should be explicitly discussed and chosen—how boundaries should be drawn (e.g., whether they are based on political or ecological criteria) may raise ethical issues. Local input is critical. *Identify stakeholders*: stakeholders include those who significantly affect or are affected by conservation plans—they have a legitimate stake in what happens. There could be feedback between this stage and almost any other stage. The text will discuss the problem of identifying legitimate stakeholders which has not received the attention that it deserves (Sarkar 2012a). *Compile and assess data*: relevant biological, ecological, and sociopolitical data must be collected, and done so in a cost-effective manner because resources are always limited and planning horizons often much less than what would be ideal. *Treat data; build models if necessary*: data treatment through statistical analysis is often required. Modeling is needed when treatment is insufficient to produce spatial data on relevant biological and socio-political factors, for instance, on the spatial distribution of species or habitat types that may be of conservation interest. *Identify and evaluate biodiversity constituents and surrogates*: stakeholders identify biodiversity constituents which requires discussion of normative commitments (Sarkar 2012a). Surrogates consist of quantitative estimators of biodiversity constituents (Margules and Sarkar 2007). *Set goals and targets*: quantitative targets for biodiversity representation must be set; other goals include spatial configuration of the conservation areas to enhance likelihood of persistence. Many of these goals must be set using discussion between stakeholders since they reflect the normative choice of what risk is acceptable, for instance, the probability of extinction in a specified future time frame. *Review existing conservation areas*: any existing conservation area network must be analyzed to determine the extent to which it already satisfies the specified goals and targets. *Prioritize areas for conservation*: new sites must be prioritized to meet the goals and targets that were set earlier. The objective is to achieve adequate representation of all biodiversity features while satisfying other desired goals. This is the most-studied part of systematic conservation planning. There are well-known algorithms and a wide variety of software decision support tools including ConsNet which was used in the example discussed here. *Assess biodiversity and site vulnerability*: prioritized areas and relevant biodiversity features must be assessed for vulnerability due to all factors. Again, the amount of risk deemed acceptable is a normative choice. *Refine networks*: if sites are vulnerable, they may be excluded from notional conservation area networks, and the selection process may be reiterated. *Incorporate additional criteria, if necessary*: additional criteria (biological, economic, cultural etc.) may need to be incorporated using multi-criteria analysis to evaluate trade-offs between them. See the discussion in the text. *Devise management plan*: management plans for the persistence of biodiversity must be developed taking into account local context, resource availability etc. This is poorly understood. *Implement conservation plan*: the management plan must be implemented for conservation to work. Consultation with local stakeholders is imperative for both ethical and practical reasons. In practice there have been very few cases in which entire conservation plans have been implemented and only a few more of partial implementation. *Monitor plan performance*: plan performance must be monitored to devise responses as necessary for adaptive management into the future. Again, this remains poorly explored (which is perhaps inevitable given how few plans have been implemented) almost a generation ago, tradeoffs must be introduced (Faith 1995; Sarkar 2005; McShane et al. 2011): for instance, attention to shape would typically dictate increasing compactness of conservation areas but alignment, for instance, along a watershed would pull in the opposite direction. Tradeoffs are ubiquitous in systematic conservation planning (Sarkar and Garson 2004).

Finally, management plans must be devised, implemented, and then monitored for performance, all three critical components of systematic conservation planning that are only beginning to be systematically studied and still remain poorly understood. At least there is now widespread recognition that implementing plans is primarily a socio-political task that requires sustained attention including the ability to communicate and negotiate with stakeholders. In retrospect what seems ludicrous is the extent to which this aspect of conservation planning was not recognized in the early decades of conservation biology (see Sarkar 2005 for more discussion).

13.3 Formal Multicriteria Analysis (MCA)

There are a wide variety of methods that have been used to incorporate multiple criteria into decisions; some of these are consistent with standard economic and decision theories; others, including the popular Analytic Hierarchy Process (Saaty 1980), are not and are, in that sense, arbitrary besides also being subject to well-known paradoxes. [However, even non-arbitrary methods cannot always avoid the last problem—see Arrow and Raynaud (1986) and Sarkar (2012a).] Those MCA methods that have been used in the context of spatial conservation planning were systematically reviewed by Moffett and Sarkar (2007); an earlier, more general, compendium by Figuera et al. (2005) provides details of all the common methods.

In what follows we will assume that conservation (and other environmental) decisions involve multiple stakeholders who act as a single decision-making unit through deliberative discussions that elucidate their (group) values (as, of instance, expressed through explicit preferences) and that this process is intentionally explicitly deliberative rather than formal. [For more on value elucidation, as I and my collaborators envision it, see Keeney (1992).] Deliberation is used to decide on the problem formulation (including opportunities and constraints), the set of feasible alternatives, the objectives and how they are structured (which will occupy the rest of this paper), and the analytic strategy to be used including whether formal analysis is either necessary or even desirable. Thus a decision to use a formal methodology (for instance, because the alternative set is too large to analyze fully through deliberation) is hardly an endorsement of formal methods *over* deliberation.

Even after deciding to use formal methods, and after choosing to restrict them to non-arbitrary ones, a wide variety of methods are available. In the early years of systematic conservation planning we explored and categorized them on the basis of the strengths of the assumptions about preferences that underlie each of them. For instance, in many circumstances alternatives (environmental plans) can be ranked on the basis of the criteria but these ordinal rankings cannot be credibly quantified into cardinal ones without introducing dubious assumptions. For instance, in the case of spatial planning units, one unit could be judged to have a more compact shape than another but there may be no acceptable measure for compactness.

We explored which multicriteria methods only assume such ordinal rankings of alternatives (and also of criteria if these are not presumed to be all of the same importance). One standard such method was Dominance (Rothley 1999; Sarkar and Garson 2004): the goal of the analysis was to eliminate all dominated (roughly Pareto-suboptimal) alternatives. While this method worked in some circumstances, the number of non-dominated alternatives (that is, those that could not be eliminated from further consideration) typically increased non-linearly with the number of criteria and a portfolio consisting of all of them became intractable for deliberative processes.

So, we were forced to take recourse in more restrictive methods that require more assumptions, and ultimately to multiattribute value theory which presumes the assignment of quantitative values to the criteria and to each element of the alternative set for each criterion (that is, a cardinal ranking of both criteria and the performance of the alternatives according to each of them). These are highly restrictive assumptions and there have been may discussions of when and how they may be justified (Sarkar 2005; Keeney 2007); these discussions deserve much more emphasis than have so far been given to them in the context of environmental decisions (see below). Multiattribute value theory then enables the attribution of an aggregated quantitative (cardinal) value to each alternative (in our case, a notional conservation area network) that can then be used as measure of its performance. (If uncertainties about outcomes are explicitly incorporated in this process, the aggregation function becomes a utility rather than a value function (Keeney and Raiffa 1993) and we would fall under the rubric of multiattribute utility theory.)

Additional complications arise when the objectives are not all independent of each other but are themselves structured, often in a complex way. Returning to the case of conservation area networks and their spatial configuration, the five criteria mentioned earlier are typically used because they are believed to be relevant to a species' persistence. Thus, there can be a hierarchy where a more important objective of persistence subsumes spatial configuration and, under the latter, the five others that all become its sub-objectives. However, some of these sub-objectives may also serve other higher-level objectives: for instance, shape and size can also be relevant to the ease of patrolling borders of conservation areas which may form part of a management strategy (for instance, of wildlife preserves under threat of illegal extraction). If we require objectives hierarchies to have simple "tree" structures (as in multiattribute value theory) then these criteria may occur more than once as sub-objectives (and we have to be sensitive about the justification of such double-counting).

Constructing the objectives hierarchy and eliciting weights for its components must precede any recourse to formal analysis in order to obtain an aggregated value for the performance of alternatives. The objectives hierarchy consists of a set of fundamental objectives with a tree of sub-objectives subsumed under each of them

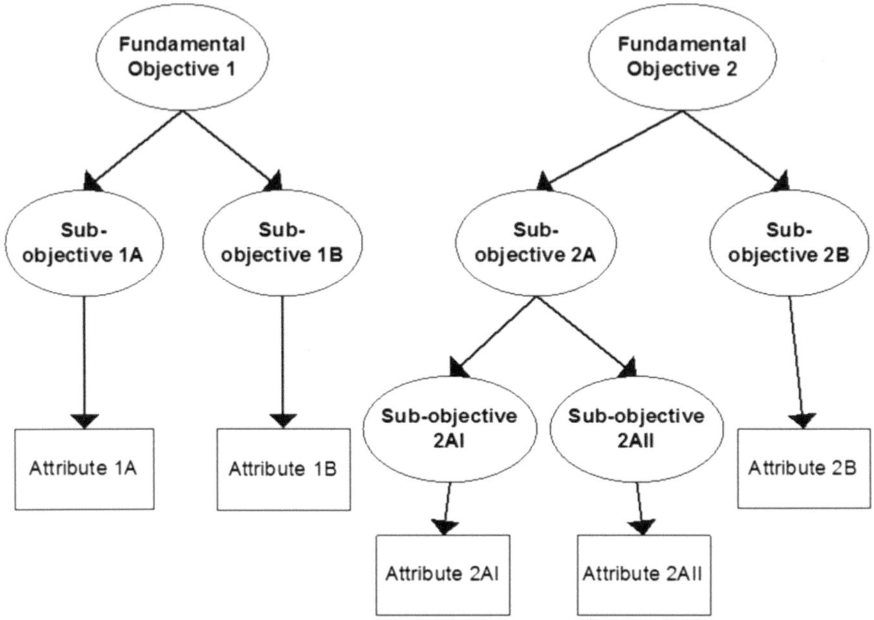

Fig. 13.2 Structure of an objectives hierarchy: the root of the hierarchy is the fundamental objective. The leaves, indicated by rectangular rather than oval shapes, are the measurable attributes corresponding to the sub-objective to which they are directly attached. The objective nodes are identified by answering the question: "What are the objectives of the decision?" The position in the hierarchy is determined by an answer to the question: "Why is this objective important?" If this question elicits an answer, the corresponding node is higher (closer to the root) of the hierarchy. If there is no answer, it is a fundamental objective. For further discussion, see the text

and with the lowest nodes (the "leaves") being associated with measureable attributes that have quantitative values (see Fig. 13.2). The development of such a structure starts with the identification of fundamental objectives that are ends-in-themselves; that is, there is no further answer to the question, "Why is this objective important?" (Keeney 1992). Objectives lower in the hierarchy (sub-objectives) are important for what they contribute to the fundamental objectives.

Construction of the objectives hierarchy is driven by elicitation from stakeholders acting as a group through deliberation. The process involves the iterated use of variants of two questions: "What are the objectives of the decision?," and the one mentioned earlier, "Why is this objective important?" The first of these provides the set of objectives for the objectives hierarchy, the second establishes the structure of the hierarchy. The elicitation process stops at the top (of the hierarchy) when fundamental objectives are reached. It stops at the bottom when the lowest–level objectives that are identified can be associated with measurable attributes. Note that the objectives hierarchy is not canonical in the sense that more than one structure can accommodate stakeholder preferences equally adequately. In practice, informal discussion may allow a first pass at a putative objectives hierarchy, which can then be subject to iterated revision and discussion. This was the strategy followed in the case study described below.

Using multiattribute value theory requires weights on each of these nodes and these must be elicited from stakeholders. Since these weights are perhaps the most misunderstood aspect of the use of most formal MCA methods, their interpretation requires remark: for multiattribute value theory the weight on a criterion is supposed to reflect its relative importance compared to other criteria within the bounds of the values of that criterion for the given set of feasible alternatives. For instance, the weight on a criterion of the size of a spatial unit is not supposed to reflect the general importance of size for conservation decisions, but its relative importance compared to the other criteria, given the range of sizes that are obtained within the alternatives set. If these sizes are very narrowly constrained, this (relative) weight may be very small even if, in general, size is a very important feature of conservation plans.

Finally, if the value function is supposed to be linear (which is the simplest case), the criteria used to construct the objective hierarchy must satisfy preference independence and difference independence conditions for the stakeholders (Keeney and Raiffa 1993). Preference independence is satisfied if preferences between two alternatives are determined only by the attributes that have different values for them and are not affected by the common values on the other attributes. Difference independence is satisfied when a preference between two alternatives differing with respect to some subset of attributes does not depend on the shared values of the other attributes.

All the complexities described in the last three paragraphs have been recently been discussed in detail by Sarkar et al. (2017) and those considerations will not be repeated here. What deserves emphasis in our context is that, so far, every stage of the decision process has required extensive deliberation between stakeholders and no mechanical formal analysis. These deliberations are influenced by the formal analysis to be performed at the next stage, which consists of an aggregation of the elicited preferences and measured values. This aggregation process imposes the requirements of the method used (multiattribute value theory in our case). Nevertheless, the decision process described here remains fundamentally deliberative and our discussion underscores the fact that the stakeholders are the decision makers and the formal method is simply a tool to be used like a calculator. In fact, the actual formal analysis is done rather trivially in most contexts by using a software tool (just like a calculator) and the results should ideally then become the locus of further discussion between stakeholders. If the stakeholders find the entire portfolio of results unacceptable, the whole process can be reiterated. The expectation is that the process will converge to an acceptable decision in at most a few iterations but there is no guarantee of such an end result.

13.4 An Example: Land Use Planning in Merauke, Indonesian New Guinea

13.4.1 Background

The example used here has been described in most detail by Ciarleglio et al. (2009b); a detailed description of the decision analysis has also been published (as noted earlier—Sarkar et al. 2017). The discussion here will be restricted to only one aspect of this effort, viz., the construction of the objectives hierarchy which will be used to illustrate the role played by deliberation to frame the problem and set it up for formal analysis.

In 2008, Conservation International (CI) contracted with the Indonesian conglomerate Medco to devise a land use plan for an industrial forestry plantation concession in Merauke in Papua Province of Indonesian New Guinea. Medco's intention was to grow trees for pulp; and its stated goals included achieving sustainability in forestry production, conservation of biodiversity, maintenance of ecosystem function, and satisfaction of community interests (Conservation International 2010).

The planning process began in Jakarta in December 2008 with a meeting between Medco and CI representatives and some of the other stakeholders (see below) in the presence of decision analysts from the University of Texas. We were not remunerated beyond our expenses and viewed our involvement as an academic research effort. We were observers at this and subsequent consultations; not being stakeholders, our involvement was limited to technical assistance in the form of responding to questions explicitly directed to us. These questions were limited to those of methodology. We did not participate in discussions of what objectives should be considered, how they should be prioritized, or what weights should be assigned. CI assumed for itself the role of a legitimate stakeholder.

Other stakeholders identified by CI were present at the Jakarta meeting except, critically, representatives of the communities affected by Medco's proposed development who (in CI's judgment) could not be included at this early stage for logistical reasons but whose values were supposed to be integrated into the spatial planning decision process once they had been fully investigated in the field by CI personnel. There were nine affected communities on the basis of the intersection of at least part of their traditional lands with the concession area. During the year after the Jakarta meeting, CI teams visited each of the communities to engage in discussion and preference elicitation. The extent to which community participation led to modifications of what was decided at the first meeting will be part of the discussion below.

The Jakarta meeting reviewed the decision context and achieved a consensus on the objectives to be incorporated. Subsequent developments only trivially modified this objectives set. A tentative objectives hierarchy was constructed at the original meeting. It also remained unexpectedly stable. In contrast, weights on criteria were not broached there and were assessed through elicitation using extensive and repeated discussions including the affected communities over the next year. Finally, the MCA was performed through a search for best-performing alternatives using the ConsNet software package (Ciarleglio et al. 2009a, 2010).

13.4.2 The Objectives Hierarchy and Fundamental Objectives

The stakeholders identified four fundamental objectives by immediate consensus at the Jakarta meeting (and no subsequent changes were necessary):

- Biodiversity representation: CI had biodiversity conservation as one of its priorities. Medco viewed incorporating biodiversity as an important contribution they expected from collaboration with CI. Stakeholders at the meeting, basing their input on what was known about community preferences, interpreted this goal to include that of maintaining ecosystem services. (Thus, in contrast to what is likely to happen in probably many other decision contexts, maintaining ecosystem services was not viewed as a fundamental objective by itself. This was a stakeholder decision achieved through deliberation at the Jakarta meeting and not brought into question during consultation with the communities at later stages.)
- Spatial Configuration: As noted in the last section, the spatial configuration of conservation area networks is expected to influence biodiversity persistence and can also be relevant to the management of conservation areas. In this case, spatial configuration was also relevant to the location of production areas for plantation crops. It was in Medco's interest to have easy access to communication links (rivers and roads, much of the latter to be constructed as part of development). Finally, spatial configuration was important for the communities. (For example, again contrary to many prior assumptions, forest resources close to villages were preferred to those that were far away so that they were easily accessible to older individuals.) These complex relationships can be credibly mapped into multiple hierarchical structures (recall that the objectives hierarchy does not have a canonical structure). The way in which the Merauke stakeholders chose to delineate their objectives hierarchy will be described in the next subsection (as shown in Fig. 13.3).
- Production suitability: Given the goals of this project, the economic viability of the proposed plantation must be taken into account. This was accepted as a fundamental objective by all stakeholders.
- Community interests: Sustainable development, as envisioned for this project, was intended to benefit not only Medco but resident communities in the concession area. Consequently, the satisfaction of community interests, as perceived by communities, was adopted as a fundamental objective; these included a

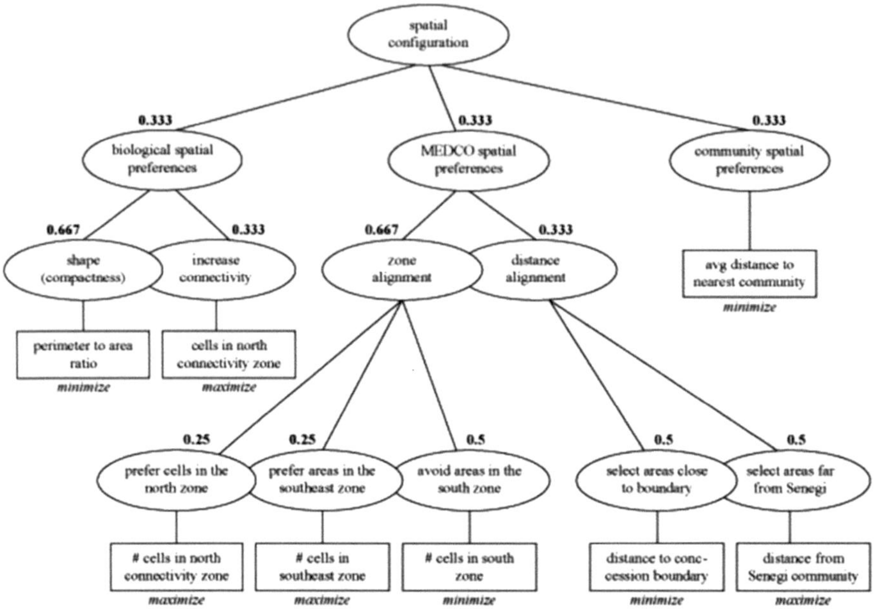

Fig. 13.3 The spatial hierarchy for Merauke (after Sarkar et al. 2017): the fundamental objective is the spatial configuration of the conservation area networks, that is, the distribution of those areas that are neither dedicated to habitation nor slated for pulp production. There are three subobectives corresponing to what is valued by the three main stakehlder groups. CI's interest in the persistence of biodiversity is decomposed into shape (with a preference for compactness) and connectivity. Both have obvious assoicated measurable attributes. Medco's spatial sub-objective is decomposed into zone and distance alignment as discussed in the text. They correspond to preferences for pulp production areas in certain parts of the concession and also close to transportation links (rivers and roads along the boundaries). The communities had one spatial sub-objective: closeness of conserved forests to their village so that they were easily accessible for resource extraction. The figure also shows one set of weights that was obtatined by elicitation and discussion

variety of goals including economic benefits as well as continued access to the natural resources of the region. Communities also demanded the setting aside of buffer areas around villages for future development, making them unavailable for both production and biodiversity conservation (Sarkar et al. 2017).

13.4.3 Sub-objectives

Identifying and structuring the sub-objectives for each fundamental objective proved to be a much more complex—and interesting—process than identifying the fundamental objectives, with multiple revisions and reiterations required to incorporate input from the affected local communities. Here attention will be restricted to the structure of the objectives hierarchy under one fundamental objective since our only purpose here is to explicate the role of deliberation in its construction rather than record all results of this exercise (which can be found in Sarkar et al. 2017). Figure 13.3 shows the structure subsumed under the fundamental objective of spatial configuration which is selected for our discussion because it is the most complex of the four structures subsumed under the four fundamental objectives.

As deliberation continued after the initial Jakarta meeting, the structure that finally stabilized as Fig. 13.3 became increasingly more complex; this development illustrates the point being emphasized here, that is, the determinative role played by deliberation even in a decision exercise that fundamentally embraced formal methods. Given what was said earlier, there were three sub-objectives at the first level: biodiversity-related spatial preferences, Medco's spatial preferences, and community spatial preferences. The first and third were included from the beginning of the discussions; the second arose after Medco reviewed some initial solutions and found them to be infeasible. (Note how the stakeholders can initiate at any stage a full reiteration of the component of the planning process that includes the use of formal methods.)

The biodiversity-related spatial preference's sub-objectives were decomposed into two others at the next level: shape and connectivity which were the only two spatial criteria that the stakeholders decided as relevant. As expected, compact shapes were preferred. Connectivity was interpreted as the selection of a contiguous body of cells connecting to existing official nature reserves that adjoined the concession. The Medco spatial preferences also had two sub-objectives at the next level, both viewed as types of alignment. The first was "zone alignment" which concerned which part of the concession areas were to be conserved and prevented from plantation (Medco preferred that these be on land less suitable for plantation farming because of soil and hydrological characteristics). The second was "distance alignment" and referred to the closeness of conservation areas to the concession boundary (easing transportation) and distance to the one of the community areas (which was considered desirable). Finally community spatial preferences had a single sub-objective, closeness of conservation areas to each community. This was believed to be important because it indicated, among other things, the ease with which older individuals could access relatively intact forests to collect resources.

13.4.4 Weights

Finally, the elicitation of weights and ensuring the satisfaction of the preference and difference independence conditions also required stakeholder deliberation. In the process, novel methodologies for weight elicitation had to be devised—for further discussion, see Sarkar et al. (2017). Final results in the form of a portfolio of plans were provided to Medco for further stages of deliberation. To the best of my knowledge none of the plans we produced have as yet been implemented.

13.5 Advantages

The discussion so far has emphasized the extent to which deliberation must be integrated into the use of formal methods when the latter are deployed appropriately. However, should the tone of that discussion give rise to the impression that the use of formal methods generates comparatively insignificant advantages over a purely deliberative approach, this section will list four advantages that make the former valuable, and probably unavoidable in many contexts. This section will also help explain why I continue to advocate the use of formal methods in environmental decisions even though I share much of Norton's skepticism about their value and recognize the potential for their misuse.

13.5.1 Large Alternative Sets

Deliberation alone becomes intractable if the number of alternatives to be considered is large unless the alternative set has some internal structure that lets large subsets be considered as modules for this purpose. There is no reason to expect that, in most contexts, the alternative set has such a convenient aggregative structure. In the case of conservation area network selection, no such structure has been identified in spite of many efforts (Sarkar et al. 2006; Ciarleglio et al. 2009a, 2010). If a land or seascape is partitioned into just a few hundred potential conservation areas, the number of feasible networks of these areas will typically be much greater than 10 million, far too many to be assessed through deliberation.

In contrast, because they are generally algorithmic, formal methods can be encoded into software programs that can generate and assay millions of alternatives in seconds, and typically find a tractable portfolio of the best of them within the time frame of a participatory planning session of stakeholders. Within systematic conservation planning this advantage led to the creation and adoption of software decision support tools as early as the 1980s (Margules et al. 1988). These original tools were restricted to the identification of notional conservation area networks on the basis of biodiversity representation alone. As the example discussed in this paper shows, tools currently in use can provide support in much more complex decision contexts.

13.5.2 Consequences of Constraints

It is a truism that all practical decisions are made under constraints. Some constraints are hard (that is, they can never be violated) and fixed (that is, they do not change during the decision process). In many—but not all—cases such well-behaved constraints can be accommodated through deliberation (with exceptions generated, for instance, when there are a large number of complex constraints). However, many constraints are soft (that is, they allow some exceptions) and flexible in the sense that they change, most notably during stakeholder discussions of tradeoffs.

Formal methods, because of their speed and the availability of computational decision support tools, are about the only way in which soft or flexible constraints can be reasonably incorporated into a decision analysis in a limited amount of time, for instance, during a planning session. These methods permit a comprehensive exploration in a limited amount of time of the effects of varying such constraints systematically.

13.5.3 Repeatability

Deliberative decision making processes often suffer from the problem that stakeholders not only negotiate preferences, which is appropriate, but also trade intuitions about facts, for instance, how a plan is likely to perform including how it may impact future preferences. Philosophical pragmatists have long been exposed to the abuse of intuitions in contemporary metaphysics and epistemology and are typically wary of intuition-based narratives which may be valuable in suggesting future inquiry but have no epistemic or ontological warrant on their own. The more practically relevant problem in our context is that reliance on intuitions, unless they get explicitly formalized, do not lead to determinate outcomes in the sense that the result of one decision making session may disagree with that of another supposed to be based on the same information and values.

Because of their algorithmic nature formal methods avoid this problem. The generation of results is repeatable; this is a close analog of the situation in experimental science where a well-designed experiment must be replicable. This aspect of formal analysis also ensures that the effects of varying any aspect of the decisionmaking process (weights, the structure of the objectives hierarchy etc.) can be studied in a controlled fashion. Once again, there is a compelling analogy to experimental science. In this somewhat unexpected way the use of formal methods provides support for Norton's emphasis on experimentalism in environmental decisions.

13.5.4 Flexible Portfolios and Further Deliberation

Formal methods allow the generation of a large number of best-performing alternatives which, ideally, become part of a portfolio returned to stakeholders for further refinement in order to arrive at a final decision. This last stage of deliberation allows the incorporation of relevant features which are not amenable to representation as formal criteria that can be used to order (or otherwise quantify) alternatives. These include many socio-political objectives, for instance, the cultural value (including religious or wilderness value) of a spatial unit. These final discussions augment the role played by deliberation and is, therefore, a critical part of the decision process as envisioned here (Sarkar 2005). (In the Merauke example we, as decision analysts, submitted such a portfolio to CI to be conveyed to Medco and that constituted the end of our role.)

Elements of the portfolio can not only be the best-performing alternatives under one set of assumptions but can also be ones generated when assumptions are varied, usually slightly, as is often done to test their robustness. In this sense, the formal decision process enables flexibility. The same point was also made earlier during the discussion of constraints on decisions.

Creating such portfolios of potential alternatives, testing robustness, and achieving flexibility using deliberation is almost always so time-consuming as to be practically impossible without the use of formal methods.

13.6 Objections and Responses

However, there are also several problems with the use of formal methods. It behooves us to recognize and explicitly address them. These will be noted next along with potential responses where available. In general, even when there is no fully adequate response to an objection, it will be seen that purely deliberative decisions fare little better.

13.6.1 Biocentric Intuitions

Within environmental philosophy there has been a longstanding and well-known dispute between anthropocentrists and biocentrists (along with ecocentrists) (Norton 1986, 1987; Sarkar 2005, 2012b). The formal approach advocated here falls squarely within the framework of anthropocentrism. Since Norton and I both endorse a weak (or tempered) form of anthropocentrism, this issue does not divide us and we could well entirely ignore the dispute with biocentrism especially since that dispute has never seemed to me of compelling importance from a pragmatist perspective.

Nevertheless, it is worth pointing out that the procedure used here at least in part captures one of the hallowed intuitions of biocentrism, that certain entities such as biodiversity have intrinsic value in the sense that they are valuable for themselves and not because of what value they contribute to some other entity (McShane 2007). The stakeholders of Merauke would agree: that is why biodiversity is a fundamental objective in their objectives hierarchy. Arguably, setting up an objectives hierarchy formalizes intrinsic value attributions when it explores and structures the values of a community: the entities that have intrinsic value are the ones that are exalted as fundamental objectives. (I doubt that many biocentrists will accept this interpretation but this remains the best sense I have been able to make of their perspective.)

13.6.2 Stakeholder Legitimacy

The procedure described above enshrines stakeholders for decision making but does not address the problem of who is a legitimate stakeholder (Sarkar 2012b). This is a serious issue especially as, in many environmental contexts including that of biodiversity conservation, there are conflicts between Northern and Southern value systems and between local and global value systems (Vermeulen and Koziell 2002)—this is a problem that has repeatedly been emphasized by political and social ecologists and anthropologists (Guha 1994; Brechin et al. 2002). It has also been extensively documented in sophisticated journalistic explorations of conservation, perhaps most famously by Dowie (2009). It remains a cause of concern that environmental pragmatists, themselves rooted in the North, have paid very little attention to his problem (and, indeed, have often been negligent of issues of environmental justice).

Turning to Merauke, why is Conservation International, based in the United States (though acting in our case through an Indonesian subsidiary) and embodying Northern values (that are controversially supposed to be global) a legitimate stakeholder? There is no easy answer. CI's role is not unusual either for it as an organization or for other such Northern organizations (Dowie 2009).

My final point though, is that insisting on deliberation alone in no way mitigates this problem. The legitimacy of those allowed to participate in deliberations—in other words, the stakeholders—remains equally open to question. Related to this point is that both formal and deliberative decision processes potentially fall afoul of being "top-down" in many contexts when external agents are involved; probably all that can immediately be hoped for short of structural reform of contemporary environmentalism is that these agents be sensitive to this problem which has no pat solution. In the long run, it behooves environmentalists to expand their agenda to embrace political challenges of this type whether they involve ranchers resisting demands of the Endangered Species Act in Texas or forest dwellers resisting onerous policies promoted by Project Tiger in India—for details of these cases, see Sarkar (2012b) and the references therein.

13.6.3 Asymmetries of Power

The last problem discussed underscores a more general one that there are asymmetries of power that our process does not address. Beyond the issues already raised, there are additional asymmetries that arise in contexts (including Merauke) in which stakeholders are entirely unfamiliar with the elicitation procedures of formal decision analysis some of which are complex (including those used to judge preference and difference independence). Enabling enough familiarity to ensure informed decisions may take more time than that available within the planning horizon. Finally, should the discussions broach the problem of the interests of future generations there is no credible way to incorporate these in the formal decision process discussed here; these are not stakeholders who can be consulted.

There should be little doubt that deliberative decision protocols have an advantage over formal ones with respect to this problem though, if these decisions are mediated by external actors, there always remains room for worry that decisions are (perhaps inadvertently) manipulated and are not truly informed. (This is yet another aspect of the "top-down" problem.) No matter what decision procedure is used, decision analysts should remain sensitive to these problems.

13.6.4 Wicked Problems

Norton (2005 and elsewhere) and others, including many of those influenced by him, emphasize that environmental problems are "wicked," that is, devoid of even a clear formulation let alone having exact criteria to determine the quality of their potential solutions. Problems of this sort cannot be fed into an algorithmic procedure. That assessment is indeed correct but the decision process described earlier begins with a deliberative stage that should be as long as is necessary to achieve a sufficiently clear and tractable formulation. It leaves open the possibility that a multiplicity of formulations may emerge to be explored from the beginning or even after formal analysis of some of them leaves stakeholders dissatisfied with outcomes.

Similarly, the process envisioned here recognizes the problems of unstable preferences and uncertain weights. Part of the advantage of having formal methods incorporated into software decision support tools (as noted earlier) is the speed of reiterated analyses when assumptions change—and that includes problem formulation besides unstable preferences and uncertain weights. In the case of the last two issues, a standard strategy is robustness analysis (mentioned earlier): tests of the stability of solutions across a range of preferences and weights. (Stable results allow us to sidestep these problems without attempting to achieve precision in many situations in which achieving precision is difficult and perhaps illegitimate, that is, the lack of precision does not reflect agents' ignorance about facts of the world but are features of their preferences as felt by them).

13.7 Conclusions

The aim of this chapter has been to argue and illustrate by example that formal decision support protocols, when deployed appropriately, do not in any way replace what should be achieved using deliberation. Thus, the use of formal methods does not in any way denigrate or de-emphasize the central role that Norton (and I) like to give to participatory processes. The trouble is that reasonably sophisticated decision analyses such as our Merauke analysis may not be representative of how formal decision support tools tend to be used. That analysis included the involvement of one of the most prominent figures in formal decision analysis (Dyer) and another (me) with a professional interest in decision theory independent of its role in environmental contexts. That led to a recognition of the subtleties of the process (for instance, in the interpretation of weights and satisfaction of the independence conditions) besides an insistence that adequate time be set aside for the deliberative elicitation process that was used to construct and refine the objectives hierarchy. To make the last point in another way: the problem of exploring and mapping the value structures of the stakeholders was taken as a critical component of the analysis and adequate resources were devoted to it. It took much more time than the formal analysis.

The contrast here is with situations in which linear value (or utility) functions are constructed by decision analysts and, at best, weights on the criteria are assigned in the abstract without paying attention to the requirement that they be sensitive to the potential range of values that can arise when they are limited to the given feasible alternatives set (and I request readers to send me examples where they are explicitly addressed). Typically, the questions of preference and difference independence are not even broached. Given the software decision support tools and other computational resources available, it is then typically trivial to compute the value functions (and GIS tools are always available to produce images that would once have been an advertiser's dream).

The trouble is that these value functions are illegitimate even though, in many circumstances the subtleties emphasized here may only make marginal differences to the ranking of alternatives (and we do not know a priori how much difference the subtleties would make). Norton has emphasized an adaptive experimentalist approach to decisions and the iterative process I have outlined fits that mold. Formal methodologies that use decision support tools mechanically violates this goal. However, what is most troubling (to me and probably to Norton) is that this process marginalizes the stakeholders who would ideally form a community united by the goal of improving their lives through improved environmental interactions. For Norton, it violates important procedural norms including a political commitment to participatory democtratic decision-making. For me, it also is a recipe for disaster when it comes to adaptive implementation, management, and monitoring because such "top-down" decisions typically generate no personal commitments to success and, to use an ex-Marxist phrase, alienates those who have the most important stake in the relevant futures. A critical goal for those of us who advocate the use of formal methods is to try to prevent their misuse. I hope that this chapter contributes to that task.

Acknowledgements The ideas reported here have been developed through discussions over several decades with a large number of collaborators: Jim Dyer, Chris Margules, and Victor Sànchez-Cordero deserve special mention. Laboratory members who have contributed to this work include Trevon Fuller, Justin Garson, Patricia Illoldi-Rangel, Alex Moffett, and Chris Pappas. For Figs. 13.2 and 13.3, thanks are due to Michael Ciarleglio. For comments on an earlier draft thanks are due to Ben Minteer and Bryan Norton.

References

Arrow, K.J., and H. Raynaud. 1986. *Social choice and multicriterion decision-making*. Cambridge, MA: MIT Press.
Borgerhoff-Mulder, M., and P. Coppolillo. 2004. *Conservation: linking ecology, economics and culture*. Princeton, N.J.: Princeton University Press.
Brechin, S.R., P.R. Wilshusen, C.L. Fortwangler, and P.C. West. 2002. Beyond the square wheel: Toward a more comprehensive understanding of biodiversity conservation as social and political process. *Society and Natural Resources* 15: 41–64.
Callicott, J.B., L.B. Crowder, and K. Mumford. 1999. Current normative concepts in conservation. *Conservation Biology* 13: 22–35.
Ciarleglio, M., J.W. Barnes, and S. Sarkar. 2009a. ConsNet: New software for the selection of conservation area networks with spatial and multi-criteria analyses. *Ecography* 32: 205–209.
Ciarleglio, M., O. Wang, and S. Sarkar. 2009b. *Area prioritization for medco concession in merauke: Report to conservation international.* Report—Technical Note 63. Austin: Biodiversity and Biocultural Conservation Laboratory, University of Texas.
Ciarleglio, M., J.W. Barnes, and S. Sarkar. 2010. ConsNet—A tabu search approach to the spatially coherent conservation area network design problem. *Journal of Heuristics* 16: 537–557.
Conservation International. 2010. *Optimizing conservation and production—a collaboration between conservation and business. Report*. Jakarta: Conservation International-Indonesia.
Dowie, M. 2009. *Conservation refugees: The hundred-year conflict between global conservation and native peoples*. Cambridge, MA: MIT Press.
Faith, D.P. 1995. *Biodiversity and regional sustainability analysis. Report*. Lyneham, ACT: Commonwealth Scientific and Industrial Research Organisation.
Figuera, J., S. Greco, and M. Ehrgott (eds.). 2005. *Multiple criteria decision analysis: State of the art surveys*. Berlin: Springer.
Guha, R. (ed.). 1994. *Social ecology*. New Delhi: Oxford University Press.
Keeney, R.L. 2007. Common mistakes in making value tradeoffs. *Operations Research* 50: 935–945.
Keeney, R.L. 1992. *Value-focused thinking*. Cambridge, MA: Harvard University Press.
Keeney, R.L., and H. Raiffa. 1993. *Decisions with Multiple objectives: Preferences and value tradeoffs*, 2nd ed. Cambridge, UK: Cambridge University Press.
Margules, C.R., and R.L. Pressey. 2000. Systematic conservation planning. *Nature* 405: 243–253.

Margules, C.R., and S. Sarkar. 2007. *Systematic conservation planning*. Cambridge, UK: Cambridge University Press.
Margules, C.R., A.O. Nicholls, and R.L. Pressey. 1988. Selecting networks of reserves to maximise biological diversity. *Biological Conservation* 43: 63–76.
McShane, K. 2007. Why environmental ethics shouldn't give up on intrinsic value. *Environmental Ethics* 29: 43–61.
McShane, T.O., P.D. Hirsch, T.C. Trung, A.N. Songorwa, A. Kinzig, B. Monteferri, D. Mutekanga, H.V. Thang, J.L. Dammert, M. Pulgar-Vidal, M. Welch-Devine, J. Peter Brosius, P. Coppolillo, and S. O'Connor. 2011. Hard choices: Making trade-offs between biodiversity conservation and human well-being. *Biological Conservation* 144: 966–972.
Moffett, A., and S. Sarkar. 2007. Incorporating multiple criteria into the design of conservation area networks: A minireview with recommendations. *Diversity and Distributions* 12: 125–137.
Moilanen, A., K.A. Wilson, and H. Possingham (eds.). 2009. *Spatial conservation prioritization: Quantitative methods and computational tools*. Oxford, UK: Oxford University Press.
Norton, B.G. (ed.). 1986. *The preservation of species*. Princeton, NJ: Princeton University Press.
Norton, B.G. 1987. *Why preserve natural variety?*. Princeton, NJ: Princeton University Press.
Norton, B.G. 2003. Defining biodiversity: Do we know what we are trying to save? Namkoong Family Lecture Series. Vancouver: University of British Columbia Department of Forestry.
Norton, B.G. 2005. *Sustainability: A philosophy of adaptive ecosystem management*. Chicago, IL: University of Chicago Press.
Pressey, R.L., and M.C. Bottrill. 2009. Approaches to landscape- and seascape-scale conservation planning: Convergence, contrasts and challenges. *Oryx* 43: 464–475.
Rothley, K.D. 1999. Designing bioreserve networks to satisfy multiple, conflicting demands. *Ecological Applications* 9: 741–750.
Saaty, T. 1980. *The analytic hierarchy process: Planning, priority setting, resource allocation*. New York: McGraw-Hill Inc.
Sarkar, S. 2005. *Biodiversity and environmental philosophy*. New York: Cambridge University Press.
Sarkar, S. 2012a. Complementarity and the selection of nature reserves: Algorithms and the origins of conservation planning, 1980–1995. *Archive for History of Exact Sciences* 66: 397–426.
Sarkar, S. 2012b. *Environmental philosophy: From theory to practice*. Malden, MA: Wiley-Blackwell.
Sarkar, S. 2014. Biodiversity and systematic conservation planning for the twenty-first century: A philosophical perspective. *Conservation Science* 2: 1–11.
Sarkar, S. 2017. Approaches to biodiversity. In *Routledge handbook of philosophy of biodiversity*, ed. J. Garson, A. Plutynski, and S. Sarkar, 43–55. New York: Routledge.
Sarkar, S., and J. Garson. 2004. Multiple criterion synchronization for conservation area network design: The use of non-dominated alternative sets. *Conservation and Society* 2: 433–448.
Sarkar, S., and P. Illoldi-Rangel. 2010. Systematic conservation planning: An updated protocol. *Natureza & Conservação* 8: 19–26.
Sarkar, S., J.S. Dyer, C. Margules, M. Ciarleglio, N. Kemp, G. Wong, D. Juhn, and J. Supriatna. 2017. Developing an objectives hierarchy for multi-criteria decisions on land use options, with a case study of biodiversity conservation and forestry production from Papua, Indonesia. *Environment and Planning B: Planning and Design* 44: 464–485.
Sarkar, S., N.C. Parker, J. Garson, A. Aggarwal, and S. Haskell. 2000. Place prioritization for Texas using GAP data: The use of biodiversity and environmental surrogates within socioeconomic constraints. *Gap Analysis Program Bulletin* 9: 48–50.

Sarkar, S., R.L. Pressey, D.P. Faith, C.R. Margules, T. Fuller, D.M. Stoms, A. Moffett, K. Wilson, K.J. Williams, P.H. Williams, and S. Andelman. 2006. Biodiversity conservation planning tools: Present status and challenges for the future. *Annual Review of Environment and Resources* 31: 123–159.

Takacs, D. 1996. *The Idea of Biodiversity: Philosophies of Paradise*. Baltimore, MD: Johns Hopkins University Press.

Vermeulen, S., and I. Koziell. 2002. *Integrating global and local values: A review of biodiversity assessment. Report*. London: International Institute for Environment and Development.

Sahotra Sarkar is Professor of Integrative Biology and of Philosophy at the University of Texas at Austin and Distinguished University Fellow, Presidency University, Kolkata. He is the author of *Genetics and Reductionism* (1998), *Molecular Models of Life* (2004), *Biodiversity and Environmental Philosophy* (2005), *Doubting Darwin? Creationist Designs on Evolution* (2007), *Systematic Conservation Planning* (2007; with Chris Magules), and *Environmental Philosophy: From Theory to Practice* (2012).

Chapter 14
Ecology, Economics and Ethics: The Three Es Required for the Sustainable Management of Wild Sentient Species

Bruce Warburton and Dean Anderson

Abstract Wild sentient species (primarily vertebrates), both introduced and indigenous, are managed for a range of reasons and often using lethal methods. Use of these tools often raises ethical challenges and uncertainty for wildlife managers, and when combined with ecological and economic uncertainties, pest control programs can fail because these uncertainties are not fully evaluated or accounted for in program plans. Bryan Norton has frequently stated that environmental management problems are often "wicked" problems, and participants in the debate bring with them a wide range of values and vocabularies. He and others have suggested that such problems might best be solved within a framework of pragmatism implemented through an empirically based, multi-criteria adaptive management system. In this chapter, we propose the adoption of a probabilistic modelling approach that could help wildlife managers frame and formalize an adaptive management approach that integrates the 3Es: ecology, economics, and ethics, one that maximizes the probability of achieving sustainable and effective wildlife management outcomes.

Keywords Adaptive management · Animal welfare · Environmental ethics Invasive species · Pest management

14.1 Introduction

Wild sentient species (primarily vertebrates), both introduced and indigenous, are managed for a range of reasons. These reasons include their undesirable impacts on indigenous biodiversity, agricultural production and infrastructure, their maintenance and transmission of disease, and management for commercial and

B. Warburton (✉) · D. Anderson
Manaaki Whenua-Landcare Research, PO Box 69040, Lincoln 7608, New Zealand
e-mail: warburtonb@landcareresearch.co.nz

D. Anderson
e-mail: andersond@landcareresearch.co.nz

S. Sarkar and B. A. Minteer (eds.), *A Sustainable Philosophy—The Work of Bryan Norton*, The International Library of Environmental, Agricultural and Food Ethics 26, https://doi.org/10.1007/978-3-319-92597-4_14

recreational harvest. Although some management strategies use nonlethal methods, most do not. As a result, wildlife managers are often challenged to justify their policies and actions, not just on the basis of ecology and population dynamics—disciplines they are familiar with—but also on their ethical acceptability. Unfortunately, most wildlife managers are poorly equipped to enter into an informed discussion on what are often complex ethical and philosophical issues. For example, choosing and defending lethal control methods poses significant ethical challenges, and defending why populations need to be managed at all can raise significant philosophical challenges.

As Bryan Norton has frequently stated, environmental management problems are often "wicked" problems, and participants in the debate bring with them a wide range of values and vocabularies. He and others have suggested that such problems might best be solved within a framework of pragmatism that endorses ethical pluralism, implemented through an empirically based, multi-criteria adaptive management system. In this chapter, we acknowledge that most decisions that wildlife managers make related to ecology, economics, and ethics have varying degrees of uncertainty. We explore whether a probabilistic modelling approach can help wildlife managers frame and formalize an adaptive management approach that integrates the 3Es: ecology, economics, and ethics. It is hoped that such formalization will encourage wildlife managers to consider the 3Es in order to maximize the probability of achieving sustainable and effective wildlife management outcomes. However, it is also necessary to acknowledge Norton's significant contribution to this topic and his caution that wicked problems (such as managing wildlife) cannot be formulated as problems of optimization nor solved using computational or algorithmic processes. So in recognition of this caution, we propose the formalization recommended in this chapter as one input to the decision making process that is needed to achieve effective wildlife management.

14.2 Managing Vertebrate Species

Vertebrate species—mammals, birds, reptiles, amphibians, and fish—are managed extensively around the world with the following objectives: (1) to re-establish or enhance the abundance of threatened species (Lloyd and Powlesland 1994; Freifeld et al. 2016); (2) to eradicate pest species (Parkes et al. 2010; Jones et al. 2016); (3) to reduce population abundance to mitigate undesirable impacts (Choquenot and Parkes 2001; Reddiex et al. 2006; Moseby and Hill 2011); or (4) to increase numbers or change the population structure to enhance recreational or commercial exploitation (Bishop et al. 2005; Perez-Espona et al. 2009). Management actions may be directed at native or introduced species, with ecological, social (including cultural), political, and economic contexts influencing the choice of action. Apart from the first reason listed above for actively managing vertebrate species (i.e., conserving threatened species), management most often requires some form of lethal intervention. Note that even though the last action—to increase numbers—

may not always directly involve lethal management, the resulting recreational or commercial exploitation does.

Whatever the reason, wildlife managers tasked with deciding what management actions are needed are challenged by a range of uncertainties related to the ecological, economic, and ethical components of their decision making (Rotmans and van Asselt 2001; Ellis and Elphick 2007; Warburton and Norton 2009).

14.3 Ecological Uncertainty

Managing wildlife can sometimes be inherently complex, because individual species are part of broader multi-species communities, which in turn contribute to higher-level ecosystem processes. Making management decisions related to these complex environments is further complicated by typically high levels of uncertainty. There are several typologies of uncertainty, which can be both ontological (i.e. inherent variability) and epistemological (i.e. level of knowledge) (Rotmans and van Asselt 2001). Knowledge uncertainty can, for example, encompass a lack of knowledge about ecological processes, model simplicity not fully representing all the components of the system being modelled, or error in measurement and therefore data. Decision uncertainty can also result from ambiguity related to the differing values of stakeholders and managers not being fully aware of such differences. Finally, linguistic uncertainty can result from language being vague, ambiguous, and (often) context dependent (Ascough et al. 2008).

There are many examples of successful management actions delivering desired outcomes. As examples, long-term suppression of ship rats (*Rattus rattus*), brushtail possums (*Trichosurus vulpecula*) and stoats (*Mustela erminea*) resulted in the significant population recovery of threatened bird species such as Mohua (*Mohoua ochrocephala*) and yellow-crowned parakeet (*Cyanoramphus auricepsreduction*) (O'Donnell and Hoare 2012). Introduction of biocontrol agents into Australia to control rabbits has been estimated to have produced an economic benefit of A$70 billion for agricultural industries over the last 60 years (Cooke et al. 2013), and significant reductions in numbers of cattle herds infected with bovine tuberculosis has been achieved in New Zealand through reducing possum numbers (Livingstone et al. 2015). Lastly, introduced species [e.g. rodents and feral cats (*Felis cattus*)] have been eradicated from many islands resulting in signifcant conservation gains (Jones et al. 2016).

There are, however, also examples of failed programs. Failures occurred because of managers having insufficient knowledge about:

(1) species interactions (Courchamp et al. 1999; Pech and Maitland 2016), resulting in the release of meso-predators whose negative impact is as big as or greater than that of the species controlled. For example, Ruscoe et al. (2011) reported that following aerial 1080 (poisoning) operations that targetted both brushtail possums and ship rats, ship rats recovered more quickly than possums

because of competitive release and recovered to twice the level found in a non-treatment site where no control had been implemented.

(2) the relationship between pest density and impacts (Choquenot and Parkes 2001; Moberly et al. 2004; Singleton et al. 2007), resulting in an inadequate reduction in pest density. For example, in New Zealand the juveniles of the flightless kiwi (Apteryx spp) are very susceptible to stoat predation, and although considerable trapping effort can be made to reduce stoat numbers, if they are not reduced to very low numbers kiwi chicks do not survive (Robertson et al. 2016).

(3) the effectiveness of the control tools used or the strategy used to apply them (Kirkwood et al. 2014; Haniza et al. 2015). For example, some rodents in Europe have developed resistance to first generation anticoagulant poisons, and although these poisons initially worked effectively their efficacy has declined (Hadler and Buckle 1992).

Although wildlife managers are confronted with significant ecological uncertainty, they often do not have the luxury of time to wait for research to ameliorate that uncertainty, because of the potentially significant opportunity costs (e.g., species extinction) associated with delaying action (Margules and Sarkar 2007; Grantham et al. 2009). Consequently, managers need tools to guide the setting of decision triggers based on the best available science (Addison et al. 2016; Cook et al. 2016).

14.4 Economic Uncertainty

Managing wildlife has associated costs that managers need to take into account when planning wildlife programs, either to decide how best to allocate limited funds across priority areas (Moore et al. 2004; McBride et al. 2005; Holmes et al. 2016) or to decide whether the funding is sufficient to undertake a program of work to achieve the desired outcome. There is a range of fixed and variable costs associated with managing wildlife (particularly invasive species management), including planning, implementation, non-target mitigation, addressing public needs, monitoring, and compliance (Holmes et al. 2016). Also, if cost–benefit analyses are required, there can be the added complexity associated with monetizing conservation assets and benefits (Naidoo et al. 2006; Shwiff et al. 2013).

Wildlife managers are faced with economic uncertainty both at the planning stage (e.g., when costing a project) and during its implementation (e.g., due to unforeseen issues that increase costs). For example, unpublished data from competitively tendered invasive brushtail possum control operations in New Zealand aimed at eradicating bovine tuberculosis from livestock (Nugent et al. 2015) show that, for one contract, managers had to select from tender prices that ranged from $NZ92000 to $NZ270,000 (a three-fold variation), and for another contract the agreed price of $NZ288,000 was increased to $NZ296,000 because of unforeseen operational issues. Similarly, for island eradication of rodents, reported costs per hectare for the implementation of aerial control ranged from $US84 to $US4,615, and for ground control from $US78 to $US12,047 (Holmes et al. 2016). As well as coping with ecological uncertainty, wildlife managers must make decisions incorporating economic uncertainty. Additionally, these two factors do not act independently, with increasing ecological uncertainty resulting in increased economic uncertainty. For example, if the planned control effort (and its associated cost) does not achieve the desired outcome then additional control and funding will be required. Conversely, if control achieves more than required then there could be significant opportunity costs (Choquenot and Parkes 2001).

14.5 Ethical Uncertainty

Wildlife management programs often generate strongly polarized dialogue because of the diverse value sets held by the various stakeholders, including wildlife managers, ecologists, philosophers, advocates for animal welfare and animal rights, and policy makers. Each have their own vocabularies, which further challenges effective communication (Bracken and Oughton 2006). Consequently, wildlife managers need to be aware of the different values that stakeholders hold, and recognize this variation in values as another level of uncertainty they need to account for.

Wildlife managers faced with manipulating vertebrate animal (i.e., sentient) species, especially if the use of lethal methods is proposed, have both animal welfare and ethical issues to address. All lethal control methods have welfare costs (i.e., they cause some level of pain or distress), ranging from fast-acting vertebrate toxic agents (VTAs) and kill traps that pose the least welfare cost (Warburton et al. 2008), to extreme welfare costs caused by some anticoagulant poisons (Mason and Littin 2003). Lethal control operations may kill high numbers of animals; for example, the number of rodents poisoned each year in the UK alone is estimated to be about 20 million (Fox and MacDonald 2012). A single aerial poison operation targeting brushtail possums and ship rats in New Zealand may kill as many as 800,000 individual animals in one or two nights. Although the number of animals killed in control operations is typically not accounted for when planning a control program, managers should consider which control strategies will affect the least number of individuals over time yet still ensure successful achievement of the desired outcome (Warburton et al. 2012).

As a result of these issues, wildlife managers often get challenged and can be the target of vociferous opposition from ethical opponents (Vucetich and Nelson 2007). Although wildlife managers are often passionate about their role in protecting conservation or production values, they are often placed on the moral back foot by animal rights advocates and moral philosophers, who claim the moral high ground by advocating nonlethal control methods (Duka and Masters 2005; Winograd 2007). Such debates have prompted an increasing focus on the welfare and ethical issues associated with wildlife management and pest control, from both those involved in the industry and externally (Dubois et al. 2017).

There has been considerably more effort put into addressing on-the-ground animal welfare concerns (i.e., practical issues related to "what" you are doing and minimizing harm) than deeper normative issues (i.e. more the "why" you are doing it), and it is these more foundational ethical issues (particularly killing) that wildlife managers are poorly trained to address. Making this more challenging is the need to consider killing as a control option and the acceptable level of pain or suffering if killing is not free of pain or distress. Most animal welfare scientists agree that killing per se, although an ethical issue, is not a welfare issue—at least if done painlessly (Broom 2011). For example, if a wildlife manager had an overabundant feral cat population to manage and had access to a cost-effective humane lethal control method and a nonlethal but less cost-effective method available, could the manager ethically justify using the lethal method? Such a situation arose when planning the eradication of feral cats from San Nicolas Island (Hanson et al. 2010). In response to demands from humane societies, the feral cats were live captured and then translocated to the US mainland to be held in captivity until they died of natural causes. The alternative, given the feral cats were live trapped, was to euthanize them humanely at their capture site and thus avoid the additional welfare costs of translocation and habituation to captivity. So why could the managers of the eradication program presumably not defend the lethal option? Norton (1995) suggested that sustaining biodiversity requires a shift from a focus on individuals to a focus on processes, and that our interactions with animals in the wild should take on a moral dimension only at the population and species level, not at the individual level (while still ensuring negative welfare impacts on individuals are minimized).

A second key issue is how much pain or suffering and what duration is acceptable if killing is not free of pain or distress. Utilitarians attempt to assess the benefits compared to the costs, but in most ecological cases, although the welfare costs might be able to be accounted for, it is often extremely difficult to measure the benefits, especially when they may be to non-sentient organisms such as plants (Warburton and O'Connor 2004). Thus for pest control to protect biodiversity values, one would have to compare the moral importance of harm to the pest, if control is undertaken, to the moral importance of biodiversity loss (of both sentient and non-sentient species). Because this requires subjective judgment of the values (including moral values) of the control program objectives, there are no clear thresholds or limits for what level of harm might be acceptable. To assist ecologists to deal with this dilemma, Parris et al. (2010) provided a framework for assessing the ethical and other trade-offs. Additionally, it has also been suggested (Warburton and Choquenot 1999) that the welfare impacts of pest control should be judged in the context of natural processes rather than against the responsibility of care that has been accepted for domestic animals. This approach aligns with that of pragmatic contextualism advanced by Minteer et al. (2004), who suggest that such an approach links environmental ethics with concrete problems of environmental practice.

14.6 Integrating Ecological, Economic and Ethical Uncertainty

Pragmatic management strategies for wildlife issues hinge on the success of the 3Es for the duration of the project; that is, the continued success of ecological outcomes, financial and institutional support, and broad acceptance of the ethics of actions undertaken. Given the inherent complexity and associated uncertainty, management should modify strategies as experience and new information are obtained. Adaptive management guided by Bayesian uncertainty modelling is an evidence-based mechanism for arriving at and revising strategies to maximize the probability that management actions will achieve their stated objectives. The concept of adaptive management (Holling 1978; Walters 1986) describes a general process for managers and scientists to collaborate to trial and learn from experience in natural resource management, and Norton (1995, 2005, 2015) has long been a strong proponent of such knowledge-based management. Applications of adaptive management have focused on successful ecological outcomes, and continued financial and institutional support (Walters and Green 1997; Bennett et al. 2005; Geremia et al. 2014), but have achieved only mixed success (Allan and Curtis 2005; Walters 2007; McCook et al. 2010). To minimize the risk of failure, it is recommended that those commissioning the work participate in some way (e.g., design, collecting data, etc.) to ensure ownership or championing of the evidence-based management approach (Walters 1986; Walters and Holling 1990; Allen 2007; Walters 2007).

Adaptive management provides a natural framework for formally and quantitatively incorporating the likely acceptance of the ethics of management actions into strategy development, especially when applied to the management of wild sentient species. Despite its conceptual appeal for wildlife management (Shea et al. 2002; McCarthy and Possingham 2007), the practical application of adaptive management is largely absent in this area (however, see Parkes et al. 2006; Ramsey et al. 2012). Conceptually it is simple and general (perhaps vague), but this allows a management-science collaboration to apply it to specific systems and problems, where it gains clarity. The problem is that it requires two-way participation and financial support. Unfortunately, these are difficult attributes to attain and maintain over the life of an environmental management program (Botrill et al. 2008; Walters 2007).

Here we describe a conceptual approach for incorporating the 3Es into an adaptive management program informed by Bayesian uncertainty modelling. The overall success of a management program for wild sentient species is contingent on the successful and sustained achievement of ecological outcomes, financial and institutional support, and the acceptance of the ethics of actions. The failure of one or more of these necessarily results in program failure. Quantitative predictions of the probability of successful outcome of each of the 3Es are made, and the probability of the overall success of the program is simply the product of each of the 3E component probabilities:

$$P(overall\,success) = P(ecology)P(economics)P(ethics) \quad (14.1)$$

where the component probabilities are conditionally independent of each other. When the probabilities are inter-related, a covariance structure can be incorporated (Gardner et al. 2000).

The individual 3E-component probabilities are derived from data-driven or expert-parameterized models. Data-driven models are preferred and should replace expert-parameterized models as a management program progresses over time. That is, data models may not be available when a program is initiated, but over time data should be collected on the ecological, economic and ethical acceptance processes, and analyzed to estimate parameter distributions. Bayesian statistics are desirable because they can take advantage of existing independent parameter estimates and expert insight of the system, and they update parameter distributions as more data are collected (Gelman et al. 2004). Parameters are incorporated using probability distributions, not point estimates, to capture uncertainty in our understanding of the processes and predictions. Models are imperfect and we need to know the level of uncertainty in model predictions, and make management decisions accordingly.

Here we explore three simple hypothetical management scenarios to illustrate how the assessment of 3E-component probabilities and the corresponding overall probability of project success can lead to informed modifications in management. Consider the case in which the mean ecological, economic, and ethical probabilities of success are 0.9, 0.6, and 0.5, respectively (Fig. 14.1a upper panel). To obtain an overall probability of success that incorporates uncertainty, we draw random variates from each component probability distribution and calculate the product (Eq. 14.1). We repeat this for 2000 iterations to obtain the resulting distribution of the probability of success (Fig. 14.1a lower panel). With two of the 3E-component probabilities equal to or slightly better than a coin toss, the overall (or 'combined') probability of success has a most likely value of 0.24. The management strategy would have to be adapted to increase the probabilities of economic and ethical success to justify financial investment. An example of this scenario could be a community-led invasive mammal control project for biodiversity benefits (Sullivan and Molles 2016). The ecological understanding and likelihood of achieving the desired outcome may be high if control is strategically deployed (Norbury et al. 2015). However, given the lack of institutional backing and the project dependence on the contributions of individual citizens with potentially high turn-over rates, there may be only moderate prospects for stable long-term support. Similarly, community-led operations may be prone to have only a moderate probability of ethical acceptance due to the lack of institutional support and oversight in the deployment of traps and toxins (Souther 2016).

In a second hypothetical case, the mean probabilities of continued economic and ethical support are 0.9 and 0.85, respectively, but the mean probability of ecological success is only 0.2 (Fig. 14.1b upper panel). Clearly the weak link is that the predicted ecological outcome for this strategy is very unlikely to be sufficient, which results in a most likely combined probability of success of 0.12

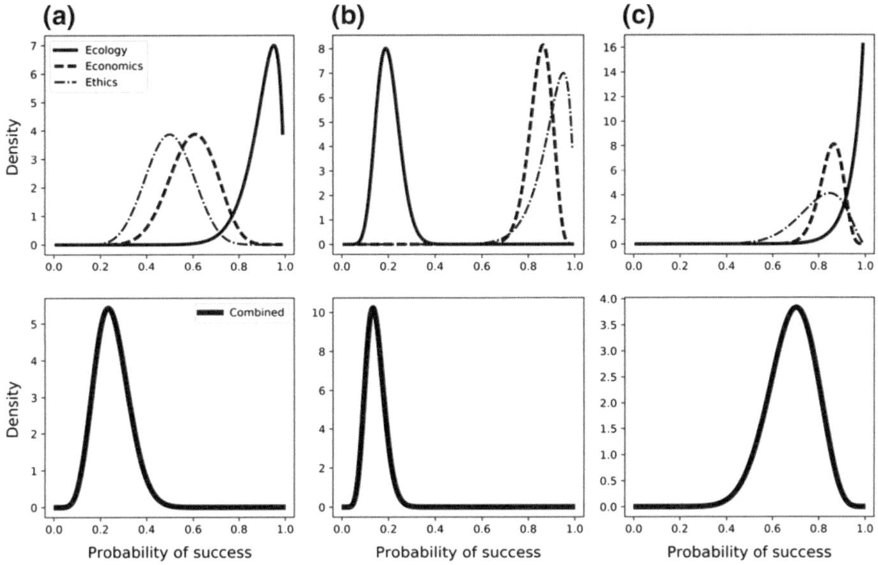

Fig. 14.1 Results of three hypothetical management scenarios illustrating how the integration of 3E-component probabilities determines the overall probability of project success. In the *upper panel* of each scenario (a, b, and c) the probabilities of ecological, economic and ethical acceptance are decomposed into individual probability distributions. The product of these probabilities represents the overall probability of program success and is shown in the corresponding *lower panel*

(Fig. 14.1b lower panel). An example of this scenario is the attempted removal of stoats from Resolution Island, New Zealand (Anderson et al. 2016). This is a Department of Conservation project that has stable financial support and is ethically accepted due to the use of animal welfare approved stoat kill-traps. However, ecological modelling shows that the current levels of wildlife control only have a 0.2 mean probability of achieving the management objectives. The probability of ecological success could be increased to 95% with a >100% increase in spending for an additional trapping session per year and deployment of improved lures. Doubling the budget, as the latter scenario, could achieve the ecological objective, but the probability of continued financial support at this level would likely be very low. These cases illustrate that the combined probability of success will always be lower than the lowest 3E component probability of success (Fig. 14.1a, b, c).

Our last hypothetical case illustrates that even with relatively high probabilities of success for the 3Es (means = 0.95, 0.85, 0.80; Fig. 14.1c upper panel), the most likely combined probability of success for the program is only a moderate 0.70 (Fig. 14.1c lower panel). While the actual probability distributions were made up to illustrate a management program with a relatively high probability of success, this scenario is represented by the aerial delivery of 1080 to suppress invasive mammals for biodiversity benefits and to eradicate bovine tuberculosis in New Zealand.

With proper application of 1080 baits possum and rat populations can be reduced with high efficacy (~95%; Nugent and Morriss 2013). For broadscale control of invasive mammals, 1080 is the most cost-effective method, which results in a relatively high probability of economic success (Warburton and Livingstone 2015). However, sustained suppression of invasive species requires repeated applications in perpetuity (Hone 2007), which increases the uncertainty of continued economic support. Similarly, while there is broad public support for the use of 1080 in environmental management in New Zealand (PCE 2011), the vocal opposition to its use increases the uncertainty in its ethical acceptance (Hansford 2016). Given a most likely probability of 0.7 and the level of uncertainty, some programs will fail.

Does this mean that all such programs are not justified? Not necessarily, as many will succeed, and adaptive management and research will aim to discover innovative ways to increase each of the 3E-component probabilities to increase the combined probability of project success.

14.7 Conclusions

The main tenet of this paper is that managing wildlife is pervaded by uncertainty, and as Norton (2005) stated, managers often do not know in advance what policies are sustainable. He also suggested that adaptive managers need to propose open-ended experimental approaches in order to learn about and reduce that uncertainty. In relation to managing invasive or over-abundant species, Warburton and Norton (2009) have proposed that an adaptive management approach is necessary when the killing of sentient species is being planned in a system with a high level of uncertainty. Failure to do so justifies the accusations of being unethical. Others have acknowledged that managing wildlife is complex and have recommended that to solve pest problems, social science and ecological science must work together (Krebs in Braysher 2017).

Here we deconstructed uncertainty into its three main components as they relate to managing wildlife: (1) ecological, (2) economic, and (3) ethical, and suggest that the uncertainty of each of these components equally influences the sustainability of a pest management program. Because the probability of each component and its associated uncertainty is multiplicative, the product of the three component probabilities results in a relatively low probability of sustainable success of the program even when each individual component probability is high. For example, if all three had success probabilities of 0.9, the success probability of the program would still be only 0.73.

Managers would benefit from adopting this simple framework, and even if their first probabilities were based on expert opinion or degrees of belief, through adaptive management these probabilities could be updated using a Bayesian model. If nothing else, recognizing that the three components have to be addressed when developing a management plan in order for that program to be sustainable would see an improvement on the status quo. In the authors' experience, most invasive species management operations do not plan for the 3Es, focus only on the tactical operational aspects, do not have any guaranteed long-term funding structures, and do not consider potential animal welfare or ethical dimensions.

Invasive species eradication plans are more likely to be sustainable through to completion because they are generally completed in relatively short time frames, have a clearly understood objective and are therefore at less risk from changing priorities or a decline or cessation of funding. Even so, Holmes et al. (2015) reported that programs to eradicate rats from tropical islands have a success rate of only 81%. Control operations that attempt to maintain animal numbers at low densities are at an even greater risk of failure because they are at continual risk of being cancelled because of changes in priorities (Parkes 1990), having funding withdrawn (Reddiex et al. 2006), or because of the loss of ongoing social license (Perry and Perry 2008). An additional risk is that many community-led programs are initiated because of a single person's passion to protect some biodiversity values, but over time this "champion" moves on for various reasons, leaving the program to decline and fail.

In this chapter, our aim was to highlight that wildlife managers need to acknowledge and consider multiple values in their decision making, including those related to ecological processes, species, and populations as well as those related to individual animal welfare, economic sustainability, professional duty, scientific integrity, and public welfare (Minteer and Collins 2008). Minteer and Collins suggested that this intersection of ecological science and values could be best addressed within the emerging field of "ecological ethics," which is both normatively pluralistic and methodologically pragmatic, ensuring animal welfare and rights questions are considered alongside environmental/ecological questions. Such an approach could be made functional by ensuring any control programs are ecologically or epidemiologically, ethically, and economically defensible (the "3Es"). That is, addressing these 3Es would be similar to the "3Rs" (replacement, refinement, and reduction) often applied to animal use in experimentation (Russell and Burch 1959). As happens for research on animals that requires animal ethics committee approval, this suggested approach for pest control and wildlife management could also be evaluated by a similar process, as suggested by Warburton and Norton (2009). Further to this, Curzer et al. (2016) suggested that there should be a fourth R (Refusal) for those experiments (in this case pest control programs) when the uncertainty is too high.

Bryan Norton (2003), as a pragmatic pluralist, suggested that to address "wicked" environmental problems it would be advantageous to operationalize environmental decision making by using a multi-criteria evaluation system. We have provided one possible framework for doing this, in which wildlife managers evaluate what impact each component of the 3Es might have on the potential success and sustainability of their pest (invasive species) control programs. Adaptive management is used to update the risk probabilities of the 3Es and improve the overall probability that programs will deliver successful and sustained achievement of ecological outcomes. It is this use of practical decision making to achieve sustainable environmental outcomes that Norton has contributed to many aspects of environmental management, work that we believe is especially valuable for addressing the wicked problems of wildlife management.

References

Addison, P.F.E., C.N. Cook, and K. de Bie. 2016. Conservation practitioners' perspectives on decision triggers for evidence-based management. *Journal of Applied Ecology* 53: 1351–1357.

Allan, C., and A. Curtis. 2005. Nipped in the bud: Why regional scale adaptive management is not blooming. *Environmental Management* 36: 414–425.

Allen, P. 2007. Commissioning research that is used: The experience of the NHS Service Delivery and Organization Research and Development Programme. *Evidence Policy* 3: 119–134.

Anderson, D.P., P. McMurtrie, K.-A. Edge, et al. 2016. Inferential and forward projection modeling to evaluate options for controlling invasive mammals on islands. *Ecological Applications* 26: 2548–2559.

Ascough, J.C., H.R. Maier, J.K. Ravalico, et al. 2008. Future research challenges for incorporation of uncertainty in environmental and ecological decision-making. *Ecological Modelling* 219: 383–399.

Bennett, J., P. Lawrence, R. Johnstone, et al. 2005. Adaptive management and its role in managing Great Barrier Reef water quality. *Marine Pollution Bulletin* 51: 70–75.

Bishop, C.J., G.C. White, D.J. Freddy, et al. 2005. Effect of limited antlered harvest on mule deer sex and age ratios. *Wildlife Society Bulletin* 33: 662–668.

Bottrill, M.C., L.N. Joseph, J. Carwardine, et al. 2008. Is conservation triage just smart decision making? *Trends in Ecology & Evolution* 23: 649–654.

Bracken, L.J., and E.A. Oughton. 2006. 'What do you mean?': The importance of language in developing interdisciplinary research. *Transactions of the Institute of British Geographers* 31: 371–382.

Braysher, M. 2017. *Managing Australia's pest animals: A guide to strategic planning and effective management*. Clayton South, Australia: CSIRO Publishing.

Broom, D. 2011. A history of animal welfare. *Acta Biotheoretica* 59: 121–137.

Choquenot, D., and J. Parkes. 2001. Setting thresholds for pest control: How does pest density affect resource viability? *Biological Conservation* 99: 29–46.

Cook, C.N., K. de Bie, D.A. Keith, et al. 2016. Decision triggers are a critical part of evidence-based conservation. *Biological Conservation* 195: 46–51.

Cooke, B., P. Chudleigh, S. Simpson, et al. 2013. The economic benefits of the biological control of rabbits in Australia, 1950–2011. *Australian Economic History Review* 53: 91–107.

Courchamp, F., M. Langlais, and G. Sugihara. 1999. Cats protecting birds: Modelling the mesopredator release effect. *Journal of Animal Ecology* 68: 282–292.

Curzer, H.J., G. Perry, M.C. Wallace, et al. 2016. The three Rs of animal research: What they mean for the institutional animal care and use committee and why. *Science and Engineering Ethics* 22: 549–565.

Dubois, S., N. Fenwick, E.A. Ryan, et al. 2017. International consensus principles for ethical wildlife control. *Conservation Biology*. https://doi.org/10.1111/cobi.12896.

Duka, T., and P. Masters. 2005. Confronting a tough issue: Fertility control and translocation for over-abundant koalas on Kangaroo Island, South Australia. *Ecological Management & Restoration* 6: 172–181.

Ellis, M.M., and C.S. Elphick. 2007. Using a stochastic model to examine the ecological, economic and ethical consequences of population control in a charismatic invasive species: Mute swans in North America. *Journal of Applied Ecology* 44: 312–322.

Fox, N., and H. MacDonald. 2012. Welfare aspects of killing wild animals in Britain. http://www.falcons.co.uk/default.asp?id=14. Accessed May 2012.

Freifeld, H.B., S. Plentovich, C. Farmer, et al. 2016. Long-distance translocations to create a second millerbird population and reduce extinction risk. *Biological Conservation* 199: 146–156.

Gardner, I.A., H. Stryhn, P. Lind, et al. 2000. Conditional dependence between tests affects the diagnosis and surveillance of animal diseases. *Preventive Veterinary Medicine* 45 (1–2): 107–122. https://doi.org/10.1016/S0167-5877(00)00119-7.

Gelman, A., J.B. Carlin, H.S. Stern, et al. 2004. *Bayesian data analysis*. Boca Raton, FL: Chapman & Hall/CRC.

Geremia, C., P.J. White, J.A. Hoeting, et al. 2014. Integrating population- and individual-level information in a movement model of Yellowstone bison. *Ecological Applications* 24 (2): 346–362.

Grantham, H.S., K.A. Wilson, A. Moilanen, et al. 2009. Delaying conservation actions for improved knowledge: How long should we wait? *Ecological Letters* 12: 293–301.

Hadler, MR., and AP. Buckle. 1992. Forty five years of anticoagulant rodenticides—past, present and future trends. In *Proceedings of the fifteenth vertebrate pest conference*. Davis, CA: University of California.

Haniza, MZH., S. Adams, and EP. Jones, et al. 2015. Large-scale structure of brown rat (Rattus norvegicus) populations in England: Effects on rodenticide resistance. Peerj 3.

Hansford, D. 2016. *Protecting paradise: 1080 and fight to save New Zealand's wildlife*. Nelson, New Zealand: Potton & Potton.

Hanson, C., J. Bonham, K. Campbell, et al. 2010. The removal of feral cats from San Nicolas Island: Methodology. In *24th Vertebrate pest conferencetimm*, ed. R. Timm, K. Fagerstone, Sacramento.

Holling, C.S. 1978. *Adaptive environmental assessment and management*. London: John Wiley and Sons.

Holmes, N.D., K.J. Campbell, B. Keitt, et al. 2016. Reporting costs for invasive vertebrate eradications. *Biological Invasions* 18: 2801–2807.

Holmes, N.D., R. Griffiths, M. Pott, et al. 2015. Factors associated with rodent eradication failure. *Biological Conservation* 185: 8–16.

Hone, J. 2007. *Wildlife damage control*. Collingwood, Australia: CSIRO Publishing.

Jones, H.P., N.D. Holmes, S.H.M. Butchart, et al. 2016. Invasive mammal eradication on islands results in substantial conservation gains. *Proceedings of the National Academy of Sciences of the United States of America* 113: 4033–4038.

Kirkwood, R., D.R. Sutherland, S. Murphy, et al. 2014. Lessons from long-term predator control: A case study with the red fox. *Wildlife Research* 41: 222–232.

Livingstone, P.G., N. Hancox, G. Nugent, et al. 2015. Development of the New Zealand strategy for local eradication of tuberculosis from wildlife and livestock. *New Zealand Veterinary Journal* 63: 98–107.

Lloyd, B.D., and R.G. Powlesland. 1994. The decline of kakapo *Strigops habroptilus* and attempts at conservation by translocation. *Biological Conservation* 69: 75–85.

Margules, C.R., and S. Sarkar. 2007. *Systematic conservation planning*. New York: Cambridge University Press.

Mason, G., and K. Littin. 2003. The humaneness of rodent pest control. *Animal Welfare* 12: 1–37.

McBride, M., K.A. Wilson, and H.P. Possingham. 2005. Incorporating investment uncertainty into the prioritisation of conservation resource allocation. In *Modsim 2005: International congress on modelling and simulation: advances and applications for management and decision making*, 2505–2511.

McCarthy, M.A., and H.P. Possingham. 2007. Active adaptive management for conservation. *Conservation Biology* 21: 956–963.

McCook, L.J., T. Ayling, M. Cappo, et al. 2010. Adaptive management of the Great Barrier Reef: A globally significant demonstration of the benefits of networks of marine reserves. *Proceedings of the National Academy of Sciences* 107: 18278–18285.

Minteer, B.A., and J.P. Collins. 2008. From environmental to ecological ethics: Toward a practical ethics for ecologists and conservationists. *Science and Engineering Ethics* 14: 483–501.

Minteer, B.A., E.A. Corley, and R.E. Manning. 2004. Environmental ethics beyond principle?: The case for a pragmatic contextualism. *Journal of Agricultural and Environmental Ethics* 17: 131–156.

Moberly, R.L., P.C.L. White, C.C. Webbon, et al. 2004. Modelling the costs of fox predation and preventive measures on sheep farms in Britain. *Journal of Environmental Management* 70: 129–143.

Moore, J., A. Balmford, T. Allnutt, et al. 2004. Integrating costs into conservation planning across Africa. *Biological Conservation* 117: 343–350.

Moseby, K.E., and B.M. Hill. 2011. The use of poison baits to control feral cats and red foxes in arid South Australia: I Aerial baiting trials. *Wildlife Research* 38: 338–349.

Naidoo, R., A. Balmford, P.J. Ferraro, et al. 2006. Integrating economic costs into conservation planning. *Trends in Ecology & Evolution* 21: 681–687.

Norbury, G.L., R.P. Pech, A.E. Byrom, and J. Innes. 2015. Density-impact functions for terrestrial vertebrate pests and indigenous biota: Guidelines for conservation amangers. *Biological Conservation* 191: 408–420.

Norton, B. 1995. Caring for nature: A broader look at animal stewardship. In *Ethics on the Ark*, ed. B. Norton, M. Hutchins, E. Stevens, et al. Washington, DC: Smithsonian Institution Press.

Norton, B. 2003. Conservation: Moral crusade or encironmental public policy? In *Recontructing conservation: Finding common ground*, ed. B.A. Minteer and R.E. Manning. Washington: Island Press.

Norton, B.G. 2005. *Sustainability: A philosophy of adaptive ecosystem management*. Chicago: IL, University of Chicago Press.

Norton, B.G. 2015. *Sustainable values, sustainable change*. Chicago: University of Chicago Press.

Nugent, G., B.M. Buddle, and G. Knowles. 2015. Epidemiology and control of Mycobacterium bovis infection in brushtail possums (Trichosurus vulpecula), the primary wildlife host of bovine tuberculosis in New Zealand. *New Zeal Vet J* 63: 28–41.

Nugent, G., and G.A. Morriss. 2013. Delivery of toxic bait in clusters: A modified technique for aerial poisoing of small mammal pests. *New Zealand Journal of Ecology* 37: 246–255.

O'Donnell, C.F.J., and J.M. Hoare. 2012. Quantifying the benefits of long-term integrated pest control for forest bird populations in a New Zealand temperate rainforest. *New Zealand Journal of Ecology* 36: 131–140.

Parkes, J.P. 1990. Feral goat control in New Zealand. *Biological Conservation* 54: 335–348.

Parkes, J.P., D.S.L. Ramsey, N. Macdonald, et al. 2010. Rapid eradication of feral pigs (Sus scrofa) from Santa Cruz Island, California. *Biological Conservation* 143: 634–641.

Parkes, J.P., A. Robley, D.M. Forsyth, et al. 2006. In my opinion: Adaptive management experiments in vertebrate pest control in New Zealand and Australia. *Wildlife Society Bulletin* 34: 229–236.

Parris, K.M., S.C. McCall, M.A. McCarthy, et al. 2010. Assessing ethical trade-offs in ecological field studies. *Journal of Applied Ecology* 47: 227–234.

PCE. 2011. *Evaluating the use of 1080: Predators, poisons and silent forests*. Wellington: Parliamentary Commissioner for the Environment.

Pech, R., and M. Maitland. 2016. Conservation of native fauna in highly invaded systems: Managing mammalian predators in New Zealand. *Restoration Ecology* 24: 816–820.

Perez-Espona, S., J.M. Pemberton, and R. Putman. 2009. Red and sika deer in the British Isles: Current management issues and management policy. *Mammalian Biology* 74: 247–262.

Perry, D., and G. Perry. 2008. Improving interactions between animal rights groups and conservation biologists. *Conservation Biology* 22: 27–35.

Ramsey, D.S.L., D.M. Forsyth, C.J. Veltman, et al. 2012. An approximate Bayesian algorithm for training fuzzy cognitive map models of forest responses to deer control in a New Zealand adaptive management experiment. *Ecological Modelling* 240: 93–104.

Reddiex, B., D.M. Forsyth, E. McDonald-Madden, et al. 2006. Control of pest mammals for biodiversity protection in Australia: I. Patterns of control and monitoring. *Wildlife Research* 33: 691–709.

Robertson, H.A., E. Craig, C. Gardiner, et al. 2016. Short pulse of 1080 improves the survivial of brown kiwi chicks in an area subjected to long-term stoat trapping. *New Zealand Journal of Zoology* 43: 351–362.

Rotmans, J., and M.B.A. van Asselt. 2001. Uncertainty management in integrated assessment modeling: Towards a pluralistic approach. *Environmental Monitoring and Assessment* 69: 101–130.

Ruscoe, W.A., D.S.L. Ramsey, R.P. Pech, et al. 2011. Unexpected consequences of control: Competitive versus predator release in a four-species assemblage of invasive mammals. *Ecology Letters* 14 (10): 1035–1042.

Russell, W., and R. Burch. 1959. *The principles of humane experimental technique*. London: Methuen.

Shea, K., H.P. Possingham, W.W. Murdoch, et al. 2002. Active adaptive management in insect pest and weed control: Intervention with a plan for learning. *Ecological Applications* 12: 927–936.

Shwiff, S.A., A. Anderson, R. Cullen, et al. 2013. Assignment of measurable costs and benefits to wildlife conservation projects. *Wildlife Research* 40: 134–141.

Singleton, G.R., P.R. Brown, J. Jacob, et al. 2007. Unwanted and unintended effects of culling: A case for ecologically-based rodent management. *Integrative Zoology* 2: 247–259.

Souther, C.E. 2016. The cruel culture of conservation country: Non-native animals and the consequences of predator-free New Zealand. *Transnational Law and Contemporary Problems* 26: 63–119.

Sullivan, J.J., and L.E. Molles. 2016. Biodiversity monitoring by community-based restoration groups in New Zealand. *Ecological Management & Restoration* 17: 210–217.

Vucetich, J., and M. Nelson. 2007. What are 60 warblers worth?: Killing in the name of conservation. *Oikos* 116: 1267–1278.

Walters, C.J. 1986. *Adaptive management of renewable resources*. New York: MacMillan Publishing Company.

Walters, C. 2007. Is adaptive management helping to solve fisheries problems? *Ambio* 36: 304–307.

Walters, C.J., and R. Green. 1997. Valuation of experimental management options for ecological systems. *The Journal of Wildlife Management* 61: 987–1006.

Walters, C.J., and C.S. Holling. 1990. Large-scale management experiments and learning by doing. *Ecology* 71: 2060–2068.

Warburton, B., and D. Choquenot. 1999. Animal welfare and pest control: The context is important. In *The use of wildlife for research*, ed. D.J. Mellor and V. Monamy. Australia: Glen Osmond.

Warburton, B., and P. Livingstone. 2015. Managing and eradicating wildlife tuberculosis in New Zealand. *New Zealand Veterinary Journal* 63: 77–88.

Warburton, B., and B.G. Norton. 2009. Towards a knowledge-based ethic for lethal control of nuisance wildlife. *Journal of Wildlife Management* 73: 158–164.

Warburton, B., and C. O'Connor. 2004. Research on vertebrate pesticides and traps: Do wild animals benefit? *Alternatives to Laboratory Animals: ATLA* 32: 229–234.

Warburton, B., N. Poutu, D. Peters, et al. 2008. Traps for killing stoats (Mustela erminea): Improving welfare performance. *Animal Welfare* 17: 111–116.

Warburton, B., D. Tompkins, D. Choquenot, et al. 2012. Minimising number killed in long-term vertebrate pest management programmes, and associated economic incentives. *Animal Welfare* 21: 141–149.

Winograd, N. 2007. *Redemption: The myth of pet overpopulation and the no kill revolution in America*. Los Angeles, CA: Almaden Books.

Bruce Warburton is a vertebrate pest biologist at Landcare Research, New Zealand. His research focusses on developing and improving aerial and ground-based strategies and tools for managing invasive mammals. He has a particular interest in animal welfare and ethics and has been a member of New Zealand's National Animal Welfare and National Animal Ethics Advisor Committees.

Dean Anderson is a quantitative ecologist at Landcare Research, New Zealand. His research focusses on the complex and interacting dynamics among environmental management, invasive species populations, and protected biodiversity or economic assets. He develops Bayesian hierarchical models to make inference on these systems, which are then converted into forecast models to explore alternative strategies to help guide management decisions.

Chapter 15
The Transformative Ark

Ben A. Minteer and Christopher Rojas

Abstract As conservationists confront an accelerating extinction crisis, zoos are emerging as potentially significant players in the effort to protect global biodiversity, a role that will likely intensify in the coming decades. It's an agenda, however, that raises a number of ethical and practical questions as zoological parks seek to balance a growing conservation mission alongside their traditional recreation and entertainment pursuits. Many of these questions were first addressed in Bryan Norton's anthology, *Ethics on the Ark*, a milestone in applied ethics and zoo conservation published in 1995. In the decades since Norton's book appeared, the function of zoos as conservation educators and as centers of public transformation has come into sharper focus, with new fields such as conservation psychology measuring the impact of the zoo visit on public perceptions, attitudes, and conservation behaviors. In this chapter, we explore some of this recent empirical work examining zoo visitors' experiences and argue that Norton's early writing in environmental ethics and conservation, particularly his notion of "transformative value," offers a philosophical grounding for understanding the ethical potential of encounters with zoo animals. We close the chapter by discussing some of the challenges and tensions that emerge when Norton's argument, which was originally presented as a justification for protecting wild biodiversity due to its ability to "transform" consumer preferences to more ecologically enlightened attitudes, is adapted to the zoo setting.

Keywords Zoos · Conservation education · Transformative value
Conservation psychology

B. A. Minteer (✉)
School of Life Sciences, Arizona State University, P.O. Box 874501,
Tempe, AZ 85287-4501, USA
e-mail: ben.minteer@asu.edu

C. Rojas
School of Life Sciences, Arizona State University, P.O. Box 874601,
Tempe, AZ 85287-4601, USA
e-mail: carojas2@asu.edu

S. Sarkar and B. A. Minteer (eds.), *A Sustainable Philosophy—The Work of Bryan Norton*, The International Library of Environmental, Agricultural and Food Ethics 26, https://doi.org/10.1007/978-3-319-92597-4_15

15.1 The Rise of Zoo Conservation

With the recognition that we are in the midst of a sixth mass extinction event on the planet (Kolbert 2015; Ceballos et al. 2017) it has become clear that protection of species in the wild will increasingly need to be supplemented by ex situ approaches; i.e., the propagation of animal populations within and by zoos, aquariums, and similar institutions (Conde et al. 2011; IUCN/SSC 2014). It's a vision that requires moving beyond the familiar understanding of zoological parks as isolated "arks" housing assurance populations to one promoting greater integration of captive and wild populations across landscapes and among conservation institutions (Redford et al. 2012; Minteer and Collins 2013; Keulartz 2015). The hope is that by coordinating planning efforts for and management of wild and captive animals along the in situ-ex situ continuum we will improve their chances in the wild and also ensure the sustainability of animal populations throughout the global zoo network (Byers et al. 2013; Traylor-Holzer et al. 2018).

The emerging zoo conservation agenda is ambitious, especially for an institution that has long embraced, for its own sake as well as out of economic necessity, its recreation function. Despite a series of high profile animal care controversies over the decades (from elephants to orcas) zoological parks remain one of our most popular cultural institutions. A frequently cited survey by the World Association of Zoos and Aquariums (WAZA) found that, globally, more than 700 million people visit a zoo or an aquarium annually (Gusset and Dick 2011). An impressive number, it's a statistic that also gives a sense of the challenge of zoos' attempts to reconfigure themselves as bona fide conservation organizations—in their budgets, programs, and operating space—while also serving as a major entertainment destination for hundreds of millions of visitors every year (Conway 2011; Grazian 2015).

Although zoos have always been entertainment venues, wildlife preservation has been part of their *raison d'etre* in the U.S. since at least the rise of the conservation movement in the late 19th century (Stott 1981). The Bronx Zoo in New York, for example, played a key role in the breeding and reintroduction of the American bison to the western plains in the early 1900s (Barrow 2009). Still, modern zoos' embrace of conservation as an explicit and significant part of their public mission arguably only became serious and systematic in the last quarter of the 20th century, a shift that was part of the institutional and policy response to public concern about animal welfare, endangered species, and environmental issues more generally (Kisling 2000). By the 1980s, many zoos began to develop Species Survival Plans (SSPs), programs that coordinated breeding and population management of threatened and endangered animals across the world zoo network (Hutchins and Wiese 1991). SSPs facilitated the recovery of the black-footed ferret and California condor, two of the great success stories in zoo conservation (WAZA 2012).

The zoo conservation agenda would gain steam in the 1990s following the publication of the World Zoo Conservation Strategy (IUDZG/CBSG (IUCN/SSC) 1993), a document that encouraged the continuing evolution of zoos into full blown "conservation centres." The message was magnified in the revised 2005 WAZA Conservation Strategy, which emphasized the importance of securing animal populations in the wild rather than in zoo enclosures (WAZA 2005). And the past decade especially has seen a dramatic surge in and intensification of zoo conservation programs and initiatives, as well as the appearance of a series of major publications, reports, and vision statements committing zoological parks to an even stronger conservation and scientific mission in the coming decades (e.g., Zimmermann et al. 2007, Fa et al. 2011; Barongi et al. 2015).

Many zoo critics, however, view these developments as little more than a cynical spin campaign to rebrand what they argue is an ethically indefensible institution—a critique common among animal rights advocates (see, e.g., https://www.peta.org/features/zoo-conservation-captive-breeding/). The distance between the professionally-run and accredited zoos of today and the sorry menageries and roadside attractions of an earlier age, however, is vast, even if legitimate animal welfare concerns remain (Clubb and Mason 2003; Maple and Perdue 2013; Mellor et al. 2015). At the very least, we can say that the recent uptick in zoo conservation efforts indicates a growing desire within today's zoo community to position these institutions as serious conservation organizations, both within the wildlife conservation professions and in the public eye.

In hindsight, one of the pivotal conversations that helped stimulate the emergence of a more articulate and reflective zoo conservation agenda was the appearance of Bryan Norton's *Ethics on the Ark,* a pioneering collection of essays on zoos, animal welfare, and conservation published by the Smithsonian Institution Press (Norton et al. 1995). The volume flowed out of a workshop Norton co-organized in Atlanta in the early 1990s that convened a diverse group of applied philosophers, zoological professionals, and wildlife scientists to ruminate on the complex and often contentious intersection of zoo animal welfare and conservation. Co-edited with several leaders in the zoo community, including Terry Maple, then-president of Zoo Atlanta, the volume quickly became a benchmark for applied environmental and animal ethics.

The treatment of zoo conservation in *Ethics on the Ark* largely focused on conservation (née "captive") breeding, wildlife reintroduction, and zoo-based conservation research; all direct activities undertaken by and within zoos to benefit wildlife health and survival in situ. Although these practices remain at the center of the zoo conservation agenda today, in the decades since *Ethics on the Ark* was published conservation *education* has become an increasingly powerful element of the zoo conservation portfolio (WAZA 2005; Sterling et al. 2007; Ogden and Heimlich 2009; Barongi et al. 2015). At the same time, the new field of conservation psychology has greatly expanded our understanding of the nature and impact of zoo education and the visitor experience, especially the perceptual, behavioral, and value dimensions of human-animal encounters and how these elements shape human interactions with species and places across wild and built landscapes (e.g., Saunders 2003; Clayton and Myers 2009). It's a body of empirical research, moreover, that is helping zoos improve their educational programs and the efficacy of their public outreach efforts (Barongi et al. 2015).

Even though these educational and psychological perspectives didn't figure directly in *Ethics on the Ark,* we believe (again, in hindsight) that Bryan Norton actually anticipated them in some of his early writing on environmental value theory and the philosophy of biodiversity conservation. In particular, Norton's notion of "transformative value," which he developed most thoroughly in his 1987 book, *Why Preserve Natural Variety?*, suggests an intriguing way to think about the impact of encounters with wild species and ecosystems on the development of environmental values and attitudes. It's an idea we believe has implications today for zoos as they seek to broaden and deepen their conservation goals.

In fact, on our reading, Norton's theory, i.e., that transformative experiences with nature can lead to changes in environmental values and preferences, is mirrored in some of the newer psychological studies of the visitor encounter with zoo animals, especially work addressing its affective dimensions. As we will see, however, several questions and concerns remain regarding this particular intersection of environmental ethics and practice, including the transferability of Norton's original understanding of transformative value to the zoo context and the uniqueness and desirability of zoos as venues of public transformation.

15.2 Norton's Transformative Value

In *Why Preserve Natural Variety?* Norton presented a sharp critique of the established economic and ethical approaches to valuing wild species and undisturbed ecosystems. By appealing narrowly to only the demand values of species, that is, the ability of plants and animals to fulfill our immediate and unreflective desires, Norton argued that the traditional economic framing was too compromised by consumptive interests to serve as an adequate foundation for nature preservation. Yet, appeal to the intrinsic value of species and ecological systems, the dominant approach within environmental ethics, was also problematic. Intrinsic value arguments were too "ethically radical," Norton believed, and the non-anthropocentric worldview supporting them too underdeveloped, to offer a compelling rationale for sound conservation policy (Norton 1987: 186–187). Dissatisfied with both approaches, he outlined and defended an alternative construct, what he dubbed "transformative value," a liberal expansion of instrumental value that Norton argued was more supportive of the species preservationist position. It was a formulation, he claimed, that avoided the philosophical liabilities ensnaring the more familiar economistic and non-anthropocentric expressions.

More specifically, Norton pitched transformative value as a dynamic form of environmental value that, although rooted in human preferences and interests, hinged on a critical distinction between different *types* of preferences, "felt" and "considered." Felt preferences were those desires or needs that could be satisfied by a specific experience; e.g., the unreflective desire for a consumer good that is sated by its acquisition. Considered preferences, however, were potential desires or needs "that an individual would express or otherwise exhibit after careful deliberation"

(Norton 1987: 9). It followed that an object or experience had transformative value if it provided "an occasion for examining or altering a felt preference rather than simply satisfying it" (Norton 1987: 10). Encounters with wild species and ecosystems, Norton believed, had this potential to trigger a questioning of our felt preferences, especially consumer ones:

> Experience of nature can promote questioning and rejection of overly materialistic and consumptive felt preferences. Appeals to the transformative value of wild species and undisturbed ecosystems thereby provide the means to criticize and limit demand values that threaten to destroy those species and ecosystems while at the same time introducing an important value that humans should place upon them. (Norton 1987: 189)

The attribution of transformative value to an object assumed that not all felt preferences are equally preferable: some felt preferences stand the test of reflective scrutiny and become considered preferences, while others wither in the gaze of reflective interrogation (Norton 1987: 11).

Furthermore, it's clear that Norton believed the transformative value of nature was premised on the acceptance of a general environmental ethos understood to be, as he put it, "objectively better" than that supporting the demand values of materialism and consumerism (Norton 1987: 210). The comparatively high status of considered preferences was ultimately the result of their comportment with what Norton depicted as a "rational" ecological worldview, a vision that embraced: (1) the human interdependence with nature; (2) a post-Cartesian, Darwinian epistemology defined by an attitude of fallibility and caution in the face of grand ecological complexity; and (3) a value system defined by ecological humility and the search for a harmonious balance with nature (Norton 1987: 204–207). The superiority of considered preferences as the end product of an encounter or experience with an object of nature possessing transformative value was therefore relative specifically to the outlook and values of nature preservationists (rather than to those of, say, real estate developers). As he wrote: "If [species preservationists] believe that the ecological world view represents a more accurate picture of the world and that the value system suggested by it is objectively better than the value system of materialism and conspicuous consumption, then they will value endangered species and natural ecosystems for their role in transforming human world views and human value systems" (Norton 1987: 210).

There is more to it, and this quick review alone raises a host of questions (chief among them being what it means to describe an environmental worldview as "objectively better," especially when conservationists today are divided over many of the values in Norton's ecological vision; see, e.g., Minteer and Pyne 2015). But for present purposes we can summarize by saying that Norton's argument for transformative value as an alternative to both narrow instrumentalist approaches and non-anthropocentric projects hinged on the potential for experiences with environmental objects to catalyze reflection and deliberation over what individuals *should* prefer or want, a process he believed would ultimately move them toward a more enlightened ecological outlook supported by a deeper co-evolutionary understanding of our place in the natural order.

It was, in many respects, a quasi-transcendentalist appeal to nature's ability to lead us to our "higher selves" as Thoreau put it; or to "think like a mountain," as Aldo Leopold directed in *A Sand County Almanac*. Both writers, in fact, served as intellectual progenitors in Norton's account of transformative value, as he acknowledged in *Why Preserve*? and would expand on in subsequent discussions (e.g., Norton 1991, 1994). Norton would also enlist the American pragmatists C. S. Peirce and John Dewey, with their predilection for dynamic and deliberative notions of value, meaning, and truth, for support as his transformative project became folded into an epistemological method of social learning, experimentation, and adaptive management in the 1990s and beyond (e.g., Norton 1995, 1999, 2005, 2015).

As we've said, the idea of transformative value was clearly a departure from the dominant normative approaches in environmental ethics that emphasized, in one way or another, the intrinsic value of nature as part of an encompassing non-anthropocentric worldview. Indeed, even though Norton stopped just short of a complete dismissal of intrinsic value in his original defense of transformative value in *Why Preserve*?, as mentioned above he made it very clear that he didn't see its necessity for species preservationists, who:

> …believe that encounters with wild species can precipitate changes in human consciousness, alterations in world views sufficient to create a new ontology, a new epistemology, and a new approach to value. If they also believe that the new, less materialistic values that are thereby created are objectively better than the materialistic, consumptive values they replace, they should value all wild species, including endangered ones, for their transformative value. On this anthropocentric basis they can argue that species should be preserved, regardless of whether they also believe that species have intrinsic value (Norton 1987: 211).

The "new, less materialistic values" Norton had in mind—those that flowed out of the ecological worldview summarized above—were in the end human values, a composite of goods and interests that terminated in the human benefit of a life lived in closer harmony with nature. As he concluded, "…If environmentalists believed that humans live more satisfying lives if they are not bound by excessive greed for material things, this belief would provide an adequate, anthropocentric support for transformative values" (Norton 1987: 238).

Given such views, it should come as little surprise that non-anthropocentric environmental philosophers greeted Norton's arguments for transformative value with what might charitably be termed deep skepticism. Holmes Rolston, a founder of modern environmental ethics and a writer known for his full-throated endorsement of an objectivist form of intrinsic natural value, discussed Norton's idea in his 1994 book *Conserving Natural Value*. There, Rolston openly wondered how a natural object that possessed only instrumental value could produce a morally significant transformation in people. "If the virtue of human character really comes from appreciating another, nonhuman form of life," Rolston asked, "then why not attach intrinsic value to this alien life?…Why praise only the virtue in the human beholder?" (Rolston 1994: 164). We wish to save species such as whales, Rolston argued, not because we seek to ennoble the human spirit, but because we recognize and seek to defend a life and a species with its own dignity and moral worth.

Fifteen years later, and referencing the opening hook and cover of Norton's 1991 book *Toward Unity Among Environmentalists* (which prominently featured a sand dollar), Rolston's view toward the concept hadn't grown much sunnier, though he now seemed to accept that transformative value could play at least *some* role in compelling environmental protection. It just wasn't the most foundational and important reason for preserving nature:

> My problem is that I do not think it is high moral ground to celebrate something else in your own self-interests, no matter how enlightened those interests (analogously to helping others in order to get a kick out of it)…The enlightened environmentalist wants photosynthesis in place, freshwater in streams, a stable climate—*and spiritual inspiration*, transformative encounter with sand dollars. Nature preservation is justified because it leads to the fostering of multiple levels of values in human life and culture worth preserving. True, we can all agree about the basic, vital, lower-order values. But the convergence is toward nobler, higher-order values, equally vital to human well-being, and these are not reducible to the familiar anthropocentric array of demand or preference values. (Rolston 2009: 111)

If Norton's argument for transformative value didn't convince many non-anthropocentric environmental philosophers in the years following the publication of *Why Preserve?*, it fared much better among conservation biologists. Transformative value, for example, appeared as a key part of the normative framework for valuing biodiversity in a popular textbook by Perlman and Adelson (1997). Norton's proposal had also received validation the year before in the science studies scholar David Takacs's book, *The Idea of Biodiversity* (1996), in which the author interviewed nearly two-dozen prominent ecologists and conservation biologists about the meaning and value of biodiversity. Takacs asked his subjects specifically about Norton's concept of transformative value and whether it resonated with them as a way to think about the value of species and ecosystems. Most of the scientists he spoke with voiced support for Norton's idea, including well-known figures like Paul Ehrlich and E. O. Wilson [although the latter told Takacs he wished Norton had grounded it more in evolutionary biology à la Wilson's own biophilia hypothesis (Takacs 1996: 235)].

Interestingly, two of the interviewees, biologists Tom Lovejoy and Peter Brussard, made an explicit connection between Norton's understanding of transformative value and the role of zoos in educating and inspiring visitors to care about the natural world. Asked if he thought biodiversity possessed transformative value as Norton described, Lovejoy replied, "Oh, I think that's perfectly true, that's perfectly true. There is a magic in living things….That's why, you know, zoos and aquaria have a greater annual attendance than all major sports events…" (Takacs 1996: 233). Brussard agreed, but took it further: "I think that lack of support for biodiversity often springs from ignorance of biodiversity. And if you can get people to zoos, if you can get people to go to natural history museums here, they're not bored but actually entranced by what they see there, that's certainly going to help…." (Takacs 1996: 232).

15.3 Zoos as Centers of Transformation

Even though conservation biologists were making the link between the notion of transformative value and the zoo experience, Norton himself didn't explicitly connect the dots between his argument for caring for biodiversity in *Why Preserve Natural Variety*? and the subsequent exploration of zoos and conservation in his 1995 *Ethics on the Ark* anthology.[1] As we've said, however, in the decades since the *Ark* volume was published the zoo community (and zoo researchers) effectively filled this breach by vigorously promoting the educational (and often, "transformative") value of the zoo experience, including the argument that zoo visits, when met with effective educational programming and progressive, "best practices" animal exhibit-craft, can encourage both a connection to animals (in the zoo and the wild) and pro-conservation behaviors.

Consider, for example, the messaging of the Association of Zoos & Aquariums (AZA), the zoological organization supporting and accrediting zoos in the United States (https://www.aza.org/conservation-education-standards-and-policies). The AZA has made a point of emphasizing the role of zoos in educating and inspiring visitors, suggesting that zoological parks provide "animal and nature experiences that engender a sense of wonder" among the public (https://www.aza.org/conservation-education-standards-and-policies). Subscribing to the general biophilic view that "the human experience requires a connection to nature," the AZA's conservation education standards state that such connections "enrich our lives and inspire our choices for future generations," and that zoos in particular seek to "promote care and positive action for the natural world."

It's an outlook shared by the larger World Association of Zoos and Aquariums (WAZA), which in its most recent conservation strategy (*Committing to Conservation*) described the purpose of zoos globally as "Instilling in all visitors a strong sense of excitement about and a desire to care for life on earth will create a solid platform for fulfilling the promise to care for and conserve wildlife" (Barongi et al. 2015: 17). By creating a "culture of conservation," WAZA argues, "zoological facilities are able to open the hearts and minds of their visitors, providing a relevant venue to convey the threats to wildlife, and to inspire, engage and guide positive environmental action" (Barongi et al. 2015: 45). The framework increasingly

[1]Norton's embrace of a more expansive pluralistic understanding of environmental values beyond the confines of the dualistic intrinsic-instrumental framework perhaps explains why he didn't continue to develop his earlier arguments for transformative value within the instrumentalist tradition, an evolution he discusses in his 2005 book *Sustainability*. But his growing emphasis on the linguistic dimensions of environmental deliberation is likely also implicated in this move away from traditional environmental value theory. In his 2015 book, *Sustainable Values, Sustainable Change*, Norton talks about "transformatives" in public debate as those "linguistic and cognitive tropes (e.g., metaphors, analogies, etc.) that can cause changes in perception and perspective regarding a situation or problem" (Norton 2015: 196). In other words, the dynamic use of language, rather than discrete encounters with wild species and ecosystems, now appears to serve as the key transformative agent in Norton's evaluative system.

adopted by zoological and other conservation organizations (including the International Union for Conservation of Nature, or IUCN) is often referred to in the shorthand of "connect, understand, act." In this framing, visitors are connected to priority species via "emotionally engaging" zoo encounters with animals, educated via interpretive displays and guided programs, and inspired to act to alleviate pressure on threatened species and habitats through a range of pro-conservation behaviors and choices (IUCN 2011).

A good illustration of this framework in action can be seen in zoo campaigns to spread awareness about the impacts of palm oil production on biodiversity in South East Asia, particularly its effect on orangutans as native forests are fragmented and destroyed (Chamberlain 2013). Melbourne Zoo's innovative conservation education program, "Don't Palm Us Off," developed a mixed-media approach to the problem, displaying video and interpretive materials in the visitor center attached to their orangutan exhibit, a program that connected information about deforestation and species endangerment with the consumption of palm oil in common consumer products (from soap to cookies). Researchers found that visitors who viewed the animals and the palm oil exhibit (which also linked to social media) reported significant increases in palm oil awareness, as well as more positive attitudes toward orangutans and greater support for labeling of palm oil products. There was also an increase in visitors' reported desire to make different consumer choices based on this experience (Pearson et al. 2014).

The palm oil exhibit is a particularly interesting case given that the program seemed to be operating with a theory of reflective experience and attitudinal/behavior change that comes close to what Norton originally had in mind with transformative value. Here, an animal encounter, shaped by the provision of contextual information about the species in the wild and the threat of habitat conversion for commercial agriculture, is believed to have triggered, at least in some individuals, a change in conservation attitude and an evaluation of consumer demand values (i.e., a distaste for products made with palm oil). The process apparently spurred the adoption of more considered preferences for products not entangled in the destruction of orangutan habitat, desires more in line with an ecological worldview supporting the protection of biodiversity and native forest sustainability. The Melbourne example therefore seems to lend some real-world support to the basic logic of Norton's theory of transformative value, at least in a generalized sense.

But how representative is it? The impact of zoo education programs on public biodiversity knowledge, attitudes, and conservation behavior remains a topic of considerable discussion and debate (see, e.g., Falk et al. 2007; Marino et al. 2010). The publication in recent years of several major studies providing empirical evidence backing up zoo conservationists' more hopeful claims along these lines, however, has bolstered the zoo-as-conservation-catalyst argument, suggesting that when zoo education programs are well-designed and well-run, visitors do indeed display greater biodiversity knowledge and at least the intention to engage in more pro-conservation behaviors (e.g., Jensen 2014).

One of the more influential and widely cited studies along these lines appeared in the journal *Conservation Biology* in 2015. There, the authors reported on a survey of more than 5600 zoo and aquarium visitors from more than two-dozen zoological parks around the globe (Moss et al. 2015). Their key finding was that their study sample demonstrated increased aggregate biodiversity understanding, as well as greater knowledge of behaviors to help protect biodiversity, over the course of their zoo visit.

In the past decade, a number of studies coming out of the emerging field of conservation psychology have greatly improved our knowledge about the dynamics of the connect-understand-act model by investigating the role of the personal connection to zoo animals in enhancing visitors' understanding of biodiversity and their motivation to engage in pro-conservation behaviors. Clayton et al. (2009), for example, surveyed more than 200 visitors at Cleveland Metroparks Zoo and found that their study participants were more inclined to report concern for zoo animals and for species as a whole when they felt a sense of connection to animals in exhibits—and when they had learned something about them and wanted to know more. More recently, Grajal et al. (2017), in a survey of nearly 3000 visitors to 10 zoos and 5 aquariums across the United States, uncovered a positive relationship between visitors' sense of connection to zoo animals and their self-reported pro-environmental behaviors (consumer choices, environmental support efforts, etc.) related to climate change, a relationship that existed regardless of participants' political orientations.

A significant thread in many of these social scientific studies of zoo visitors' connection to animals at zoological parks is the significance of emotion and a positive response to viewing zoo animals, including building a social identity that incorporates a relationship to animals and to nature generally (Swanagan 2000; Fraser et al. 2007; Clayton et al. 2011). In an important paper that parsed out some of the key elements in this process, Ballantyne, Packer, and Sutherland (2011) explored visitor experiences at four wildlife tourist venues in Australia, including a public aquarium and a marine park. In their post-visit surveys, the researchers uncovered four levels of response to animal encounters: sensory impressions (seeing and hearing), feelings (emotional attraction), reflection (cognitive processing of the event), and behavior. The interplay of emotional connection and reflection on the animal encounter by visitors proved especially significant in the study. As the authors put it, "It was the combination of emotional affinity with a reflective response that appeared to have the most powerful impact on visitors, leading to a concern and respect not only for the specific individuals encountered in the wildlife tourism experience, but the species as a whole" (Ballantyne et al. 2011: 774). The direct encounter and emotional connection with wildlife in their sample appeared to make conservation issues more personal and consequently more salient for people, which led the authors to suggest that such experiences may indeed be important for inspiring pro-environmental attitudes and actions in zoological parks and other wildlife viewing venues.

This general conclusion has been borne out in other work exploring the relationship among direct animal experiences in the zoo setting, emotional affinity, and behavioral intentions. Powell and Bullock (2014) studied public perceptions at three animal exhibits at the Bronx Zoo and found that zoo visitors who experienced strong (positive) emotional responses to viewing the animals in their study (tigers, African wild dogs, and spotted hyenas) reported greater conservation mindedness, and that visitors' emotional connection was shaped by the sense of having a general "encounter" with animals, making eye contact with them, observing active animal behaviors, and other factors. Similar results have been reported by Luebke et al. (2016) and Hacker and Miller (2016), who found that having such up-close interactions with animals in zoos across a range of species (from spotted hyenas to elephants) and across different institutions produced a more intense positive emotional connection between visitors and zoo animals, one they believe to be a critical mediator among observation, reflection, and the desire to engage in behaviors that have positive conservation implications.

Obviously, much more remains to be learned about the meaning and impact of animal encounters in zoos on visitor perceptions and, especially, on post-visit attitudes, values, and behaviors relevant to conservation. The point we simply want to make here is that these and similar other studies of zoo visitors and their experiences with animals in zoological parks coming out of the field of conservation psychology seem to provide promising empirical evidence supporting, in a general way, Norton's theory of transformative value. That is, based on this work it seems possible that many professionally-run zoos offer a venue that facilitates, and in some cases actively encourages, the kind of reflection Norton argued is an important function of direct experiences with species in moving people toward a more ecologically enlightened outlook—one that prizes biodiversity and places great store in efforts to preserve it.

15.4 Concerns and Open Questions

If this association between Norton's notion of transformative value and zoos (as the context for such transformations) is plausible, we believe it suggests a compelling ethical and practical justification of zoos as "instruments" for pro-conservation value transformation. Drawing on empirical studies from conservation psychology and referencing Norton's argument surrounding value transformation, zoos can defend their identity as conservation organizations by appealing to a sound philosophical and experiential understanding of the value of public encounters with their animal collections. Moreover, the attraction of zoos as entertainment destinations, it could be argued, only enhances zoos' opportunity to connect large numbers of people to animals, and to get them to think and act in a way more congruent with wildlife conservation goals. This idea of zoos as agents of public transformation therefore joins other conservation-based justifications for zoological parks in the modern era, taking its place alongside the more traditional population management, research, education, and field support functions.

Yet, as mentioned above, Norton's argument for transformative value also raises several questions, including concerns stemming directly from its extension to the zoo setting. Indeed, the application of transformative value to zoo animals (and to zoos as a whole) introduces a specific set of issues surrounding the scope, direction, and ethical implications of "enlightening" experiences in zoological parks, considerations that will need to be addressed before any final conclusion can be reached about the suitability and significance of Norton's theory for defending zoos as centers of pro-conservation transformation.

For starters, there is a potential problem with the context. It is clear from Norton's discussion of transformative value in *Why Preserve?* that the species and conditions he has in mind that trigger transformation are primarily "natural" or "wild" environments rather than the comparatively artificial and contrived ex situ environment of the zoo. As he put it, "*Wild species and pristine ecosystems…* provide the occasions for forming and criticizing our values, as felt preferences are measured against the evolving [ecological] world view. In this way they have transformative value" (Norton 1987: 212; emphasis added). Zoo animals obviously aren't truly or fully "wild" given the condition of captivity, and zoo landscapes do not qualify as "pristine landscapes" in anything but a cosmetic sense. So, there is a lingering question about the suitability of Norton's value concept in the zoo case and its portability outside of the in situ environment.[2] We will return to this point below.

Another worry is the impact of zoo visitors on the animals themselves. Studies have shown that visitor encounters, especially the more interactive experiences that produce strong (positive) emotional affiliations between the public and zoo animals, can result in increased stress on the latter, prompting a set of animal welfare concerns. (Davey 2007; Fernandez et al. 2009). In such cases the value and legitimacy of the transformative zoo experience will be rightly called into question if it is gained at the expense of duties to promote animal wellbeing. These are responsibilities the zoo community, despite the protestations of animal rights critics, takes very seriously (Mellor et al. 2015: 72).

Furthermore, there is a potentially troubling issue surrounding the ethical valence of the outcome of transformational experiences in zoos; namely, the possibility that transformative experiences in zoos may actually reinforce or introduce *negative* emotional or cognitive attitudes toward animals. There is some evidence, for example, that zoo education programs for children can promote negative associations with animals if not properly run (Jensen 2014).

---

[2]There is a potential further problem with the extension of transformative value in the zoo context, at least for achieving conservation goals. The personal identification with zoo animals and emotional affinity described in social scientific studies of visitor responses, in addition to (or rather than) catalyzing concern for the plight of wild populations and species, could reinforce anthropomorphic tendencies at the level of the individual animal that end up making population management (in both zoos and the wild) more contentious, as was the case with the controversial killing of Marius the Giraffe in the Copenhagen Zoo in 2014. Marius was deemed a "surplus animal" in the parlance of zookeeping and euthanized for zoo population control purposes (Minteer 2014). At the very least, such cases suggest that the emotional and ethical connection to animals in the zoo setting is a complex affair, and that more individualistic ethical responses are possible alongside the population- and species-level attitudes and outcomes described in the conservation psychology literature and surveyed in this chapter.

The potential for Norton's transformative process to result in negative as well as positive value outcomes has been highlighted by Sahotra Sarkar, one of the few philosophers to have explored Norton's original argument in depth and to have carefully weighed its appeal as a justification for biodiversity protection. Sarkar describes this "directionality problem" besetting transformative value in his 2005 book, *Biodiversity and Environmental Philosophy*:

> Even experience of biodiversity can have potentially negative transformative value. Imagine an individual visiting a wildlife preserve for the first time. An accidental encounter with a dangerous wild animal, such as a venomous snake, can potentially be terrifying. It may even lead to a lack of sympathy for that animal's conservation. What is important in this context is that an experience of this sort may result in a change in the felt preferences of an individual in a way that affects biodiversity conservation negatively even more generally. (Sarkar 2005: 97)

Although Sarkar refers to the context of the wildlife preserve, as mentioned above negative responses to animal encounters are also possible in zoos. These typically do not entail safety concerns, though as the recent case involving Harambe the gorilla in the Cincinnati Zoo reminds us, they can (McPhate 2016). Rather, the adverse reaction can be an aesthetic response to perceived sub-par zoo conditions or to certain taxa (e.g., reptiles), or simply visitor disappointment in encountering sleeping, out-of-sight, and/or inactive animals (Cushing and Markwell 2011; Godinez et al. 2013).

Sarkar identifies a further problem with Norton's argument, however: the challenge of delimiting the entities claimed to possess transformative value. That is, if objects other than wild species and ecosystems can trigger an evaluation of our felt preferences then the value of biodiversity in prompting this shift is considerably reduced. In the zoo context, we believe this "boundary problem," as Sarkar terms it, presents a two-pronged challenge.

First, the zoo case would seem to undermine protectionist arguments premised on the uniqueness of encountering wild species in their indigenous habitats and experiencing the more "pristine" environmental conditions described by Norton. If zoos (or more precisely, encounters with zoo animals) also have transformative value and the potential to help convert thoughtless consumers into thoughtful conservationists, then such arguments for protecting wildlife in situ, preserving nature from destructive human influence, and so forth, would seem to lose much of their normative force. At the very least, a case would have to be made that there is a tight and mutually supportive connection between the values and choices encouraged by the zoo experience and those relating to species and habitat protection in the field.

The boundary problem afflicting transformative value reveals another dimension in the zoo context, however, one that potentially undermines arguments defending the distinctive transformative value of zoological parks. If there is good reason to believe other experiences and encounters that do not require the exhibition of live animal collections can elicit a similar evaluation of preferences (and a positive attitudinal shift toward conservation), then the power of the zoo-as-transformer argument is, once again, weakened considerably. This is especially true in light of persistent ethical objections to the practice of keeping animals in captivity and the presumption that, all things being equal, doing so reduces their freedom (Jamieson 1985). Consider, for example, the impressive work of wildlife photographers such as National Geographic's Joel Sartore, whose indelible portraits of endangered animals in his "photographic ark" project are some of the most visually arresting and emotionally moving images of wildlife ever created (Sartore 2017). Although Sartore photographs many of his animals in zoos, the photographs quickly come to possess their own value; the appreciation of them, in other words, does not require any indexing to the original photographic subjects in zoological parks. And then there is the ubiquitous video documentations of wildlife captured by the Discovery Channel, Animal Planet, and the steady stream of Imax documentaries that provide a degree of pictorial detail and immersion in animal worlds once thought impossible. The best of these do a masterful job of encouraging a sense of wonder and inspiration, and perhaps even transformation.

Finally, we believe the zoo context also courts a unique hybrid version of the directionality and boundary problems confronted by transformative value. As we've said, by transposing Norton's notion of transformative experience to the zoo we are shifting the environmental context in question from (as least comparatively) wild-and-free to captive-and-controlled, a process that could be criticized as changing the subject of what constitutes a proper biodiversity ethic by reinforcing an attitude of dominion over animals rather than a respect for nature—a worry that evokes Holmes Rolston's earlier criticisms of transformative value. As a result, we have either weakened the ethical content of the transformative experience by normalizing the human power over animals and nature in the zoo setting, or we've undercut the original transformative value-based argument for preserving biodiversity in situ by showing that other, i.e., less wild and more artificial, conditions can serve as catalysts for reflection and environmental value reassessment. Perhaps we've even managed to do both at once.

These are all serious concerns. Although we can't possibly address them all in the remaining space we have here, we can at least offer a few preliminary thoughts that may help to soften a couple of them.

Regarding the objection that zoo encounters with animals, while they may be sufficient for transformative experiences, are not necessary (especially since virtual encounters can deliver the same results without running afoul of potential animal rights concerns related to captivity), we've discussed above how psychological research on zoo visitors has revealed the importance of direct interaction with zoo animals in forming emotional connections and encouraging pro-conservation perspectives. The intimate nature of the encounter and a personal identification with the animal, in fact, seem to be key in this process, including activities like making eye contact, which appears to play an important role in the positive affective response of zoogoers (Powell and Bullock 2014). Even in the digital age there is still a wide experiential gulf (albeit a narrowing one) between seeing an orangutan up close and personal and viewing a picture of an orangutan in a book or an image on a screen, however high the definition. Of course, one could still insist that these kinds of interactive and personal animal encounters are ethically and aesthetically more desirable when they happen in the wild via safaris or other in situ wildlife viewing experiences; that is, in contexts where the animal is relatively free and the conditions are predominantly natural. But for those without the means or inclination to take part in these experiences (which describes a considerable segment of the population), the local zoo offers a far more accessible and democratic form of direct animal encounter, especially in an urban setting. Therein lies at least part of the promise and potential of zoos' transformative value.

The worry that the zoo experience could end up reinforcing an ethic of animal dominion, however, is real and is probably an unavoidable concern as long as humans run the zoo. At the same time, zoos' ongoing efforts to enhance the naturalism of zoo exhibits and to provide various exhibit enrichments that allow animals to engage in more natural and characteristic species behaviors remind us how far today's professionally-run and accredited zoos have come from the unpleasant concrete-and-bars era (Hanson 2002; Hancocks 2010). One of the more innovative and provocative visions in naturalistic zoo design to emerge in recent years is "Zootopia," a planned expansion of Denmark's Givskud Zoo expected to launch its first phase in 2019. A hybrid of the immersive zoo and safari park, Zootopia is envisioned as a (nearly) wall-less and cage-free zoo landscape in which the animals roam relatively freely in multispecies habitats, a design driven by the goal of significantly reducing and concealing the human presence. Visitors to Zootopia will be sequestered in an ingenious series of hidden viewing bunkers and moved through the air via mirrored pods suspended on a wire track. Elsewhere they'll use bicycles and boats to get close to the zoo's elephants and zebras, which will be separated by an inventive array of natural and undetectable barriers (Minteer 2018). Although the initial media response to the Zootopia concept was fairly overheated—e.g., one journalist went so far as to suggest that the Denmark plan "reverses the role of captor and captive" (Wainwright 2014)—Zootopia represents a potentially dramatic revolution in zoo design and a great step forward in the immersive zoo model. If the full plan is built out as envisioned and the new expansion is successful (at least by its own lights), it will raise the aesthetic and ecological bar for naturalism and wildness in zoological parks.

Ironically, however, the more zoos move in this direction (which we can either call "radical immersion" or "extreme naturalism"), the less they will resemble zoos as we traditionally understand them. They'll be closer to small, highly managed wildlife parks; venues capable, perhaps, of greater visitor transformation and inspiration but places that will still never be as natural or as wild (however we define these vexed terms) as the nature parks and reserves they emulate. Nevertheless, Zootopia and its institutional progeny may end up marking not only a new phase in zoo design, but a milestone in the ethical evolution of the zoo toward a less patently artificial and more ecologically inspired facility, one that could push the moral and ontological notion of "captivity" close to its breaking point.

15.5 An Enduring Legacy

As this volume ably demonstrates, Bryan Norton's career is distinguished by his deep and sustained engagement with a string of fields lying beyond philosophical shores, from conservation biology and ecological economics, to environmental management and sustainability science. More than any other environmental philosopher Norton deserves credit for pushing a field that has often struggled with its own intellectual insularity into a more interdisciplinary, experimental, and collaborative space. In the process, he's shown how environmental philosophers can thrive outside the hothouse of traditional philosophy departments and contribute something unique and even useful to programs in public policy, natural resources, environmental studies, sustainability, and the life sciences.

And Norton's work is still opening up avenues for interdisciplinary research and collaboration, including, as we've discussed in this chapter, a novel area of study bridging environmental philosophy and conservation psychology focused on how zoo animal encounters can shape environmental values and promote a desire to engage in conservation-supporting behaviors. Although we need to learn more about the nature and impact of these experiences on the values, attitudes, and especially, the behaviors of zoo visitors, we believe that Norton's work on transformative value provides a useful philosophical anchor for this emerging body of empirical and experimental research. We hope Norton's work will keep inspiring and pushing our thinking about the responsibilities and possibilities that emerge from our complex ethical and experiential relationships with other species wherever we find them: in zoos, in the wild, or in the places in-between.

Acknowledgements We thank Sahotra Sarkar for his helpful comments on and suggestions for an earlier version of this chapter.

References

Ballantyne, R., J. Packer, and L.A. Sutherland. 2011. Visitors' memories of wildlife tourism: Implications for the design of powerful interpretive experiences. *Tourism Management* 32: 770–779.

Barongi, R., F.A. Fisken, M. Parker, and M. Gusset (eds.). 2015. *Committing to conservation: The world zoo and aquarium conservation strategy*. Gland, Switzerland: World Association of Zoos and Aquariums Executive Office.

Barrow Jr., M.V. 2009. *Nature's ghosts: Confronting extinction from the age of Jefferson to the age of ecology*. Princeton, NJ: Princeton University Press.

Byers, O., C. Lees, J. Wilcken, and C. Schwitzer. 2013. The one plan approach: The philosophy and implementation of CBSG's approach to integrated species conservation planning. *WAZA Magazine* 14: 2–5.

Ceballos, C., P.R. Ehrlich, and R. Dirzo. 2017. Biological annihilation via the ongoing sixth mass extinction signaled by vertebrate population losses and declines. *PNAS* 114(30): E6089–E6096, published ahead of print 10 July 2017. https://doi.org/10.1073/pnas.1704949114. Accessed on 14 May 2018.

Chamberlain, G. 2013. Orangutans fight for survival as thirst for palm oil devastates rainforests. *The Guardian*. December 14. https://www.theguardian.com/world/2013/dec/15/orangutans-fight-for-survival. Accessed on 31 July 2017.

Clayton, S., and G. Myers. 2009. *Conservation psychology: Understanding and promoting human care for nature*. Oxford, UK: Wiley-Blackwell.

Clayton, S., J. Fraser, and C.D. Saunders. 2009. Zoo experiences: Conversations, connections, and concern for animals. *Zoo Biology* 28: 377–397.

Clayton, S., J. Fraser, and C. Burgess. 2011. The role of zoos in fostering environmental identity. *Ecopsychology* 3: 87–96.

Clubb, R., and G. Mason. 2003. Animal welfare: Captivity effects on wide-ranging carnivores. *Nature* 425: 473–474.

Conde, D.A., N. Flesness, F. Colchero, O.R. Jones, and A. Scheuerlein. 2011. An emerging role of zoos to conserve biodiversity. *Science* 331: 1390–1391.

Conway, W. 2011. Buying time for wild animals with zoos. *Zoo Biology* 30: 1–8.

Cushing, N., and K. Markwell. 2011. I can't look: Disgust as a factor in the zoo experience. In *Zoos and tourism*, ed. W. Frost. Bristol, UK: Channel View Publications.

Davey, G. 2007. Visitors' effects on the welfare of animals in the zoo: A review. *Journal of Applied Animal Welfare Science* 10: 169–183.

Fa, J.E., S.M. Funk, and D.M. O'Connell. 2011. *Zoo conservation biology*. Cambridge: Cambridge University Press.

Falk, J.H., E.M. Reinhard, C.L. Vernon, K. Bronnenkant, N.L. Deans, and J.E. Heimlich. 2007. *Why zoos and aquariums matter: Assessing the impact of a visit*. Silver Spring, MD: Association of Zoos and Aquariums.

Fernandez, E.J., M.A. Tamborski, S.R. Pickens, and W. Timberlake. 2009. Animal–Visitor interactions in the modern zoo: Conflicts and interventions. *Applied Animal Behaviour Science* 120: 1–8.

Fraser, J., S. Gruber, K. Condon. 2007. Exposing the tourist value proposition of zoos and aquaria. *Tourism Review International* 11(3): 279–293.

Godinez, A.M., E.J. Fernandez, and K. Morrissey. 2013. Visitor behaviors and perceptions of jaguar activities. *Anthrozoös* 26: 613–619.

Grajal, A., J.F. Luebke, L.-A. DeGregoria Kelly, J. Matiasek, S. Clayton, B.T. Karazsia, C.D. Saunders, S.R. Goldman, M.E. Mann, and R. Stanoss. 2017. The complex relationship between personal sense of connection to animals and self-reported proenvironmental behaviors by zoo visitors. *Conservation Biology* 31: 322–330.

Grazian, D. 2015. *American zoo: A sociological safari*. Princeton, NJ: Princeton University Press.

Gusset, M., and G. Dick. 2011. The global reach of zoos and aquariums in visitor numbers and conservation expenditures. *Zoo Biology* 30: 566–569.

Hacker, C.E., and L.J. Miller. 2016. Zoo visitor perceptions, attitudes, and conservation intent after viewing african elephants at the San Diego zoo safari park. *Zoo Biology* 35: 355–361.

Hanson, E. 2002. *Animal attractions: Nature on display in American zoos*. Princeton, NJ: Princeton University Press.

Hancocks, D. 2010. The history and principles of zoo exhibition. In *Wild mammals in captivity. Principles and techniques for zoo management*, 2nd ed, ed. D.G. Kleiman, K.V. Thompson, and C. K. Baer. Chicago: University of Chicago Press.

Hutchins, M., and R. Wiese. 1991. Beyond genetic and demographic management: The future of the species survival plan and related AAZPA conservation efforts. *Zoo Biology* 10: 285–292.

IUCN. 2011. Connect-understand-act model engages zoo visitors in fighting extinction. https://www.iucn.org/content/connect-understand-act-model-engages-zoo-visitors-fighting-extinction. Accessed on 31 July 2017.

IUCN/SSC. 2014. *Guidelines on the use of ex situ management for species conservation*. Version 2.0. Gland, Switzerland: IUCN Species Survival Commission.

IUDZG/CBSG (IUCN/SSC). 1993. *The world zoo conservation strategy: The role of zoos and aquaria of the world in global conservation*. Brookfield, IL: Chicago Zoological Society.

Jamieson, D. 1985. Against zoos. In *In defense of animals*, ed. P. Singer. Oxford: Basil Blackwell.
Jensen, E. 2014. Evaluating children's conservation biology learning at the zoo. *Conservation Biology* 28: 1004–1011.
Kisling Jr., V.N. 2000. Zoological gardens in the United States. In *Zoo and aquarium history: Ancient animal collections to zoological gardens*, ed. V.N. Kisling Jr. Boca Raton, FL: CRC Press.
Keulartz, J. 2015. Captivity for conservation? Zoos at a crossroads. *Journal of Agricultural and Environmental Ethics* 28: 335–351.
Kolbert, E. 2015. *The sixth extinction: An unnatural history*. New York: Henry Holt.
Luebke, J.F., J.V. Watters, J. Packer, L.J. Miller, and D.M. Powell. 2016. Zoo visitors' affective responses to observing animal behaviors. *Visitor Studies* 19: 60–76.
Maple, T.L., and B.M. Perdue. 2013. *Zoo animal welfare*. Heidelberg: Springer.
Marino, L.S., O. Lilienfeld, R. Malamud, N. Nobis, and R. Broglio. 2010. Zoos and aquariums promote attitude change in visitors? A critical evaluation of the American zoo and aquarium study. *Society and Animals* 18: 126–138.
McPhate, M. 2016. Gorilla killed after child enters enclosure at Cincinnati Zoo. *New York Times*, May 29, 2016. https://www.nytimes.com/2016/05/30/us/gorilla-killed-after-child-enters-enclosure-at-cincinnati-zoo.html?_r=0. Accessed on 31 July 2017.
Mellor, D.J., S. Hunt, and M. Gusset (eds.). 2015. *Caring for wildlife: The world zoo and aquarium animal welfare strategy*. Gland, Switzerland: World Association of Zoos and Aquariums.
Minteer, B.A. 2014. Can zoos save the world? *The Conversation*. https://theconversation.com/can-zoos-save-the-world-32356. Accessed on 31 July 2017.
Minteer, B.A. 2018. The parallax zoo. In *The ark and beyond: The evolution of zoo and aquarium conservation*, ed. B.A. Minter, J. Maienschein, and J.P. Collins. Chicago: University of Chicago Press.
Minteer, B.A., and J.P. Collins. 2013. Ecological ethics in captivity: Balancing zoo and aquarium research under rapid global change. *ILAR Journal* 54: 41–51.
Minteer, B.A., and S.J. Pyne (eds.). 2015. *After preservation: Saving American nature in the age of humans*. Chicago: University of Chicago Press.
Moss, A., E. Jensen, and M. Gusset. 2015. Evaluating the contribution of zoos and aquariums to Aichi biodiversity target 1. *Conservation Biology* 29: 537–544.
Norton, B.G. 1987. *Why preserve natural variety?*. Princeton, NJ: Princeton University Press.
Norton, B.G. 1991. Thoreau's insect analogies: Or, why environmentalists hate mainstream economics. *Environmental Ethics* 13: 235–251.
Norton, B.G. 1994. Thoreau and Leopold on science and values. In *Biodiversity and landscapes: A paradox of humanity*, ed. K.C. Kim and R.D. Weaver. New York: Cambridge University Press.
Norton, B.G. 1995. Integration or reduction: Two approaches to environmental values. In *Environmental pragmatism*, ed. A. Light and E. Katz. London: Routledge.
Norton, B.G. 1999. Pragmatism, adaptive management, and sustainability. *Environmental Values* 8: 451–466.
Norton, B.G. 2005. *Sustainability: A philosophy of adaptive ecosystem management*. Chicago: University of Chicago Press.
Norton, B.G. 2015. *Sustainable values, sustainable change: A guide to environmental decision making*. Chicago: University of Chicago Press.
Norton, B.G., M. Hutchins, E.F. Stevens, and T.L. Maple (eds.). 1995. *Ethics on the ark: Zoos, animal welfare and wildlife conservation*. Washington DC: Smithsonian Institution Press.
Ogden, J., and J.E. Heimlich. 2009. Why focus on zoo and aquarium education? *Zoo Biology* 28: 357–360.
Pearson, E.L., R. Lowry, J. Dorrian, and C.A. Litchfield. 2014. Evaluating the conservation impact of an innovative zoo-based educational campaign: 'Don't Palm Us Off' for orangutan conservation. *Zoo Biology* 33: 184–196.
Perlman, D.G., and G. Adelson. 1997. *Biodiversity: Exploring values and priorities in conservation*. Malden, MA: Blackwell.

Powell, D.M., and E.V.W. Bullock. 2014. Evaluation of factors affecting emotional responses in zoo visitors and the impact of emotion on conservation mindedness. *Anthrozoös* 27: 389–405.
Redford, K.H., D.B. Jensen, and J.J. Breheny. 2012. Integrating the captive and the wild. *Science* 338: 1157–1158.
Rolston, H. III. 1994. *Conserving natural value*. New York: Columbia University Press.
Rolston, H. III. 2009. Converging versus reconstituting environmental ethics. In *Nature in common? Environmental ethics and the contested foundations of environmental policy*, ed. B.A. Minteer. Philadelphia: Temple University Press.
Sarkar, S. 2005. *Biodiversity and environmental philosophy: An introduction*. New York: Cambridge University Press.
Sartore, J. 2017. *The photo ark. One man's quest to document the world's animals*. Washington, D.C. National Geographic.
Saunders, C.D. 2003. The emerging field of conservation psychology. *Human Ecology Review* 10: 137–149.
Sterling, E., J. Lee, and T. Wood. 2007. Conservation education in zoos: An emphasis on behavioral change. In *Zoos in the 21st century: Catalysts for conservation?*, ed. A. Zimmermann, M. Hatchwell, L. Dickie, and C. West. Cambridge, UK: Cambridge University Press.
Stott, R.J. 1981. Historical origins of the zoological park in American thought. *Environmental Review* 5: 52–65.
Swanagan, J.S. 2000. Factors influencing zoo visitors' conservation attitudes and behavior. *The Journal of Environmental Education* 31: 26–31.
Takacs, D. 1996. *The idea of biodiversity: Philosophies of paradise*. Baltimore, MD: Johns Hopkins University Press.
Traylor-Holzer, K., K. Leus, and O. Byers. 2018. Integrating ex situ management options as part of a one plan approach to species conservation. In *The ark and beyond: The evolution of zoo and aquarium conservation*, ed. B.A. Minteer, J. Maienschein, and J.P. Collins. Chicago: University of Chicago Press.
Wainwright, O. 2014. Denmark's cage-free zoo will put humans in captivity. *Guardian,* August 5. http://www.theguardian.com/artanddesign/architecture-design-blog/2014/aug/05/denmark-cage-free-zoo-will-put-humans-in-captivity. Accessed on 31 July 2017.
World Association of Zoos and Aquariums (WAZA). 2005. *Building a future for wildlife—The world zoo and aquarium conservation strategy*. Bern, Switzerland: WAZA Executive Office.
World Association of Zoos and Aquariums (WAZA). 2012. Fighting extinction. *WAZA Magazine* 13.
Zimmermann, A., M. Hatchwell, L. Dickie, and C. West (eds.). 2007. *Zoos in the 21st century: Catalysts for conservation?* Cambridge, UK: Cambridge University Press.

Ben A. Minteer is Professor of Environmental Ethics and Conservation in the School of Life Sciences and the Arizona Zoological Society Endowed Chair at Arizona State University. His work has appeared in *Science, Nature, PNAS, Conservation Biology*, *Environmental Ethics*, and many other journals ranging across the sciences and environmental humanities. Minteer has also published a number of books, including most recently *The Ark and Beyond: The Evolution of Zoo and Aquarium Conservation* and *After Preservation: Saving American Nature in the Age of Humans* (both with the University of Chicago Press).

Christopher Rojas is a Ph.D. candidate in Biology and Society in the School of Life Sciences at Arizona State University. His dissertation focuses on the intersection of environmental pragmatism and adaptive co-management approaches to conservation.

Chapter 16
The Pragmatist's View: A Conversation with Bryan Norton

Curt Meine

Abstract In a wide-ranging interview conducted in July 2016, Bryan Norton reviewed his career path and contributions to environmental ethics and policy over more than four decades. Norton discusses the origins of his interest in environmental philosophy; the emergence and evolution of environmental ethics; his engagement with the ideas of Aldo Leopold and Baird Callicott; his work as an applied ethicist in the policy arena; his exploration of the core concepts of sustainability, convergence, and ecosystem service in his thinking; and future directions in environmental ethics and policy.

Keywords Bryan Norton · Environmental ethics · Sustainability
Convergence hypothesis · Aldo Leopold

In July 2016 the 3rd Biennial North American Congress for Conservation Biology convened in Madison, Wisconsin. In anticipation of Bryan Norton coming to Madison to present at the conference, I asked him if he would make time to film an interview, to be shared through a video series of the Center for Humans and Nature called "Conversations around the Green Fire." Bryan graciously agreed and, in an hour-and-a-half conversation stretching over two sessions, we were able to explore at least a portion of his background, career, influences, and ideas.[1] Bryan opened our conversation with an introduction to his academic interests and career.

I have been associated with the School of Public Policy at the Georgia Institute of Technology since 1988 and I remain associated with them. I'm an emeritus professor at this point, but I still work with graduate students. I grew up in

[1]I have edited the text of our conversation for clarity and have provided references. I also asked Bryan Norton to review my editorial changes. The filmed conversation is available on the website of the Center for Humans and Nature, www.humansandnature.org.

C. Meine (✉)
Aldo Leopold Foundation and Center for Humans and Nature,
P.O. Box 38, Prairie du Sac, WI 53578, USA
e-mail: curtmeine@gmail.com

S. Sarkar and B. A. Minteer (eds.), *A Sustainable Philosophy—The Work of Bryan Norton*, The International Library of Environmental, Agricultural and Food Ethics 26, https://doi.org/10.1007/978-3-319-92597-4_16

Michigan and went to the University of Michigan where I studied first political science and then philosophy, and I have a degree in philosophy, working mainly in the philosophy of language and the philosophy of science.

My main concern became: how do concepts change through time? This was back when Thomas Kuhn's work *The Structure of Scientific Revolutions* was first published, and it was all about these shifting paradigms (Kuhn 1962). What I noticed was that you could play this paradigm game historically as he did, or play it simultaneously and say that each field has its own paradigm. I started to say, "Okay, what Kuhn says about paradigms is pretty applicable to conservation and environmental ethics"—the field that I gradually got into. Because what we must do in conservation is bring together insights from many different disciplines, and that means you are sort of trying to hop from one paradigm to another.

I see a certain unity to all of that in my history, which started with the philosophy of language and how we learn and how science changes, but that provided the underpinnings of my work in conservation.

You did not start out as an environmental ethicist. In fact, the field of environmental ethics and philosophy had not yet fully emerged at the beginning in your career. So how did you make that transition—from a more traditional academic career in philosophy to environmental ethics?

Well, I had a long-standing interest in the environment, so that was always there. When I got my degree in philosophy, I studied the formal aspects of philosophy, logic, semantics, and so forth. I wrote my dissertation on a positivist named Rudolf Carnap (Norton 1970). Carnap was very special in that he started out as a sort of a hardcore positivist; everything had to verifiable and so on. But at the end of his career he came to realize that we really have to step back from science and make decisions about how we are going to talk about the world and about the linguistic aspects of the science. He came to the conclusion near the end of his life that we needed to become more pragmatist in our approach. Positivism held that hardcore science provided the only kind of knowledge we need. Carnap—amazingly, in at least his 70s, I guess—said that this is not working; that we really need to step back from observation and understand the language we use. That is, we also need to start talking about language as he, as an empiricist, could.

I finished my degree in 1970 at the University of Michigan and got my first job teaching philosophy at a small college, New College, in Sarasota, Florida. We had three philosophers. I was responsible for analytic philosophy. That turned out to be an excellent laboratory for me to work out my ideas because we had excellent students in small classes and I had the chance to really gradually develop my ideas. And then, just as a stroke of luck, I received a grant from the National Endowment for the Humanities to develop a curriculum in ethics and the environment. I joined a biologist and a literature professor and we taught these classes jointly on humanities and the environment.

And just as that was happening, as we were working on that grant, I found an opportunity through Mark Sagoff of the Institute for Philosophy and Public Policy at the University of Maryland. Mark had applied for multiple grants and got two at the same time. For one of them he said, "Well, we'll hire somebody to do the second one." That second grant was a terrific opportunity for me. It was a project funded by the National Science Foundation to ask two questions: What is the strongest rationale for the Endangered Species Act? And, secondly, given the understanding of the rationale, how should we prioritize our efforts in order to protect species?

Well, that was an intellectual windfall, because that was exactly what I was interested in. I started working on that grant and the result was two books on what we today call "biological diversity"—that term came into use right after I had published my own monograph on the subject "Why Preserve Natural Variety?" (Norton 1987). That became the foundation on which I built my career.

At that point I was at the University of Maryland doing this research, but I did not want to give up tenure. So I went back to New College for another two or three years, then moved on to Georgia Tech because I was looking for an institution that would be more supportive of research. When I arrived there I was to be a member of the school's social sciences program, which was mainly a service teaching unit. But then, another stroke of luck: the entire non-engineering part of Georgia Tech was reorganized to create, among other things, the School of Public Policy. That was just right down my alley. I was now able to do philosophy in the context of public policy. Along with two or three other major faculty members, we created the School of Public Policy. It's now a functioning and highly successful school. I retired several years ago, but I'm still active with students. I can't give up! I'm working now with two graduate students and continue to write and attend conferences and so forth.

Let's backtrack a bit now. There are some details I want you to fill in for me. I'm struck by the fact that you got your first academic position in 1970.

Yeah.

So your career begins essentially with Earth Day.

That's right.

And the full flowering of the environmental movement. We have to remind younger folks that the field of environmental ethics and philosophy did not even exist then—certainly not in the sense that it does now. There weren't many resources to draw upon, and anyone who was teaching at that time was a pioneer in the field. So do you look back on that as something of an historic coincidence or serendipity?

Of course, in the first five or six years when I was at New College I was concentrating on getting my research done, improving my teaching, getting tenure. Once that had happened I realized, having written my first book—which was called *Linguistic Frameworks and Ontology* (Norton 1977)—that if I continued in this path I would in the future be writing for a very small audience. I made a decision that I'd like to expand my audience, in two senses. One, I would speak to more people and have a larger readership than I did in my first book. But, I also wanted to address real social issues. In that way I expanded my career doubly. That was the mid- to late-1970s. And then this opportunity at the Institute for Philosophy and Public Policy came along in 1979 and into the 1980s.

Help us understand what that moment was like for that first wave of environmental ethicists. What was the atmosphere of the time? And how did traditional philosophy handle this upstart new field and people like you?

Well, we should distinguish the field and *me*. I think people who started contributing to this field were either historians of philosophy, or they were either historians or ethicists, and that's one of the reasons they started calling the field environmental *ethics*. I would prefer to call it environmental *philosophy*. The reason is that I come at it from the philosophy of science, and I'm interested in all the philosophical issues that get involved in, first of all, making judgments about what's right and wrong in a society but, also, putting something into policy form.

So those things led me from a more formalistic logical approach to philosophy to a more social philosophy approach. Of course, ethics was involved; but I stood out in two ways, I think, among that early group. First, I was one of just two or three who came from a background in the philosophy of science and was offering those insights. But, also, I was involved in that project I spoke about before, the project that Mark Sagoff originally formulated on the Endangered Species Act. At that time Henry Shue was the Director of the Institute for Philosophy and Public Policy, and he encouraged me from the day I arrived to get into policy. He said, "Take advantage of the fact that we're just a subway ride away from Washington, D.C. That's where policy is happening. We want you to be active."

And so I joined groups and spent a lot of time talking with people who were actually *doing* conservation. For example, I joined a group that was interested in strengthening plant conservation. Back then we were mainly looking at big animals. So I met that group. And through those conversations we talked about a lot of stuff that was interesting and important policy, who's doing what and so forth. But the concept of intrinsic value and that kind of argument—[that] never came up. These people knew what they had to do [for plant conservation]—they were doing excellent work—but they weren't asking why. They thought it was just obvious that we should save plants from extinction.

So that set me apart in two ways. First, I came more from a scientific point of view, and less from the ethicist's point of view. And, second, as soon as I started doing this I was thrown into the middle of policy. I was on several policy boards. For example, I was one of the original members of the EPA's Environmental Economics Advisory Committee. That was just one of the things I did, and other philosophers sort of looked at me and said, "Is Norton still a philosopher?" And I said, "Yes, I am! I'm a practical philosopher." People took quite a while to come around to that. But as we talk about the future of environmental ethics it seems to me my approach is attracting more and more of the younger people coming into the field. So I'm feeling somewhat like it worked out better than a lot of people thought it would.

That's amazing. We lay philosophers tend to think of "formal" philosophy as divorced from real-world issues, not wanting to deal with messy politics, not wanting to get enmeshed in that. But for you it was exactly the opposite. You did not start in philosophy and then get into policy. From the beginning you were working in both worlds and making the connections.

That's right. And what I found was that I asked a lot more interesting questions when I came at it from the practical end of the story. My colleagues, whom I greatly respect and have learned from, like Holmes Rolston and Baird Callicott—they came at it more from the point of view of "How can we make a philosophical argument?" And of course that's important, and it helps us I think to sort things out. It's important that we do pay some attention to the "why?" questions. Why is it important to save species? I mean that's very important. I think those are important questions, and I don't mean to stay away from those questions. But I think we get better purchase on them if we start from a practical point of view. Look at the Endangered Species Act as an example from my own career, and think about how you would explain and justify the Endangered Species Act to people—to non-philosophers and policy people and the general public.

That brings us to our own point of connection, and that is the influence of Aldo Leopold on you, your writing, your thought, and your work. Leopold has always been, at least in my reading of your work, a touchstone for you. Help us understand what Leopold's value has been for you as you were working through these ideas—how his tradition, his contribution as a "proto-environmentalist" (if you will), comes in. There has long been this desire in environmental ethics to "get Leopold right," to explain Leopold in a way that helps us advance our ways of thinking. How did Leopold begin to enter into your way of putting pieces together?

Well, I became friends with Baird Callicott. A lot of people can't believe that when they see how we write about each other, but we became friends. And, of course, around that time he was working on his book, *In Defense of the Land Ethic* (Callicott 1989). He was publishing papers in environmental ethics on that subject, most of which ended up, I guess, in that anthology.

And so I found it necessary to bounce my ideas, which are very different ideas, off Baird. I was introduced to Leopold by reading *A Sand County Almanac* because Baird said I must do so and I did. And then I read Baird's interpretation of this and I said, "But this is all wrong!" And why? Well, I think I reacted against Baird's approach of treating Leopold as the first and foremost defender of intrinsic value in nature. I realized that I looked at it more from the point of view of Leopold as a scientist—a scientist who was trying to do something; but who also, when he realized how difficult what he was trying to do was, could step up to the more philosophical level and think philosophically, even though he had little training.

Leopold had more training in philosophy than we think. We know, for example, that he and Estella used to read William James to each other. So there is evidence that Leopold was thinking more philosophically. But the fact is that he was bringing scientific concepts into the fold. At first I was just sort of reacting, saying "I don't think this intrinsic value thing is as important to Leopold as most people are saying." That was my first step. But then a second step was to notice Leopold's way of thinking as presented in two very short little essays in *A Sand County Almanac*: first *Marshland Elegy* and later *Thinking Like a Mountain*.

What Leopold came to see in thinking about cranes was that they are, as he said, "the trumpets in the orchestra of evolution." Consequently he saw them as having great value; they have a right to exist because of their long tenure. That seems to me be Leopold's core idea in the essay *Marshland Elegy* (Leopold 1949). And I thought, "Yeah, he's onto something there." In that essay he talks about how the cranes were integrated into an ecological system since the Ice Age. He also noted that their evolutionary history went way past the Ice Age, and that these "trumpets" are still with us telling us something. And what they're telling us in that essay, I think, is that humans have to live in three different scales of time. We live in *human* time, which Leopold introduced at the beginning of that essay, being impatient as he heard cranes coming toward the marsh. He introduces the passage of time as humans experience it. Then he pointed out that we also live in *ecological* time and, more deeply, in *evolutionary* times. And these different processes all "condition the daily affairs of birds and men," to quote one of the passages.

Knowing a little bit about science, I realized Leopold had created, forty years before it was defined, the importance of hierarchy theory in ecological science. Hierarchy theory says that systems in nature are generally organized in hierarchies, with different scales of space and time. Larger units enfold smaller ones. The larger units change quite slowly, whereas the smaller units that make up the large units change much more rapidly. Leopold really had an extraordinary, interesting concept of ecosystems and gave us a sort of frame in which to understand how humans fit in.

What you just explained to us, Bryan, was how Leopold anticipated the importance of understanding time scales and the temporal dimension in thinking about how the world around us works and how our decisions within it have to fit that. How about the spatial scale?

Well, what hierarchy theory does for us is to relate time and space. As large systems change slowly, they become sort of the background for what we do. Smaller systems—like our yard, for example—change much more rapidly. I noticed that Leopold had this very sophisticated conception of time and space, even though he introduced it metaphorically. And then I read the essay *Thinking Like a Mountain* (Leopold 1949), in which Leopold described how he had destroyed the wolf populations [in the American Southwest] and how he later came to regret that after the deer population he was trying to increase started to actually decrease from overpopulation and over-browsing of the mountainside. Leopold wrote about the famous incident where, having killed an old wolf, he watched the *green fire die in her eyes*. What he learned there, I think, was that wolves function in a larger and slower system than humans do.

Humans want to get a large amount of goods from animals and plants; that is, from nature. At that point we're impatient. We think in terms of economics only. Consequently, we can come to hate something like the wolf, even though, once you start to see the wolf's role in longer-term developments, the wolf becomes almost like a savior, right?—someone, something of great value. In that essay, which is generally described as Leopold's *mea culpa* essay, he says "I am guilty."

Having destroyed the wolves and changed the ecological system, he then regretted it. "He fell back on that hierarchical framework to say, 'I was thinking only like a human'. I was only thinking in terms of increasing the deer herd for hunters. But then I realized that my activities on that level spilled over to affect the usually slower-scale changes." In fact he saw the impacts on the ecological scale, the scale at which deer and wolves interact. The deer populations growing out of control were a result of his removing the wolves. So his human thinking destroyed a very complex system. He goes on to also talk about erosion of the mountainside, which would essentially be geological time.

The way I read those two essays is: in one Leopold creates this extraordinarily complex and interesting picture of ecology and ecological systems; in the second, he uses that to explain how he made a terrible mistake. So what's interesting about that is that Leopold starts to see the world on three different levels: the human, the ecological, and the evolutionary. And he felt that we can see positive values on each of those three levels. The positive level for humans is generally economic and developmental. The level of interactions among species (wolves and deer, in particular) would be the ecological scale. And what he realized was that, however important our economics are, if we destroy the ecological system, it's going to come back and bite us even at the human level. So his explanation of why he went wrong was very much based in a scientific model, which he showed through the metaphor of "thinking like a mountain." The wolf was thinking like a mountain, and it was interacting on an ecological level.

As you're explaining it now, Bryan, this gives me some new insight into one of the important themes in all of your work. And let's contextualize it a little bit. In the early years of environmental ethics, a lot of the debate in the field involved anthropocentric and non-anthropocentric perspectives, and instrumental and intrinsic values. I'm making the connection now perhaps that I hadn't made quite as explicitly before, that it's not one or the other, it's about *contextualizing* these views. And how Leopold is not on one side or another, per se, but holds this vision of being embedded in systems. Can you say more about your efforts to work through that debate, to try to reconcile those perspectives? Help us understand how you have tried to work through that classic dualism in environmental ethics.

Maybe I found myself in that position because I came at it from [a] public policy [perspective]. I was doing public policy by the time I was doing environmental ethics. I think that had a lot to do with it. I mean, I was forced there. After I joined the Environmental Economics Advisory Committee of the EPA, for example, I would have been laughed out of the room if I had started in saying, "Oh, you don't need to worry about these supply and demand curves. Nature has intrinsic value." Well, it wouldn't have worked. So I had to figure out ways of contributing in contexts where philosophers and environmental ethicists had seldom tread. I had to create something. This sense of scale, the importance of temporal and spatial scale, are all tied up in that.

Having come at it from a science point of view, I realized the relevance of Leopold's very sophisticated scientific model. And I realized that, yeah, I can make a contribution to policy. But I couldn't simply take the writings of someone like Baird Callicott, as much as I respect his work, and drop it into a conversation about supply and demand curves. It doesn't work for those people, because their whole model involves trying to solve every problem on the first level, the human interaction level. And Leopold is saying, "No, you've got to pay attention to these larger scales." I didn't succeed in convincing economists to be Leopoldians at all. But I think I was able, at points along the way, to say, "Well, listen, you have to pay attention to scale here." You don't have to invoke the whole structure in order to recognize the importance of scale issues.

That became my theme and maybe my distinctive contribution (or one of my contributions): that, coming at it from a scientific point of view, but trying to deal with public policy people, I had to come up with somewhat different approaches than were typical in environmental ethics. And when I would explain this to my colleagues, I very quickly became the ogre. I was in an environmental ethics conference once, sitting in the audience. There was a discussion going on, and a young woman whom I'd never met and didn't know spoke up and said, "Well, Bryan Norton would say... *if we think Bryan Norton is still a philosopher*...." And, you know, my friends in the audience of course started clapping.

But that was sort of the attitude, right? Because I didn't follow the same ethical conversation that they were engaged in, I was considered an outsider. Well, I think today I'm no longer thought to be an outsider. In fact, there are a lot of people who would love to claim that they were the ones who originated this sort of perspective from within public policy. And now we have people setting up centers that are designed to increase our impact in public policy. For me, that was never in doubt. That was more important to me than whether it was useful to talk about intrinsic value or not. But of course I got branded an anthropocentrist. And so, I played the game.

I'm really a pluralist. I'm willing to let people have the values they feel and express them. But I thought that somebody needs to stand up for the point of view of anthropocentrism. It's also true that, even though I have some non-anthropocentric leanings, it just seems more important to me that we see the world from a human perspective, because we're the ones who are making the decisions, right? Even if I think that nature has great value, I'd rather come at public policy problems from a point of view that asks how humans have gotten into this big problem, and how are they going to get out of it? I tended to look at the problem from a policy point of view, as essentially a set of problems about human values—economic versus various other kinds of values, what at that time I called "transformative values."

In *Why Preserve Natural Variety?* I argued that nature has value to humans because we can learn from it. And I've cited people like Thoreau and Muir and Leopold, all of whom have somewhere in their writings a sort of conversion experience, right? Leopold had it with the wolves. Muir had it when he was wandering in Canada, avoiding fighting in the Civil War. Muir found a tiny, a beautiful orchid and fell down and wept. And Thoreau, of course, is all about transformation.

So what I saw there was a sort of trope, shared by these three great conservationists. There was value involved, and it wasn't economic value, but a kind of human value in increasing our understanding, finding a better place for ourselves. So the learning process has value itself. I guess that makes me an educator; I mean, I care about people learning. And I think that the process of learning, the process of doing biology and doing ecology, is one that transforms people. I tried to find a position that wasn't all the way committed to intrinsic value, but that introduced values which could not be fully explained within an economic system.

I thought that I made that case very clearly, but people who are already deeply committed non-anthropocentrists, like Baird Callicott, couldn't buy it. It would have detracted from the purity of their position. To them it sounds instrumental; they're working with a distinction between intrinsic value and instrumental value. What I'm saying is, nature is instrumental to humans, in terms of food and fiber and so forth; but nature is also useful to human beings—*instrumental*—in their becoming truly human, in learning who they are and how they fit into the world. And so I've never thought of myself as a narrow anthropocentrist.

In the paper that I published in 1984, "Environmental Ethics and Weak Anthropocentrism" (Norton 1984), I tried to show that the people who are arguing against anthropocentrism are actually arguing against a very narrow approach to anthropocentrism, that is, the view that all values are economic. Well, I've studied enough economics that I know that there's a better answer here. What I was pointing out was that there are human reasons that are not economic reasons. These involve human values that are of great importance in terms of human maturing, growing up, learning who we are; and stop thinking of ourselves as God's children and start thinking of ourselves as developing better understanding and becoming more human by understanding our real role in these systems.

It's interesting to think that this discussion was unfolding within the environmental ethics community even as the concept of sustainability was gaining acceptance in the policy world, in the larger world, and in public awareness. These things were happening simultaneously. Yet, we've been having this conversation and the term sustainability has not even come up yet—perhaps because it was not yet a viable concept in the policy arena.

No, it was of course the mid-1980s before the concept of sustainability got a great deal of traction. Sustainability really is about thinking over long scales of time, right? So, in my case, starting from the problem of why we should save a species, I find Leopold and realize how important his work is on ecological organization. And then I realize that an important part of that is looking at decision-making with a very long time frame. It seemed a natural move, for me, having started with biology and biological diversity.

Some of the things that I learned from that journey would carry over, because in both cases the key is to gain a longer sense of time—so that when we're making decisions, we're not paying attention only to what's going to happen in the next five minutes or the next five years, but we also have to think in terms of centuries and millennia. I saw the power of the concept of sustainability, and think it's a really interesting concept in the following sense: sustainability came on the scene and became very important in policy and politics before it was ever defined or given a clear meaning.

What we had at that point was a battle going on: important people disagreeing about sustainability and how to understand it. Some people see economics as something that occurs within a very short time frame (like Leopold killing all the wolves and maximizing deer herds). Those kinds of interventions become more and more dangerous when you realize what impacts they may have on the larger and slower-scaled systems of an ecologically-organized, hierarchical system. But I really saw this as an opportunity to say, "Okay, there's a policy issue going on. And people are going back to their original disciplines and saying, 'Well, we can define sustainability.'" I worked with economists and, in fact, published papers with economists about how economists and ecologists think about sustainability differently.

I've mentioned that when I came into philosophy, I became intrigued with Thomas Kuhn's ideas on the structure of scientific revolutions, because he introduced the concept of paradigms. He was talking historically about how science changed across time. But here I'd walked into a problem where different paradigms were trying to capture a very important public policy concept. The economists had a concept of sustainability, what we call "weak sustainability." Ecologists had a much stronger view. Some philosophers tried to relate it to the rights of future generations and so on. So I was right in the middle of a conflict over paradigms. And it just was the most natural move in the world for me to move from thinking through biological issues and, with Leopold, to extend that to broader issues about everything from clean water to long-term issues we face today, like climate change.

Let's think about where we are at this moment. Now you're not the only philosopher on the block thinking about the philosophy of sustainability. You've *taught* the other kids on the block, and they're taking it in new directions with new realities. We have climate change coming at us. We have problems of justice and equity, globally now, and awareness of this in a way we didn't have maybe twenty-five years ago, when you were first helping us to

think through this. Give us a sense of where you think we are now and where this is going to take environmental ethics as a field.

Where we are now, I think, is that two important things are going on in philosophy and at the edges of philosophy, and in the intellectual world generally. One of them is that philosophy needs to make a contribution to policy. The second one is that when we're talking about policy, we're talking about very long-term problems. Those two things are bedeviling us at this point. We've had quite a few discussions about how conservation biologists can have an impact on policy makers. Those same conversations go on in environmental ethics.

It seems to me that what's happened is that, as people in environmental ethics have developed the ideas of intrinsic value, they're finding it harder and harder to apply them when we have to talk to policymakers who probably don't think that nature has intrinsic value. Or, if they do, they won't legislate on that basis. And we need to think about longer term.

I hesitate to predict this, but it looks to me like the field of environmental ethics and environmental philosophy is moving away from the controversy over intrinsic value versus instrumental value. We're now recognizing, I think, that most values are relational, that they depend on the context. I mean, you can talk about nature having intrinsic value, but in some contexts, that just isn't going to be very plausible. In other contexts, people think of nature and value nature in ways that simply can't be reduced to a simple conception of instrumentality. I've already argued for transformation as an important value that we derive from nature.

And that's what we need: we need a transformation in the way we see the world and our place in it. I don't know how to bring that about. To build on your point, I think some of my students out there will carry this forward; and when I say "my students," I don't mean the small number who studied with me, but the students who have responded to my writings. I don't want to build up my own contribution too much, except to say that I did start this trend, I think, and it's changed and developed over time.

Now we're at a point where the difficulty of the problems we face and the bifurcation of our population into the people who are pro-environment and anti-environment have brought us into a new era. In this era, what's going to be really important is that we expand our conception of time and not think of ourselves in terms of what can we accomplish in the short run. We really need to be able to accomplish things that have long-term impacts. And you and I are probably equally confused and concerned: it's not easy to see how that goes forward.

What I am saying, though, is that I think it has a better chance of going forward if we start from the problems, try to develop processes where people address problems in fruitful ways. That means developing more a *procedural* understanding of how we make environmental decisions, which is really the focus of my most recent book. That's what I think the future holds. And just one more smaller point about that: we really have a small group of people who are trying to develop this idea of values being relational and contextual. That seems to be very important and

I'm really hoping that I'll be able to contribute in a small way to that development, as it's being picked up by the next generation.

You've been one of the stalwarts in the field, Bryan, and have helped train the first generation—maybe two generations now—of environmental philosophers who are active in policy circles. Let's look ahead and ask: What advice should we give the next generation of philosophers who want to contribute to these discussions? And what kind of qualities will they need?

One thing that seems to be really important at this point is that we move our arguments away from what I call "ideology." Ideology would be a commitment to a view such as "all values are economic values" or "all environmental values are intrinsic values." That to me is ideology. I do not see how we could come up with any kind of empirical test to tell whether those statements are true. And I only believe in empirical science. So I think we're going to move away from that. Those values are fine; they're out there and people are expressing them. What I want to figure out is how we can move forward as a diverse society.

I think the early years of environmental ethics were sort of colonial, in the sense that we tried to take world problems and colonize them within philosophy. What I'm saying is what we really need to do is to think about better processes for making decisions—not just once we've all agreed on all the values and all of our assumptions, because we'll never get there. We need to be able to make better decisions in a diverse society, where people express very different values.

In my most recent book (Norton 2015), and in the work that I expect to do in the future—and what I hope others will follow me in doing—I pay a lot of attention to how people make decisions. That might be: How do you form a watershed management group? How do you manage a watershed management group? How do you keep it on target? How do you avoid people splitting up because of ideologies?

And so my most recent book is all about *procedural rationality*. How can we make decisions in an appropriate way? This leads us, again, to relationships: What kind of relationships do we want to have with nature in the future? I think that's going to involve a lot of different local communities starting to develop a sense of their own place, recognizing the context in which they face environmental problems and coming together as groups that can really struggle with these very difficult questions.

To me it's okay that we have Mr. X saying "Economics is the most important thing"; and Mrs. Y saying "Oh, but I think nature has intrinsic value." I'd love to have those people in my conversation. But we should be able to make a decision even as they differ, because there certainly are some things we can do. And this goes back to my convergence hypothesis: that there are things that we can do, despite difference, that are going to be helpful to nature, as well as to human beings. And that seems to me to be the sweet spot. As we go forward we want to have processes whereby groups form, recognize their divergences and disagreements, but say, "Can't we find something, some action, that we can take that will have a positive impact on our economics and a positive impact on the integrity of nature (or whatever we want to say)?"

So my idea is to shift toward what is called procedural rationality. And in my book I offer ten heuristics of how organizations and groups can make better decisions—what to concentrate on, what to avoid, and so on. So I think the philosopher's opportunity here is to be the pragmatist, to say that there's always going to be some opportunity that we can move forward with. And if we can move forward, then we'll develop trust and become a more cohesive community.

We have examples of this happening. The tree-planting programs in Africa and India are my favorite example. They plant trees. They provide small loans to women, usually, to plant the trees. They plant them too close together for them to thrive, so that within two or three years they have to start thinning. Well, suddenly they have an economy for selling firewood. They are going to be developing that economy and, if they're using degraded land, it's also going to improve the land ecologically.

In these kinds of community efforts, you don't have to determine whether you think trees have intrinsic value or instrumental value to recognize that a hillside devastated with erosion is better off with trees on it, right? That's not a philosophical question, but philosophy can help us to raise those questions and to improve our decision-making process. And I think that's the best we can hope for. I don't think we're going to find somebody making a discovery and winning a Nobel Prize for discovering that nature has intrinsic value, or that nature doesn't have intrinsic value. There are no Nobel prizes to be won there. What we can do is to start interacting with each other in ways that are more likely to find common ground, and I think that's the best we can do.

We've been tracking the development of your ideas through your writing and career. When did you start thinking about the theme of *convergence* in your work?

It started when I was in Washington D.C. at the Center for Philosophy and Public Policy, now the Institute for Philosophy and Public Policy. As I mentioned earlier, I met up with some groups in Washington and began to have conversations with them. Interestingly, I was sort of a naive environmental ethicist. I guess everybody was at that point. We're talking about 1979 and '80. When I was meeting with these people (who were, again, for the most part doing plant conservation), there would be a lull in the conversation at lunch sometimes and I would try out the idea of intrinsic value. And I got dead silence. I mean, nobody was interested in the subject. What I surmised from that was that these people are committed to a relatively clear objective, which is to save plant species. But they didn't think very deeply about how their values were different from other values, or anything like that. And they weren't very interested in the subject. Well, that grated on me for a while. I mean, *why aren't these people interested in all this important work*?

As a pragmatist, I started thinking, "Well, you know, who's wrong here? And maybe I, as an outsider to this kind of thing, am making assumptions that really don't hold." The other thing that I noticed was that I would see diversity in perspectives. Some people would say, "Oh, I think it's a terrible thing that we don't attribute intrinsic value to plants." That would seldom come out, but once in a while. And then there were people, of course, who were talking about how important plants are in the ecological system and so forth. So there were a lot of explanations of their motivations.

What I realized was that we're all doing the same thing. We were all concerned about achieving the same goals. I thought, "Well, you know, if you think about it, in the very long run, humans can be damaged by species loss. The longer we stretch our timeframe, the more we start to emphasize protecting plants as a foundation of ecosystems." On the other side of that, I started to realize that the humanists, the anthropocentrists, were as committed to the same goals as the non-anthropocentrists. With quite a few environmental problems, I thought, what if we did everything that we could to protect a human value that's involved in this system, how would that differ from what we would do if we said we're in it to save nature and humans? We're the problem, right?

From that I came to the conclusion that there a lot of policies out there that we could start with that don't lead to any kind of conflict about values, because both the anthropocentrists and the non-anthropocentrists were equally committed to the same objectives. Sure, we could have an interesting conversation, but, aside from that conversation, we were doing the same thing. That practical experience led me eventually to say, "Well, you know, we should choose those problems that are so important to humans and so important to endangered species that we should be able to find policies that will serve both of those ends."

I came up with this idea as I was writing the book *Why Preserve Natural Variety?* I first proposed this in the last few paragraphs of a paper in *Environmental Ethics* called "Conservation and Preservation: A Conceptual Rehabilitation." (Norton 1986). And then I developed it in *Why Preserve Natural Variety?,* which had a whole chapter devoted to what I called "the convergence hypothesis." Again, if we look at the broadest human values, the most long-range human values, and compare them with the values of ecological systems, these are really headed in the same direction. If we damage nature, we damage ourselves in the long run.

What surprised me when I proposed the hypothesis was how controversial it was. I was basically thinking, "Well, sure, we can go a long way with this convergence hypothesis. Anything that we are able to do today, it seems to me, would probably fall into one of those two categories. And our objectives would not shift because of our values." But what was interesting was that the non-anthropocentrists became livid. I was beaten about the head and the shoulders for years because, being an anthropocentrist and "stuck in anthropocentrism," I would never do anything important to protect the environment.

What's interesting is that when Ben Minteer published the book *Nature in Common* (Minteer 2009), which was really focused on the convergence hypothesis, you could go through its table of contents and sort the people into anthropocentrists and non-anthropocentrists, into policy people and philosophers. What we find is that all the policy people said, "This is obvious." And all of the philosophers like Laura Westra and Baird Callicott got furious at the idea, because it debunked their claim that they were real radicals, and that thinking about intrinsic value is a radical change in human history.

So it's interesting that the people who actually tried to get something done, the policy people, were very comfortable with the convergence hypothesis. It holds together what they're doing, and forestalls a great number of criticisms and disagreements. But the people who thought we need to reject anthropocentrism and open up a whole new way of thinking, what they called "deep ecology"—it took on other names over time—they were so committed to that idea that they found it damaging to their viewpoint for somebody to come along and say, "Well, a lot of the things you want to do could also be justified by anthropocentrism." What they wanted was to have a sharp bifurcation, because that made them sort of the stars of the show, if I could be a little nasty about it.

I mean, I was cutting into their thunder, right? Their thunder was that anthropocentrism was causing every problem. And I'm coming back saying: "Well, there's short-term anthropocentrism, economic anthropocentrism; but there's also long-term anthropocentrism, where we can even talk about the rights of future humans and so on. And it seems to me that the rights of future humans cannot be secured without saving a lot of nature and vice versa."

So I have thought of the convergence hypothesis as sort of the keystone of my pragmatism. I feel like the convergence hypothesis helps us to get things done and to work together without a lot of extraneous argumentation about what our motivations are. And since I've published that, I feel that, to my way of thinking, it pretty much resolves the anthropocentrism/non-anthropocentrism debate in favor of what I today would call maybe "administrative anthropocentrism." That is, when we're trying to decide on policy, and we're relying on values, they're going to be human values. It's going to be humans expressing those values. In that sense, humans administer the world, but for a variety of values. I think of administrative anthropocentrism as not really controversial. It's just that we're human and it's our values that we're pursuing. And solutions to problems and resolutions of arguments are ultimately going to come down to people's values.

As you discussed earlier, Bryan, one of the key themes in your work over the years has involved the challenge of translating philosophical values into the economic values that get traction in the policy arena and in the "real world." You've written extensively about the different ways we need to think about economic valuation. And you have seen, over the course of your career, the crystallization of this under the rubric of *ecosystem services*. How does this fit together? How does the ecosystem services approach help or hinder progress on these issues that you have raised over the years?

I need to give a contextual answer to that question. That is, in some contexts I think ecosystem services can be useful. In others, I think they're extremely misleading and damaging.

Let me start with the ways in which I think it can be damaging. In certain contexts, people come to the ecosystem services argument, I think, by noticing that there are values that can't be measured economically, or by normal methods. That is, using supply and demand economics, you really can't capture certain values that people feel very deeply about. And the ones I would emphasize are those transformative values, the ways that nature can make us better people.

I think there is fairly broad agreement that cost-benefit analysis is, at this point, quite incomplete and needs to be supplemented if we're ever going to get to where we can do a whole cost-benefit analysis and get to the bottom line. I think what happened with the initial introduction of ecosystem services was that quite a few people saw it as a patch on cost-benefit analysis; that is, there's this missing part of cost-benefit analysis, and if we start talking about ecosystem services and start putting dollar values on that, we can bring the cost-benefit analysis to a better conclusion. But I think that's just a category mistake, you might say, in that the methods that you use in economics—measuring preferences and so forth—are just different methodologies than people are using with ecosystem services. Economic values are marginal values, values at the margin. Ecosystem service concepts are totalistic values. You add them together and you get mush; nothing. And so people who have tried to reduce environmental values problems to ecosystem services are, I think, doing a real disservice. Values are diverse and you're never going to make them fit into one mold.

So do I throw the concept out? No. Because, in my philosophy, there's a crucial point at which humans have to recognize how the things that they care for and value are worth saving, but people don't understand how to connect their values back to the physical system.

And so I introduced what I think can be a useful term, "opportunity." Opportunity is a hybrid concept. On the one hand, it requires certain kinds of ecological functioning or natural functioning. It also implies that there's a good thing you can do with it, right? So I've found that if *ecosystem services* is properly used, it can show people how changes to the ecological systems can really matter to them.

Here's a simple example: skiing. There are a lot of people who love to ski; it's an excellent sport, etc. But you can't ski without snow. So if somebody is saying, "Basically what we're concerned about is if snow goes away in New England, the slopes will no longer be very attractive. They'll still be making snow, but they'll not have real snow." And so this is a value that is really quite selfish in a sense ("I love skiing"). Well, that's tough; you're a loser. So, skiing has a great appeal to human beings, but global climate change could totally wipe out that whole option, that *opportunity*. There's an example, I think, where opportunity requires a certain physical system, one that's cold enough to foster snow. And that connects to a real human value, recreation.

So it seems to me that ecosystem services can be helpful if used in that way, as a way of pointing out to people, "Oh, you like X. But if you like X, then you should care about Y, Z, and so on"—back into history and into the physical system and not the purely mental system. So you can think of these two different ways that people have used ecosystem services. The first, as a sort of patch on cost-benefit analysis, I think is a big mistake. It's full of confusions and it's reductionist. It misses a lot of values. But if you see it as a way of getting people to trace a path from the things they really care about back to specific aspects of nature, then we're doing something. We're telling people what they need to care about, what they need to do if they really care about skiing or any other human value.

Bryan, when I think of your work, the image of a stream comes to me, and I see channels braiding. You start out working on endangered species, which flow into scale issues, which flows in and out of sustainability philosophy and pragmatism. And looking forward, now that you've earned your retirement, I know that you're not going to stop and the river will keep flowing. And your students and those who've read and found value in your work will continue to develop your ideas as we face new circumstances and develop new concepts. I'd like to conclude with an open-ended question about the next generation of philosophers who want to contribute to environmental sustainability discussions. What will our next generation, our next wave of thinkers and doers, need to help us sort through these challenges going forward?

Okay. To comment first on all of the different threads that have, in some ways, come together and gone apart. The core of that is pragmatism. Pragmatism is a forward-looking philosophy. The truth is that which emerges over time. We're always going to make mistakes, but science, through many, many iterations, gradually gets further away from error and maybe even a little closer to the truth. So that's the pragmatist view—that we'll never have complete knowledge, because it's constantly changing. That's what really holds the whole thing together, I think, as far as my thought goes.

Looking forward, and putting all those things together, I think the most important thing that we face today is the need to make better decisions. In *Sustainable Values, Sustainable Change*, (Norton 2015), I shift away from what most philosophers seek, which is substantive rationality (where a decision is defined as rational if it aids us in getting to a stated objective). But we can contrast that with the view of Herbert Simon, the artificial intelligence genius. He pointed out that it's probably more interesting in most circumstances to talk about procedural rationality, which says that a decision is rational if it was arrived at by an appropriate method or process (Simon 1976). So I hope that the people who have read my work, including especially this most recent book, will see the opportunity here to shift away from many of their arguments about who's right about what values and so forth, and toward a position that says: "Let's allow diversity. Let's encourage some people to defend intrinsic value, other people to defend transformative value, and others instrumental value. Let's be sure that they're interacting in a way that's fair, where all the voices are being heard, and where rationality is based on appropriateness."

I think this will unify, on the one hand, the policy work that I've done by shifting our attention to how do we make decisions and away from what's ultimately valuable. What I hope for the future is that we'll get better and better at designing policy, designing processes by which communities can, with all voices heard, fairly propose and defend policies, which can then in turn become experiments. And that brings us to adaptive management, which we haven't talked about very much, but adaptive management really is, from my point of view, the on-the-ground representation of pragmatism.

That line of thought leads me to consider the current political reality, Bryan. I fully appreciate what you're saying about process—that we need open, transparent, fair, inclusive processes to help us get through these dilemmas. Yet we find ourselves in a time when ideology rules and political divisions seem harder to overcome than ever. Do you see some hopeful signs out there? That perhaps people are tiring of this kind of situation? That this alternative, more positive approach can take hold?

In thinking about this positive approach, as you put it, is, we have to talk about the context. At the federal level in the U.S., we are not ready to make rational, environmental decisions. It's obvious, right? We've been able to hold the line on the Endangered Species Act and in a number of other ways, but we're not getting the kind of support from the federal government that we absolutely need. On the other hand, if you look at the local level, you see a totally different story. Social science researchers have studied hundreds of watershed management projects. Some have been going on for ten or twenty years, but many of them are just popping up as local people, from the ground up, start to discuss what are they concerned about, and what they can do about it.

So I think the hope for the future is that we'll get a lot better at making better local decisions. And then somehow—and this is sort of where the miracle comes in—that gets expanded and generalized so that it applies at larger scales, at the state level and the federal level. That's a serious and difficult goal to reach, but that's what I see us aimed at: start locally and have local communities discussing, face-to-face as much as possible, what they love and what they care about and what they see as a threat. And in many cases they can come up with something they can agree on doing. What we find in these groups is that, once they start working together, even on quite neutral things—even if it's nothing more than monitoring, for example—they start to trust each other more. If they trust each other more, then they're able to try more serious experiments and adaptive management can really take hold at that point.

I suppose that's all very visionary, but I think we see it happening on the local level. The question is, are we going to be able to scale it up to the federal level and have that same kind of respect for open discussion and choice of policies that support a whole diverse set of values? If that were to happen, I would be very happy. I don't suppose I'll be around to see it, but I would love to peek back into the world and see how it goes. Because if it doesn't, if it goes badly, there may not be a world to peek back into.

I have a lot of faith in people, in people's opinions and abilities when they really feel like they're operating in a democracy. When they feel that there are people behind the scenes manipulating decisions, when they find that they've run into dead-ends all the time, this is what turns people away. So what we need, what we hope for, is that more and more local groups are able to achieve a procedural rationality, whereby they are making better decisions than they would have made without it. And then the hope is that can be scaled up and that we will eventually, as a nation, start acting like a democracy, which we're not doing right now.

Bryan, obviously we can't cover all your work, which is vast—

The secret to that is that I've been at it a long time!…

—but thank you for all you've done. Thanks so much for sitting down with us, and for what you'll continue to do, because I know that you are hardly done. We'll look forward to continuing the conversation.

Okay, great. And I enjoyed it also.

References

Callicott, J.B. 1989. *In defense of the land ethic: Essays in environmental philosophy*. Albany: State University of New York Press.

Kuhn, T. 1962. *The structure of scientific revolutions*. Chicago: University of Chicago Press.

Leopold, A. 1949. *A sand county almanac and sketches here and there*. New York: Oxford University Press.

Minteer, B. (ed.). 2009. *Nature in common?* Philadelphia: Temple University Press.

Norton, B.G. 1970. *Linguistic frameworks and ontology: A study in the meta-ontology of Rudolf Carnap*. Dissertation, University of Michigan.

Norton, B.G. 1977. *Linguistic frameworks and ontology: A re-examination of Carnap's metaphilosophy*. The Hague: Mouton Publishers.

Norton, B.G. 1984. Environmental ethics and weak anthropocentrism. *Environmental Ethics* 6 (2): 131–148.

Norton, B.G. 1986. Conservation and preservation: A conceptual rehabilitation. *Environmental Ethics* 8 (3): 195–220.

Norton, B.G. 1987. *Why preserve natural variety?* Princeton, NJ: Princeton University Press.

Norton, B.G. 2015. *Sustainable values, sustainable change: A guide to environmental decision making*. Chicago: University of Chicago Press.

Simon, H.A. 1976. From substantive to procedural rationality. In *25 years of economic theory: Retrospect and prospect*, ed. T.J. Kastelein, S.K. Kuipers, W.A. Nijenhuis, and R.G. Wagenaar, 65–86. Leiden: Martinus Nijhoff Social Sciences Division.

Curt Meine is a conservation biologist, historian, and writer who serves as Senior Fellow with the Aldo Leopold Foundation and the Center for Humans and Nature, and as Associate Adjunct Professor at the University of Wisconsin-Madison. He has written several books, including *Aldo Leopold: His Life and Work* and *Correction Lines: Essays on Land, Leopold, and Conservation*, and served as the on-screen guide in the documentary film *Green Fire:Aldo Leopold and a Land Ethic for Our Time*. He is coeditor most recently, of the bioregional anthology. *The Driftless Reader* (2017).

Appendix
Reflections on the 1970 Norton-Carnap Correspondence

Bryan G. Norton

As I worked on my dissertation in the late 1960s, I found a few points in my readings on Carnap that left me puzzled. I mentioned this to one of my committee members and he surprised me by saying, "Why don't you write to him?" This suggestion struck this shy and quiet graduate student as requiring hubris I could not muster: Why would Carnap, an internationally known and respected philosopher find time to respond to a mere graduate student at another institution?

We thought Carnap had, by this time, returned to Germany and, since we knew he had been recently at UCLA, I tried mailing him there in hopes they would forward it. So I cranked up the Smith-Corona typewriter, put a piece of paper—plus carbon—and typed out my questions, making countless erasures and smudging the carbon almost beyond reading. As it turned out, I was able to formulate two questions that were puzzling to me and, once I thought I'd made my inquiries as clear as possible, I sent the letter off, hardly expecting anything to come of it. It turned out that Carnap was still at UCLA and only a couple of weeks later, I received a detailed and carefully annotated response to my two questions, which included corrections he would like to make to "Empiricism, Semantics and Ontology" (Carnap 1950).

My first question had to do with many-sorted languages, a technical device that indexes properties according to the type of entities they may be applied to. For example, one who expects to speak of both material objects and numbers in one language, one might introduce the properties "heavy" and "prime" but restrict their applications so that the former could only be coupled with a variable ranging over physical objects and the latter would be limited to application to numbers. On this understanding, the sentences "7 is heavy" and "physical object x is prime" are not false, but meaningless. In a one-sorted language—which Quine favored—these sentences would be false. I was exploring whether Carnap's approach might support a pluralistic ontology, and symbolize this pluralism by using a many-sorted language. In the letter, Carnap confirmed this idea and he emphasized that such semantic rules are best thought of as "a matter of choice." This was encouraging, in that he confirmed that his system could be used to explicate an argument, for

S. Sarkar and B. A. Minteer (eds.), *A Sustainable Philosophy—The Work of Bryan Norton*, The International Library of Environmental, Agricultural and Food Ethics 26, https://doi.org/10.1007/978-3-319-92597-4

example, between a monist and a dualist, even though the designation of objects to sorts is merely a technical device.

Carnap, however, did not provide me with a telling argument against Quine's belief that many-sorted languages involve only a technical detail, to be excised according to Occam's razor. Quine thought anything that can be said in a many-scored language can be said in a one-sorted language, so he favored the simplest alternative. In the end, I concluded that this controversy had more to do with the two philosophers' different viewpoints on the role of logical exegesis. Quine favored development of systems of language and logic that, by passing tests of simplicity, etc. *replace* prior, less improved systems. Carnap, on the other hand, favored the completion of many alternative languages and he experimented with elements of systems, hoping to clarify specific points by introducing languages developed to clarify specific aspects of more inclusive systems.

Perhaps surprisingly, I have found many-sorted systems of logic to be useful in my environmental work. That idea is embodied in my mature work in that, emphasizing the importance of scale, I, in essence, treat scale-sensitive assertions as examples of "many-sorted" technical languages (i.e., a many-sorted language might elucidate the intuition that, say individual specimens, populations, and ecological processes, should be given different treatment in that each "sort" can be modified by characteristics appropriate to that sort, but not characteristics appropriate to other sorts). My response to this point has been to accept Quine's assertion that many-sortedness is a technical detail, but then argue that it is "useful" to adopt a language with such a separation in order to highlight the scalar and complex nature of ecological systems. Hence, while "System x is more resilient than system y" has variables ranging over ecosystems, "Animal, a, is dehydrated," ranges over individual animals. So, while Quine is correct that this divided ontology can easily be accommodated in a one-sorted language by creating hypotheticals that limit the applicability of predicates, we can argue that—in particular situations where science is describing objects that exist on and change on different scales—many-sorted languages can help to make multiscalar descriptions of nature more perspicacious.

I also asked Carnap about a seeming contradiction—or at least lacuna—in his treatment of external sentences. In "Empiricism, Semantics, and Ontology," he had explained that he was comfortable developing languages, the semantics of which require that he accept abstract entities as the referent of assertions like, "There are classes" (…"numbers", etc.), because such sentences, when considered internally, are made trivially true by the conventional choices made in specifying semantic rules. My puzzlement arose because it would seem that a perfectly parallel conventionalistic argument would, if applied to a language that refers to material objects as one of the sorts it describes, the existence of material objects would be true "by virtue of the rules of language."

In this case, I think Carnap avoided the issue. He correctly pointed out that, whenever he discusses external questions, he was mainly concerned with re-interpreting and rehabilitating languages of mathematics and languages referring to abstract entities such as classes and relations. His response that one might resolve this difficulty by avoiding rules specifying material objects by developing an ontology of space-time points, and build up an ontology of objects from these trivially existing points, is pertinent as a possible work-around. A problem remains. If a philosopher *does* wish to develop a materialistic language, complete with semantic rules allowing the internally interpreted sentence, "There are material objects," then this sentence (as explained by the internal-external distinction), would be analytic—true by the semantic rules of the language. So, as Carnap said in his reply letter, "One might wish to regard this statement as synthetic, because it is for the observer an empirical fact that he has seen such objects." So, he says, "I left open the question whether [this statement] is analytic or not. Fair enough. But it is also true that, if the internal-external distinction is to provide a new paradigm for linguistic philosophy, one would hope that its implication that questions of existence, viewed as internal to a system, are analytic, should apply in general.

So, when Carnap responded to my question about such sentences, he said in his reply that he was comfortable saying that internal existence claims of the existence of abstract entities were trivial, but he "hadn't decided" whether the particular sentence, "There are material objects," would be analytic." This sounds like a significant lacuna in his philosophy, but in fact he was able to resolve the question to some degree by introducing a new epistemological category: sentences made true because they allow us to more efficiently and effectively describe the world.

Leaving aside this relatively restricted point, I have understood my work in environmental philosophy and policy as engaging with scientists and activists in a process of making "external" decisions—choices to develop language and concepts that capture both the urgency of human values and aspirations, and possibilities for better understanding and communication regarding environmental problems.

Reference

Carnap, R. 1950. Empiricism, semantics, and ontology. *Revue Internationale de Philosophie* 4: 20-40. Reprinted in the Supplement to Carnap. R. 1956. *Meaning and necessity: A study in semantics and modal logic,* enlarged edition. Chicago: University of Chicago Press.

Bryan G. Norton
Dep't of Philosophy
University of Michigan
Ann Arbor, Mich. 48104
June 1, 1970

Professor Rudolf Carnap, Professor Emeritus
University of California, Los Angeles
Department of Philosophy
Los Angeles, California

Dear Professor Carnap:

Greetings! I am a graduate student in philosophy at the University of Michigan, and am writing my Ph. D. thesis on some of your work. Specifically, I am explicating, in more detail, your distinction between internal and external questions of existence, as it was set out in the essay "Empiricism, Semantics, and Ontology." And, I am attempting to defend your distinction against the attacks of Professor Quine in his paper, "On Carnap's Views on Ontology." I am finding my work on your essay both interesting and fruitful, and I feel that I owe you a debt of gratitude for all that I have learned from it.

There are two points, however, upon which I feel that I have not understood you completely, and would greatly appreciate a comment from you which I am sure would clarify my thinking.

First, concerning your use of a many-sorted language, I do not understand how you intended to separate the objects in the domain into the proper categories. That is, how is one to know whether a given object is to be within the range of one particular style of variables, or another? In The Logical Syntax of Language, the only discussion of this point is concerned with the theory of logical types. This is fine for separating material objects from properties, say; because the theory of types divides the domain into individuals, properties of individuals, properties of properties, etc. But, I do not see how the theory of types could be of any help in separating, say, material objects from spatial points, which would be on the same level of the theory of types, but which should, it seems, be ranged over by different styles of variables. Perhaps you deal adequately with this point somewhere, but I have been unable to find the reference. If you have, could you point out where, and if you haven't, I would greatly appreciate a brief comment from you.

Second, I am confused about the status of questions of existence which might be external, (i.e. they contain a "universal word", as you describe them in section 76 of Logical Syntax) but which are treated as internal. I interpret "Empiricism, Semantics, and Ontology" as saying that all statements in this class are analytic and trivial. This seems correct, that you adopt the first order predicate calculus, (including '(Ex)(Fx -Fx)' as a theorem, for each style of variables).

However, in Logical Syntax, you attempted to avoid this result (see section 37a), appealing to the intuitive truth that questions of existence are empirical questions, and should not have their answers entailed by the acceptance of a logic. Then, again, in "My Views on Ontological Problems of Existence," in the Schilpp volume, you said, "I have called existential statements of this kind, formulated within a given language, internal existential statements. They are usually analytic and trivial." (P.A. Schilpp, The Philosophy of Rudolf Carnap, p. 871.) Why do you say "usually analytic"? I am greatly puzzled as to how to interpret these seeming changes in position. Perhaps I have misinterpreted "Empiricism, Semantics, and Ontology." At any rate, I would greatly appreciate a comment from you on this point, also.

I hope you will not find my writing to you impertinent, but, I know of no other way to straighten out my confusion on these points. Both points have turned out to be fairly important in my attempts to explicate your exact position.

I hope, also, that this letter finds you in good health and good spirits. If you answer my letter, I will be very happy and grateful. Incidentally, if you should be interested in having a copy of my thesis, just mention this, and I would be more than happy to send one to you. As I said, I already feel indebted to you, and will feel this even more acutely if you answer my letter.

Very sincerely

p.s. I am sorry about having to send this letter through UCLA, but not knowing your address in Germany, I had no other way of getting it to you.

RUDOLF CARNAP
11728 DOROTHY ST., APT. 304
LOS ANGELES, CAL. 90049
PHONE 826-6377

June 18, 1970

Mr. Bryan G. Norton
Department of Philosophy
University of Michigan
Ann Arbor, Michigan 48104

Dear Mr. Norton:

I was glad to see from your letter of June 1 that you are writing a Ph.D. thesis on some of my philosophical views. I can understand very well that you find it difficult to reconcile the various formulations which I used at different times to express my views. I have a bad memory; when I write a new article or a new book, then I have only a clear picture of my views at that time. But I do not remember in detail what I said at earlier times. Sometimes a discrepancy is caused by my use of a changed terminology. But in some other cases there is an actual change in my view. In a case of that kind, if I am aware of the change, I usually make it explicitly clear.

I presume that you are reading my article "Empiricism, Semantics, and Ontology" in the appendix of my book <u>Meaning and Necessity</u>. Today I would like to make the following changes in the formulations for the sake of greater clarity:

(1) In the last sentence of section 1, delete "abstract entities" and insert behind "other field" the following: "a language in which they can speak about abstract entities;".

(2) In the first paragraph of section 2 (about 2/3s down the page), replace "within the framework" by "within the system of entities represented by the linguistic framework".

(3) Page 214 near the middle: (a) behind "theoretical justification" insert "of the kind just indicated"; (b) behind "assertion of reality" insert "or existence of the system of entities in question".

Maybe you could find some clarification of my views in these questions of existence in the Schilpp volume on my philosophy under "Replies" section 4 "My views on ontological problems of existence".

On many-sorted languages, see my book <u>Introduction to Symbolic Logic</u>, p. 83 last paragraph (there is a printing error; instead of "D19 through D22" it should be "D14* through D17*." You find an example for a three-sorted language in section 47a. There I explain also, in the paragraph included in square brackets, the practical difference between the use in geometry of a one-sorted language and a three-sorted language.

This is a matter of choice. We should not ask: "Are points and lines of the same kind or of different kinds?". Let b be a point and m be a number. Then in a one-sorted language L_1, the expression "b = m" is meaningful but false, In a two-sorted language L_2 that expression is not meaningful, that is to say, we make the syntactical rules of formation in such a way that it is not a sentence. In the preliminary deliberation about the choice of the language form we take into consideration whether we wish to take the same one- and more- place predicates as meaningful for the two kinds of objects or not. For example, do we wish to make the predicate "positive" applicable not only for numbers but also for points (as meaningful, irrespective whether true or false).

Some remarks on questions of existence. You quote from the Schilpp volume page 371: "these [internal questions of existence] are usually analytic". I was here chiefly thinking of the questions about the existence of "higher" entities, for example classes of objects, classes of such classes, classes of numbers, and the like. Existence statements of this kind, concerning abstract entities, I regard as analytic. In this context I left open the question whether the statement "there are objects (observable bodies)" is analytic or not. One might wish to regard this statement as synthetic, because it is for the observer an empirical fact that he has seen such objects. My main interest in this discussion was the problem, mostly discussed by philosophers, of the existence of abstract entities (classes, relations, etc.). [By the way, I have made the attempt (in Meaning and Necessity, page 86) to interpret also such statements of individuals ~~problems~~ as analytic, by taking as individuals (entities of order 0) not objects, but positions in the space-time system. My main interest here was to overcome the difficulty which Russell found in the construction of his system of arithmetic, where he needed the assumption that the number of individuals is infinite.]

Thank you for offering to send me a copy of your thesis when it is finished. But as a consequence of cataract operations of both eyes a year ago, my reading is rather slow and therefore I'm afraid that I shall not be able to find the time for reading the thesis. But I am certain that the Department of Philosophy at UCLA would appreciate receiving a copy for their library. They are always anxious to receive anything on my philosophy, and I might then also have a look at it.

At present I am still living in Los Angeles, but I might move for good to Germany, maybe in the summer of 1971. With best wishes for good progress in your work and the thesis,

Sincerely yours,

Rudolf Carnap

Rudolf Carnap

RC:jbc

Transcripts of Letters

Bryan G. Norton
Dep't of Philosophy
University of Michigan
Ann Arbor, Mich. 48104
June 1, 1970

Professor Rudolf Carnap, Professor Emeritus
University of California, Los Angeles
Department of Philosophy
Los Angeles, California

Dear Professor Carnap:

Greetings! I am a graduate student in philosophy at the University of Michigan, and am writing my Ph.D. thesis on some of your work. Specifically, I am explicating, in more detail, your distinction between internal and external questions of existence, as it was set out in the essay "Empiricism, Semantics and Ontology." And, I am attempting to defend your distinction against the attacks of Professor Quine in his paper, "On Carnap's Views on Ontology." I am finding my work on your essay both interesting and fruitful and I feel that I owe you a debt of gratitude for all that I have learned from it.

There are two points, however, upon which I feel that I have not understood you completely, and would greatly appreciate a comment from you which I am sure would clarify my thinking.

First, concerning your use of a many-sorted language, I do not understand how you intended to separate the objects in the domain into the proper categories. That is, how is one to know whether a given object is to be within the range of one particular style of variables, or another? In The Logical Syntax of Language, the only discussion of this point is concerned with the theory of logical types. This is fine for separating material objects from properties, say; because the theory of types divides the domain into individuals, properties of individuals, properties of properties, etc. But, I do not see how the theory of types could be of any help in separating, say, material objects from spatial points, which would be on the same level of the theory of types, but which should, it seems, be ranged over by different styles of variables. Perhaps you deal adequately with this point somewhere, but I have been unable to find the reference. If you have, could you point out where, and if you haven't, I would greatly appreciate a brief comment from you.

Second, I am confused by the status of questions of existence which might be external, (i.e. they contain a "universal word", as you describe them in section 76 of Logical Syntax) but which are treated as internal. I interpret "Empiricism, Semantics and Ontology" as saying that all statements in this class are analytic and trivial. This seems correct, that you adopt the first order predicate calculus, (including that '(Ex)(Fx v -Fx)' as a theorem, for each style of variables).

However, in Logical Syntax, you attempted to avoid this result (see section 37a), appealing to the intuitive truth that questions of existence are empirical questions, and should not have their answers entailed by the acceptance of a logic. Then, again, in "My Views on Ontological Problems of Existence," in thc Schilpp volume, you said, "I have called existential statements of this kind, formulated within a given language, internal existential statements. They are usually analytic and trivial." (P. A. Schilpp, The Philosophy of Rudolf Carnap, p. 871) Why do you say "usually analytic"? I am greatly puzzled as to how to interpret these seeming changes in position. Perhaps I have misinterpreted "Empiricism, Semantics and Ontology." At any rate, I would greatly appreciate a comment from you on this point, also.

I hope you will not find my writing to you impertinent, but, I knew of no other way to straighten out my confusion on these points. Both points have turned out to be fairly important in my attempts to explicate your exact position.

I hope, also, that this letter finds you in good health and good spirits. If you answer my letter, I will be very happy and grateful. Incidentally if you should be interested in having a copy of my thesis, just mention this, and I would be more than happy to send one to you. As I said, I already feel indebted to you, and will feel this even more acutely if you answer my letter.

Very Sincerelyx
Bryan G. Norton

p.s. I am sorry about having to send this letter through UCLA, but not knowing your address in Germany, I had no other way of getting it to you.

RUDOLF CARNAP
11728 Dorothy St, APT. 304
LOS ANGELES, CAL. 90049
PHONE: 826-6377

June 18, 1970

Mr. Bryan G. Norton
Department of Philosophy
University of Michigan
Ann Arbor, Michigan 48104

Dear Mr. Norton:

I was glad to see from your letter of June 1 that you are writing a Ph.D. thesis on some of my philosophical views. I can understand very well that you find it difficult to reconcile the various formulations which I used at different times to express my views. I have a bad memory; when I write a new article or book, then I have only a clear picture of my views at that time. But I do not remember in detail what I said at earlier times. Sometimes a discrepancy is caused by my use of a changed terminology. But in some other cases there is an actual change in my view. In a case of that kind, if I am aware of the change, I usually make it explicitly clear.

I presume that you are reading my article "Empiricism, Semantics, and Ontology" in the appendix of my book Meaning and Necessity. Today I would like to make the following changes in the formulation for the sake of greater clarity:

(1) In the last sentence of section 2, delete "abstract entities' and insert behind "other field" the following: "a language in which they can speak about abstract entities;".
(2) In the first paragraph of section 2 (about 2/3s down the page), replace "within the framework" by "within the system of entities represented by the linguistic framework".
(3) Page 214 near the middle: (a) behind "theoretical justification" insert "of the kind just indicated"; (b) behind "assertion of reality" insert "or existence of the system of entities in question".

Maybe you could find some clarification of my views on these questions of existence in the Schilpp volume on my philosophy under "Replies" section 4 "My views on ontological problems of existence".

On many-sorted languages, see my book Introduction to Symbolic Logic, p. 83 last paragraph (there is a printing error; instead of "D19 through D22" it should be "D14* through D17*." You find an example for a three-sorted language in section 47a. There I explain also, in the paragraph included in square brackets, the practical difference between the use in geometry of a one-sorted language and a three-sorted language. This is a matter of choice. We should not ask: "Are points and lines of the same kind or of different kinds?". Let b be a point and m be a number. Then in a one-sorted language L1, the expression "b = m" is meaningful but false. In a two-sorted language L2 that expression is not meaningful, that is to say, we make the syntactical rules of formation in such a way that it is not sentence. In the preliminary deliberation about the choice of the language form we take into consideration whether we wish to take the same one- and more-place predicates as meaningful for the two kinds of objects or not. For example, do we wish to make the predicate "positive" applicable not only for numbers but also for points (as meaningful, irrespective whether true or false).

Some remarks on questions of Existence. You quote from the Schilpp volume page 371: "These [internal questions of existence] are usually analytic". I was here chiefly thinking of the questions about the existence of "higher" entities, for example classes of objects, classes of such classes, classes of numbers, and the like. Existence statements of his kind, concerning abstract entities, I regard as analytic. In this context I left open the question whether the statement "there are objects (observable bodies)" is analytic or not. One might wish to regard this statement as synthetic, because it is for the observer an empirical fact that he has seen such objects. My main interest in this discussion was the problem, mostly discussed by philosophers, of the existence of abstract entities (classes, relations, etc.). [By the way, I have made the attempt (in Meaning and Necessity, page 86) to interpret also such problems[1] as analytic, by taking as individuals (entities of order 0) not objects, but position in the space-time system. My main interest here was to overcome the difficulty which Russell found in the construction of his system of arithmetic, where he needed the assumption that the number of individuals is infinite.]

Thank you for offering to send me a copy of your thesis when it is finished. But as a consequence of cataract operations of both eyes a year ago, my reading is rather slow and therefore I'm afraid that I shall not be able to find the time for reading the thesis. But I am certain that the Department of philosophy at UCLA would appreciate receiving a copy for their library. They are always anxious to receive anything on my philosophy, and I might then also have a look at it.

At present I am still living in Los Angeles, but I might move for good to Germany, maybe in the summer of 1971. With best wishes for good progress in your work and the thesis.

Sincerely yours,
Rudolf Carnap

RC:jbc

---

[1]Carnap inserted a hand-written correction here, crossing out "problems" and inserting in the margins "statements about individuals."